Springer Series on Naval Architecture, Marine Engineering, Shipbuilding and Shipping

Volume 22

Series Editor

Nikolas I. Xiros, University of New Orleans, New Orleans, LA, USA

The Naval Architecture, Marine Engineering, Shipbuilding and Shipping (NAMESS) series publishes state-of-art research and applications in the fields of design, construction, maintenance and operation of marine vessels and structures. The series publishes monographs, edited books, as well as selected PhD theses and conference proceedings focusing on all theoretical and technical aspects of naval architecture (including naval hydrodynamics, ship design, shipbuilding, shipyards, traditional and non-motorized vessels), marine engineering (including ship propulsion, electric power shipboard, ancillary machinery, marine engines and gas turbines, control systems, unmanned surface and underwater marine vehicles) and shipping (including transport logistics, route-planning as well as legislative and economical aspects).

The books of the series are submitted for indexing to Web of Science.

All books published in the series are submitted for consideration in Web of Science.

Alexander Arnfinn Olsen · Fidaa Karkori

Containerized Cargo Handling and Stowage

Principles and Procedures

Springer

Alexander Arnfinn Olsen
Southampton, UK

Fidaa Karkori
Southampton, UK

ISSN 2194-8445 ISSN 2194-8453 (electronic)
Springer Series on Naval Architecture, Marine Engineering, Shipbuilding and Shipping
ISBN 978-3-031-63328-7 ISBN 978-3-031-63329-4 (eBook)
https://doi.org/10.1007/978-3-031-63329-4

This Springer imprint is published by the registered company Springer Nature Switzerland AG
The registered company address is: Gewerbestrasse 11, 6330 Cham, Switzerland

In loving memory of Mohamed Karkori

Limitless and immortal, the waters are the beginning and end of all things on earth.

Heinrich Zimmer

Preface

When the first container vessel docked in the German port of Bremen on 6 May 1966, it caused, as does any novelty, vast excitement. But back then, nobody could imagine the extent of the revolution that had been set in motion. Initially, people smiled condescendingly at the "boxes," but soon it was the potential for rationalisation that became the central focus of attention. Labour-intensive and damaging cargo handling processes at the ports were no longer necessary. Transshipment times were cut to a fraction of those previously required on general cargo ships thanks to the possibility of moving entire truck loads on to ships in the form of containers. That said, the real revolution, however, only began as an increasing number of ports developed facilities for the transshipment, storage and clearance of containers. In parallel to this, the slot capacities for containers on ships rose from just a few hundred to the massive 23,000 we are seeing on the latest classes of ships, with even larger shipping units now being planned. Ship speeds, which have almost doubled since containerisation first got under way at the end of the 1960s, are also a significant factor that has enabled containerisation to grow exponentially. Ever faster and more economical transportation methods employing logistics systems that span the entire globe have made it possible to achieve the economies that we know today. These are synergy processes that will continue to lead to above average growth rates in container transport.

Many recurrent problems associated with traditional general cargo transport, such as petty theft, load damage as a result of repeated transshipment, and so forth, appeared to have been solved at once with the introduction of the "container" transport system. Its supporters were soon proclaiming a new age of damage-free transportation thanks to these huge metal boxes. These unrealistic hopes were soon dispelled in the face of new types of damage which were specific to containers. Organised crime emerged in place of petty theft, with entire containers being stolen to order. Professional cargo care, as provided by an experienced and well-trained crew during transportation on general cargo ships, is not carried out for containers, and the cargo is instead left "to its own devices." Professional stowage of the cargo by dock workers on board the general cargo ships now takes the form of stuffing (packing) the containers inland, either at "stuffing centres" or by the freight forwarder. This resulted in the emergence of a new and typical pattern of damage. Moisture, mould and corrosion are equally responsible for damage as the failure to secure cargo correctly. Indeed, there is close correlation between the latter and accidents related to cargo securing on large container ships, which are responsible for the loss of hundreds of containers every year. This can also threaten the very existence of entire container ships and their crew. Employing untrained personnel to load containers can also have a negative effect on the quality of transport operations, as does reductions in packaging, which was initially welcomed on environmental grounds, but soon fell afoul of an exaggerated cost awareness.

This book aims to provide the professional cargo officer with a source of guidance and information based on accepted industry best practice. The global economy is dependent on the smooth exchange of goods. Any damage results in a waste of resources. Loss prevention measures are not only economically necessary but also directly protect the marine environment. Surveys of transport practice have revealed that almost seventy percent of all packed containers, swap bodies, road and rail vehicles, or other cargo transport units, exhibit shortcomings in packing and load securing which can result in damage. It should be recognised, however, that even a book which has been edited and updated over many years cannot take account of every variant which may be encountered in practice; therefore, this book does not make any claim to completeness, being less than comprehensive or perfect in many parts. Nevertheless, it is hoped that this book will serve as a source of guidance for the professional and will make at least some contribution to preventing cargo damage and loss. The authors hope that it will be possible gradually to update this book as new best practices emerge so that it will continue to be a useful source of advice to anyone responsible for packing and securing containers.

Southampton, UK
December 2023

Alexander Arnfinn Olsen
Fidaa Karkori

Acknowledgements It is with grateful thanks that the authors acknowledge the assistance and support of Dr. Dieter Merkle at Springer Nature; Prince Nelson, Boopalan Renu and the editing and production team at Straive; and the numerous individuals who have contributed both their time, expertise and support to this book. To you all, we offer our gratitude and sincerest thanks.

Contents

Part I Development of Containerisation

1 The Development of Containerisation ... 3
1.1 Introduction to the Container Vessel ... 5
1.2 Container Ship Architecture and Design ... 10
1.2.1 All-Container Ships ... 15
1.2.2 Development of the 'Hatch-Less' Container Ship ... 16
1.2.3 Refrigerated and Reefer Ships ... 21
1.2.4 Container Passenger Ships ... 23
1.2.5 Feeder Container Ships ... 23
1.2.6 Ore Bulk Container Carrier ... 25
1.3 Container Flows ... 30

2 Positioning and Securing of Containers ... 33
2.1 General Stowage on Ships ... 33
2.2 General Securing Information ... 35
2.2.1 Securing in Vessel Holds by Cell Guides Alone ... 35
2.2.2 Securing Containers in Vessel Holds by Cell Guides and Pins ... 35
2.2.3 Securing Containers in the Vessel Hold by Conventional Securing and Stacked Stowage ... 37
2.2.4 Securing Containers in the Vessel Hold by Block Stowage and Stabilisation ... 37
2.2.5 Securing Containers on Deck Using Container Guides ... 39
2.2.6 Securing Containers on Deck Using Block Stowage Securing ... 40
2.2.7 Securing Containers on Deck Using Stacked Stowage Securing ... 42
2.3 Container Stowage Plans ... 44
2.4 Bay Plan ... 48

3 Responsibility for Container Packing and Cargo Insurance 53
3.1 Intermodal Carriage of Containers 54
3.1.1 Carriage by Road 54
3.1.2 Carriage by Railway 55
3.1.3 Carriage on Inland Waterways 56
3.1.4 Carriage by Sea 59
3.1.5 Transfer of Containers at Inland Terminals 59
3.2 Packing of Hazardous Materials 60
3.3 Container and Cargo Insurance 61
3.4 Container and Hull Insurance 63
3.5 Insurance Liabilities 63
3.5.1 Containers Provided by the Shipper 63
3.5.2 Containers Provided by the Carrier 63

Part II Causes of Damage and Loss During Transit

4 Damage and Loss During Transit Caused by Mechanical Stress ... 69
4.1 Examples of Inadequate Securing 70
4.2 Loss Prevention Through Training 71
4.3 Mechanical Stresses in Maritime Transport 76
4.3.1 Avoidable Shipping Stresses 77
4.3.2 Unavoidable Shipping Stresses 79
4.3.3 Static Mechanical Stresses 81
4.3.4 Stacking Crush Pressure 83
4.3.5 Dynamic Mechanical Stresses 86

5 Mechanical Stress During Transport 91
5.1 Mechanical Stresses During Maritime Transport 91
5.2 Mechanical Stresses During Inland Waterway Transport 100
5.3 Mechanical Stress in Land Transport 100
5.4 Mechanical Stresses During Railway Transportation 104
5.5 Mechanical Stress During Cargo Handling 107

6 Non-mechanical Stress During Transport 115
6.1 Climatic Stresses 115
6.2 Biotic Stresses 117
6.3 Chemical Stresses 119

Part III Container Design and Construction

7 Container Technical Aspects 125
7.1 Container Dimensions and Weights 128
7.2 Container Operational Markings 135
7.2.1 Stackability Markings 137
7.2.2 Height Markings 140
7.3 Optional Markings 141
7.4 Other Markings 142
7.5 Arrangement of Obligatory and Optional Markings 143

8 Container Design 149
8.1 Container Types 155
8.1.1 General Purpose Containers 155
8.1.2 Dry Bulk Containers 160
8.1.3 Special Dry Bulk Containers 161
8.1.4 Thermal Containers 162
8.1.5 Open-Top Containers 164
8.1.6 Half-Height/Open-Top Containers 165
8.1.7 Coil Containers 167
8.1.8 Tank Containers 169
8.1.9 Half-Height Tank Containers 170

9 Container Structural Testing Regimes 177
9.1 Racking Test 179
9.2 Container Maintenance and Oversight 179
9.3 Approved Continuous Examination Programme (ACEP) 180
9.4 Approval of New Containers by Design Type 180

10 Container Identification Systems and Marking 185
10.1 Container Size and Type Codes 190
10.1.1 Examples of Container Size and Type Coding Conventions 191
10.2 Marking of Containers Carrying Dangerous Cargo 207
Annex 213

Part IV Container Load Packing

11 Impact of Packaging and Packing Aids on Load Security 221
11.1 The Influence of Packaging Receptacles and Packaging Aids on Load Securing 223

12 Unitisation and Palletisation 227
12.1 Securing Goods in Packaging Receptacles 252

13 Marking of Goods 255
13.1 Labelling in Accordance with the Standards 256
13.2 Approved Marks and Labels 258
13.3 Meanings of the Pictorial Symbols According to DIN, ISO and IEC 259
13.3.1 Keep Dry 259
13.3.2 'Protect from Heat' and 'Keep Out of Direct Sunlight' 261
13.3.3 Do Not Use Hooks 261
13.3.4 Centre of Gravity 262
13.3.5 Top Heavy 263
13.3.6 Clamp Here 263
13.3.7 Sling Here 264

13.3.8 Fragile Goods . . . 264
13.3.9 Keep Upright . . . 266
13.3.10 Do Not Roll . . . 268
13.3.11 Tear Off Here . . . 269
13.3.12 No Hand Truck Here . . . 269
13.3.13 Do Not Destroy Barrier . . . 269
13.3.14 Electrostatic Sensitive Device . . . 270
13.3.15 Permitted Temperature Range . . . 270
13.3.16 Temperature-Controlled Load . . . 271
13.3.17 Do Not Use Forklift . . . 271
13.3.18 Safe to Use Forklift . . . 271
13.3.19 Stacking Limitation . . . 272
13.3.20 Protect from Heat and Radioactive Sources . . . 274
13.3.21 Do Not Expose to Magnetic Fields . . . 274
13.4 Tags and Detachable Labels . . . 274

14 Packing and Stowing Methods . . . 277
14.1 Preparatory Work . . . 278
14.2 Selecting and Checking Container Conditions and Standards . . . 283
14.3 Using Canvas Covers . . . 284
14.4 Before Packing . . . 286
14.5 Chassis Supports . . . 286
14.6 Swap-Bodies . . . 288
14.7 Driving into The Container . . . 288
14.8 Stowage Planning . . . 289

15 Segregation, Dunnage and Stowage . . . 299
15.1 Using Segregating Materials . . . 299
15.2 Dunnage . . . 301
15.3 Basic Stowage Methods . . . 306

16 Packing Safety and Prohibited Practices . . . 317
16.1 On Completion of Packing . . . 325
16.2 Final Work in the Door Area . . . 328

Part V Container Load Securing

17 General Load Securing Methods . . . 333
17.1 General Load Securing Methods . . . 333
17.2 Achieving a Tight Fit Using Container Components and Special Components . . . 336
17.3 Achieving a Tight Fit by Filling in Gaps . . . 339
17.4 Achieving a Tight Fit by Bracing . . . 341
17.5 Achieving a Tight Fit by Lashing . . . 343
17.6 Friction Securing . . . 352
17.7 Securing Against Tipping and Other Hazards . . . 357
17.8 Securing by Nailing . . . 359

18 Load Securing, Filling Materials and Use of Friction 361
18.1 Tie Down Lashing 363
18.2 Direct Lashing 366
18.3 Wood 372
18.4 Filling Materials and Airbags 378
18.5 Artificial Tight Fits 380
18.5.1 Gang Nails 380
18.5.2 Chocks 381
18.6 Friction and Friction Enhancing Mats 382
18.7 Calculating the Friction Coefficient μ For Bulk Cargo 385
18.8 Units of Measurement When Using the Transformed Formulas 386
18.9 Sliding Friction Coefficients 386
18.10 Factors Which Alter the Friction Coefficients 387
18.11 Examples of Different Friction Enhancing Materials 388

19 Lashing Materials and Equipment 393
19.1 Natural Fibre and Manmade Material Ropes 393
19.2 Wire Clamps (Clips) 401
19.3 Anchor Shackles and Chain Shackles 403
19.4 Turnbuckles 404
19.4.1 Determining the Diameter of a Lashing Point: 20 mm 405
19.5 Pre-fabricated Lashing Belts 410
19.6 Packing Straps 412
19.7 Belt Tensioners 414

Glossary 415

Abbreviations and Acronyms

‘	Foot/feet
“	Inch/es
°	Degree(s)
°C	Degrees Centigrade
°F	Degrees Fahrenheit
∞	Infinity
ABS	American Bureau of Shipping
ACEP	Approved Continuous Examination Programme
BACO	‘Ba’rge ‘co’ntainer carrier
BGV D 29	German Federal Accident Prevention Regulations
BIC	International Container Bureau
BV	Bureau Veritas
CAP	Common Agricultural Policy
CO_2	Carbon dioxide
CONRO	Container Roll-on/Roll-off
CPC	Cellular Palletwide Containers
CS	Calculation Strength
CSC	International Convention for Safe Containers
CSS Code	Code of Safe Practice for Cargo Stowage and Securing
CTU	Cargo Transport Unit
daN	Decanewton
DNV	Det Norske Veritas
DWT	Deadweight
f	Force
FCL	Full Container Load
FEU	Forty-foot Equivalent Unit
FIS	Freight Information System
FloFlo	Float-on/Float-off

FN	Normal force
Ft	Foot/feet
g	Gravity
g.n.t	Gooseneck tunnel
GL	Germanischer Lloyd
GP	General Purpose Container
GRP	Glass-fibre Reinforced Plastic
hPa	Hectopascals
ILO	International Labour Organisation
IMDG	International Maritime Dangerous Goods Code
IMO	International Maritime Organisation
ISO	International Standardisation Organisation
ITU	International Telecommunication Union
Kg	Kilograms
kN	Kilonewton
LASH	Lighters Aboard SHip
lbs	Pounds
LC	Lashing Capacity
LCL	Less Than Full Container Load
LoLo	Load-on/load-off
LR	Lloyds Register of Shipping
M	Metres
Mbar	Megabar
Mm	Millimetres
MPa	Megapascal
MS	Motor Ship
MSL	Maximum Securing Load
MT	Motor Tanker
MV	Motor Vessel
OBC	Oil-Bulk-Container Carrier
OS	Open-Side Container
OTOS	Open-Top/Open-Side Container
PA	Polyamide
PE	Polyethylene
PES	Polyester
PP	Polypropylene
RoRo	Roll-on/Roll-off Vessel
StoRo	Stowable Roll-on/Roll-off Vessel
TEU	Twenty-Foot Equivalent Unit
TIS	Transport Information System
TS	Turbine Ship
UK	United Kingdom
UN	United Nations

US	United States
UV	Ultra-Violet
VCI	Volatile Corrosion Inhibitors
VDI	Verein Deutscher Ingenieure
WoWo	Walk-on/walk-off Vessel

List of Figures

Fig. 1.1 Photograph of the *SS Ideal X* 4
Fig. 1.2 Photograph of the *SS Maxton* 5
Fig. 1.3 Photograph of the first containers to be loaded onboard the *SS Ideal X* 6
Fig. 1.4 Photograph of the river barge *Aladin*, navigating the Seine, France, undated 8
Fig. 1.5 Photograph of the *Grimaldi America*, departing Falmouth, England, 2014 11
Fig. 1.6 Photograph of a top spreader with an ISO container and tank container 12
Fig. 1.7 Photograph of the *Mairangi Bay*, Rotterdam, Holland, 1983 13
Fig. 1.8 Photograph of the *Bremen Express*, Hamburg, Germany, 2008 14
Fig. 1.9 Photograph of the *Emma Mærsk*, Aarhus, Denmark 2019 14
Fig. 1.10 Photograph of the *MOL Triumph*, arriving Port of Rotterdam, Holland, 2017 21
Fig. 1.11 Photograph of the *CSCL Indian Ocean*, Port of Felixstowe, England, 2018 22
Fig. 1.12 Photograph of the reefer, *Dole Ecuador*, Puerto Barrios, Guatemala, 2009 23
Fig. 1.13 Photograph of the passenger RoRo *Xiang Xue Lan*, Singapore, 2008 24
Fig. 1.14 Photograph of the feeder vessel *Frederik*, Hamburg, Germany, 2019 24
Fig. 1.15 Photograph of the *Herald of Free Enterprise,* Port of Zeebrugge, Belgium, undated 26
Fig. 1.16 Photograph of the *Viking Sally* (later renamed *Estonia*), departing Stockholm, Sweden, 1989 27
Fig. 1.17 Photograph of the *Repubblica Argentina*, departing Casablanca, Morocco, 2022 28

Fig. 1.18 Photograph of the *Ever Ace*, Port of Hamburg, Germany, 2022 28
Fig. 2.1 Containers stowed lengthwise from fore to aft 34
Fig. 2.2 Containers stowed athwartship (i.e., athwartships stowage) ... 34
Fig. 2.3 Cell guides in an all-container vessel 36
Fig. 2.4 Guide rails of two adjacent slots 36
Fig. 2.5 **a** Example of stacked stowage with conventional securing. **b** Example of stacked stowage with conventional securing 38
Fig. 2.6 **a** Fastening containers together. **b** Fastening containers together 39
Fig. 2.7 Fastening containers together 40
Fig. 2.8 **a** Example of block stowage method with stabilisation. **b** Photograph of block stowage method with lower stabilisation. **c** Photograph of block stowage method without upper stabilisation 41
Fig. 2.9 **a** Example of block stowage securing on deck. **b** Photograph of crisscross lashing on different container stacks. **c** Photograph of crisscross lashing on different container stacks 43
Fig. 2.10 Securing of a 3-tier stack on board a semi-container ship 44
Fig. 2.11 Row numbering where there is an even number of rows 46
Fig. 2.12 Row numbering where there is an odd number of rows 46
Fig. 2.13 Numbering of the port rows of a ship 46
Fig. 2.14 Numbering of the starboard rows of a ship 47
Fig. 2.15 Row numbers of the aft bay of a ship 47
Fig. 2.16 Numbering of horizontal container layers, or tiers 48
Fig. 2.17 Bay plan 48
Fig. 2.18 Colour-labelled containers in a bay plan 49
Fig. 2.19 Typical bay plan (Blank) 49
Fig. 2.20 Identification of container codes 50
Fig. 2.21 Typical bay plan (Completed) 51
Fig. 3.1 Photograph of empty containers at a container depot 56
Fig. 3.2 Photograph of a container transfer using a stack loader 57
Fig. 3.3 Photograph of a container loaded on a skeleton chassis 57
Fig. 3.4 Photograph of a container loaded on a flatbed trailer 58
Fig. 3.5 Photograph of containers loaded on a railway carriage 58
Fig. 3.6 Example of an inadequately secured container load 61
Fig. 3.7 Example of a properly stowed and secured container 62
Fig. 3.8 Generic leased GP container provided by the shipper 64
Fig. 3.9 Carrier's container with the carrier's identifying mark clearly legible 65
Fig. 4.1 Non-regulation complaint dangerous cargo unit 70
Fig. 4.2 Non-regulation compliant dangerous cargo unit 71
Fig. 4.3 Machine without securing means (Not fit for shipping) 72
Fig. 4.4 Earth borer without securing means (Not fit for shipping) 72

Fig. 4.5 Properly secured container load 73
Fig. 4.6 Properly secured container load 74
Fig. 4.7 Container ship *Seatrain Bennington*, circa 1980 77
Fig. 4.8 *MSC Nederland* collision, Bayport, Texas, 29 October 2011 78
Fig. 4.9 Collapsed container 78
Fig. 4.10 Example of a poorly stacked container 79
Fig. 4.11 Example of a well stacked container 80
Fig. 4.12 Overview of shipping stresses 81
Fig. 5.1 Demonstration of a ship motions (Yawing is motion around the ship's vertical axis) 94
Fig. 5.2 Demonstration of ship motions (Heaving is motion along the ship's vertical axis) 95
Fig. 5.3 Rolling is the movement of a ship around its longitudinal axis, the rolling angle in this case being 10° 96
Fig. 5.4 Demonstration of a rolling angle of 45° 98
Fig. 5.5 Photograph of bulges in containers as a result of home-grown acceleration forces 98
Fig. 5.6 Photograph showing a container "ripped apart" as a result of home-grown acceleration forces 99
Fig. 5.7 Acceleration in longitudinal direction according to VDI guidelines 102
Fig. 5.8 Acceleration in transverse direction according to VDI guidelines 102
Fig. 5.9 Acceleration in longitudinal direction according to VDI guidelines (rail) 106
Fig. 5.10 Acceleration in longitudinal direction according to VDI guidelines (rail) 106
Fig. 5.11 Photograph of the Port of Cape Town container terminal, South Africa 109
Fig. 5.12 Photograph of handling a container with a reach stacker and top spreader, Torshavn, Faroe Islands 110
Fig. 5.13 Photograph of a skeleton chassis 110
Fig. 5.14 Photograph of a mobile container gantry crane, Port of Greenock, Scotland 111
Fig. 5.15 Photograph of an unmanned container shunter, Port of Antwerp, Belgium 111
Fig. 5.16 Photograph of quayside gantry cranes, Port of Liverpool, England 112
Fig. 5.17 Photograph of a damaged tank container with hazmat kit 112
Fig. 5.18 Photograph of a damaged container at the Birzebugga Triq Kalafrana Freeport, Malta, 2008 113
Fig. 6.1 Photograph of the Sirex wasp 118
Fig. 6.2 Photograph of lumber with anti-Sirex wasp treatment 118

Fig. 6.3 Photograph of the *SS Marilu* carrying prefabricated houses for the Queensland Housing Commission from Genoa, Italy, 1952; the shipment was infected with Sirex wasp 119
Fig. 6.4 Tank container with various HAZMAT labels clearly visible 120
Fig. 7.1 Maximum mass for a 20′ container according to DIN ISO 688 130
Fig. 7.2 Maximum mass for a 40′ container according to the previous standard or optionally for a more recent 20′ 131
Fig. 7.3 Photograph of a warning affixed to a 2.5 m wide container 132
Fig. 7.4 Examples of container volume data 132
Fig. 7.5 Photograph showing the volume in litres on a tank container 133
Fig. 7.6 Photograph showing the minimum information required on a half-height open-top 134
Fig. 7.7 Photograph showing markings on platform containers 135
Fig. 7.8 Photograph showing weights and volumes on 20′ reefers and 40′ general purpose containers 136
Fig. 7.9 Photograph showing an inappropriately positioned sticker 136
Fig. 7.10 Example of a CSC plate giving full information 137
Fig. 7.11 Example of a separate plate giving full information 138
Fig. 7.12 Photograph showing standard operational markings (gross and tare weights) 138
Fig. 7.13 Examples of the different ways of writing 2.6 m and the imperial customary dimension 139
Fig. 7.14 Examples of the different stackability markings for air transport and air/sea transport 139
Fig. 7.15 Symbol warning against the danger posed by overhead power cables 140
Fig. 7.16 Example of the different methods of writing 2.9 m and the imperial customary dimension 140
Fig. 7.17 Examples of how yellow and black stripes constitute part of the height marking 141
Fig. 7.18 Photograph of a container with a full set of operational markings 142
Fig. 7.19 Example of weight and volume markings on 40′ containers 143
Fig. 7.20 Photograph of partially hidden or covered weight and volume data 143
Fig. 7.21 Example of how the maximum gross weight may deviate from the standard values 144
Fig. 7.22 Example of M.G.W. = Maximum Gross Weight ... data 144
Fig. 7.23 ... and in Chinese characters too! 145
Fig. 7.24 Example warning about folding doors on a half-height open-top container 145

Fig. 7.25 Photograph of operating and safety instructions on a thermal container 145
Fig. 7.26 Example of a statement of compliance with structural regulations from Bureau Veritas 145
Fig. 7.27 Example of desirable arrangements of black and yellow warning markings 146
Fig. 7.28 Photograph of a (UK) van with correctly painted red and white warning 146
Fig. 7.29 Selective enlargements 147
Fig. 7.30 Arrangement of obligatory and optional markings 147
Fig. 7.31 Typical arrangement of acceptable container markings 148
Fig. 7.32 Selective enlargements 148
Fig. 8.1 Diagram illustrating the essential components of a standard ISO container 150
Fig. 8.2 Diagram illustrating the part names of the standard ISO container floor 150
Fig. 8.3 Photograph showing ISO corner castings (Vertical and horizontal) 154
Fig. 8.4 Photograph showing the securing of containers on deck 154
Fig. 8.5 Photograph of a typical 20' container 156
Fig. 8.6 Photograph of a typical 40' container 157
Fig. 8.7 Photograph of a typical 45' container 157
Fig. 8.8 Photograph of a typical high-cube container 158
Fig. 8.9 Photograph of an **a** open-sided and **b** open-side and open-top container 159
Fig. 8.10 Photograph of a typical dry bulk container 161
Fig. 8.11 Photograph of a refrigerated (reefer) container 163
Fig. 8.12 Photograph of a thermal container 164
Fig. 8.13 Photograph of an open top container 165
Fig. 8.14 Photograph of a half-height/open top container 166
Fig. 8.15 Photograph of an open-side/open-end platform 168
Fig. 8.16 Photograph of an open-side/closed-end platform 168
Fig. 8.17 Photograph showing the detail on a standard P8 platform 169
Fig. 8.18 Photograph of a closed side flat rack container 169
Fig. 8.19 Photograph of a frame container 170
Fig. 8.20 Photograph of a coil palette 171
Fig. 8.21 Photograph of a typical tank container 172
Fig. 9.1 Photograph of a typical CSC plate 178
Fig. 9.2 Example of the KOR container class mark 183
Fig. 10.1 Example of a typical horizontal container identification mark 186
Fig. 10.2 Example of a typical vertical container identification mark 189
Fig. 10.3 Side view markings on containers 189
Fig. 10.4 Markings on the door of a container 191
Fig. 10.5 Typical container markings (tank container) 192

Fig. 10.6 Disadvantage of check digit [0] 192
Fig. 10.7 Correct identification using check digit [0] 192
Fig. 10.8 Container identification including country, size and type codes 193
Fig. 10.9 Flatbed container 197
Fig. 10.10 20′ long refrigerated container 198
Fig. 10.11 Size, and type codes additionally stated in accordance with current standard 198
Fig. 10.12 Norfolk Line container with container codes clearly identifiable 204
Fig. 10.13 Correct marking on a container side 208
Fig. 10.14 Correct marking on a container side 208
Fig. 10.15 Non-regulation affixing of a placard 209
Fig. 10.16 Previous variant of the primary and subsidiary risk placards 209
Fig. 10.17 New regulations: primary and subsidiary risk placards bear class number 210
Fig. 10.18 Examples of primary and subsidiary risk placards plus UN Number 211
Fig. 10.19 Correct: 4,000 kg consignment with hazardous materials, therefore UN number stated 211
Fig. 10.20 Incorrect: < 4,000 kg consignment with hazardous materials, therefore UN number not stated 211
Fig. 10.21 Warning sign for containers in which dry ice or other expendable refrigerants are used for refrigeration purposes 212
Fig. 12.1 Example of palletisation which will actually increase freight costs 228
Fig. 12.2 Example of goods that are not safe for carriage 228
Fig. 12.3 Example of an accident waiting to happen due to poorly loaded pallets 229
Fig. 12.4 Example of stowing two cartons too many! 230
Fig. 12.5 Example of incorrectly packed pallets containing boxed goods 230
Fig. 12.6 Example of poor palletisation 231
Fig. 12.7 Example of a load unit that is not suitable for transportation ... 231
Fig. 12.8 Example of a load unit that is not suitable for transportation ... 232
Fig. 12.9 Example of deficiencies when forming load units 232
Fig. 12.10 Example of unacceptable packing of pallets with drums 233
Fig. 12.11 Example of a unit load with serious deficiencies 235
Fig. 12.12 Example of a unit load with serious deficiencies 236
Fig. 12.13 Example of a unit load with serious deficiencies 236
Fig. 12.14 Example of a unit load with serious deficiencies 237

Fig. 12.15 **a** Pallets that are appropriately dimensioned for transportation by lorry. **b** Pallets that are not designed for the container. **c** Telescope cartons with pallet-like bases designed for use in the container 238
Fig. 12.16 Load dimensioned to suit the internal measurements of the container 239
Fig. 12.17 Example of poor unitisation 239
Fig. 12.18 Example of acceptable unitisation 240
Fig. 12.19 Diagonal shifting of the "lid" of a unit load 240
Fig. 12.20 Pallet and lid with "lozenge" can guarantee a tight fit 241
Fig. 12.21 Side view of the strapped pallet 241
Fig. 12.22 Example of very poorly stowed and strapped loads 242
Fig. 12.23 Example of crates fit for writing off due to insufficient diagonal bracing 243
Fig. 12.24 Example of a unit load that is not fit for transportation 244
Fig. 12.25 In accordance with the IMDG Code (Load unit that contravenes the regulations) 244
Fig. 12.26 Example of poor unitisation hindering the correct packing of containers 245
Fig. 12.27 This unit load does not comply with the IMDG Code 246
Fig. 12.28 No firm bond to the pallet; pallet not loaded flush, unit cannot be stacked 246
Fig. 12.29 Example of a load unit that does not comply with the regulations 247
Fig. 12.30 Example of where the projection of the load beyond the edges of the pallet may present a potential source of damage 248
Fig. 12.31 Example of where the projection of the load beyond the edges of the may present a potential source of damage 248
Fig. 12.32 Example of where the projection of the load beyond the edges of the pallet may present a potential source of damage 249
Fig. 12.33 Example of faulty formation of unit loads in violation of the loading guidelines 250
Fig. 12.34 Example of incomplete strapping 250
Fig. 12.35 Example of a unit load that does not comply with the IMDG code 251
Fig. 12.36 Example of a unit load that does not comply with the IMDG code 252
Fig. 12.37 Example of poorly constructed case 253
Fig. 12.38 Example of a poorly constructed case designed to accommodate two disassembled tractors for maritime transport 254
Fig. 13.1 Example of a marking that conforms to the current standards 257

Fig. 13.2 Example of non-compliant and incorrect markings and design ... 258
Fig. 13.3 Example of markings that do not comply with the current standards ... 259
Fig. 13.4 Photograph of a tarpaulin-covered case with the umbrella symbol ... 260
Fig. 13.5 Graphic showing the marking of the centre of gravity on three sides of a load, in accordance with the current DIN standard ... 262
Fig. 13.6 Example of a package showing the outdated "top-heavy" symbol ... 263
Fig. 13.7 Graphic of case with a centre of gravity in the middle of the load ... 265
Fig. 13.8 Graphic of a case where the centre of gravity is not central ... 265
Fig. 13.9 Graphic of a case whose centre of gravity is not central and with asymmetric slinging instructions ... 265
Fig. 13.10 Example of signage showing additional slinging information using special markings ... 266
Fig. 13.11 Example of a pallet box and the appropriate handling symbols ... 267
Fig. 13.12 Example of the use of incorrect handling symbols ... 268
Fig. 13.13 The recommended use of the symbol should be clearer than any proprietary labels ... 272
Fig. 13.14 Cargo is at risk if the handling symbols are missing ... 273
Fig. 14.1 Example safety notice inside a typical shipping container ... 280
Fig. 14.2 Example of a typical CSC plate ... 285
Fig. 14.3 Examples of dangerously inadequate support of a trailer chassis ... 287
Fig. 14.4 Example of a correctly supported chassis ... 287
Fig. 14.5 Photograph of a swap-body standing on its support legs ... 288
Fig. 14.6 Examples of not packed securely versus packed securely containers ... 289
Fig. 14.7 Example of the typical information displayed on 20′ and 40′ containers ... 290
Fig. 14.8 Photographic representation of a flat rack with overhang on both sides in the cell guides of a ship ... 293
Fig. 14.9 Photographic representation of a flat rack with overhang on both sides in the cell guides of a ship ... 294
Fig. 14.10 Photograph of an over height flat rack ... 294
Fig. 14.11 Example of the prescribed centre of gravity for loaded containers ... 294
Fig. 14.12 Ideal load distribution in a container with a payload of 18,000 kg * *not to scale* ... 295
Fig. 14.13 Graphic of a blank form for a packing plan ... 296
Fig. 14.14 Graphic of a blank form for a packing plan with dm^2 grid ... 296

Fig. 14.15 Graphic of a blank form for a packing plan with dm^2 or cm^2 grid ... 296
Fig. 15.1 Photograph of interlayer dunnage on a load of pails ... 307
Fig. 15.2 Example of cases with class 1.4 G air bags, stowed for ready access ... 307
Fig. 15.3 Example of correct stowage suitable for handling with forklift trucks in a box container ... 308
Fig. 15.4 Example of horizontal and vertical tiers ... 309
Fig. 15.5 The lower a package lies in the gap, the greater the opening angle and therefore the greater the forces transmitted ... 312
Fig. 15.6 Load forces on other cargo and components by the means of transport caused by the opening angle ... 313
Fig. 15.7 Principle of the locking coil ... 313
Fig. 15.8 Maximum protrusion for goods stowed with the bilge and cant-line method ... 314
Fig. 15.9 Photograph of nested I beams ... 316
Fig. 16.1 Avoid concealing CAP products ... 322
Fig. 16.2 Stackreacher ... 324
Fig. 16.3 Incorrectly loaded container with dangerous goods ... 330
Fig. 17.1 The sliding loading method—side view ... 334
Fig. 17.2 The sliding loading method—plan view ... 334
Fig. 17.3 Compact tight-fit securing of a power substation using specially shaped stanchions crosswise on a flat rack ... 337
Fig. 17.4 Example of stanchion pockets on a specially designed flat rack ... 338
Fig. 17.5 Graphic of a power substation secured using removable and adjustable stanchions ... 338
Fig. 17.6 Graphic of power substation secured using removable and adjustable stanchions—top view ... 338
Fig. 17.7 Example of incorrect securing by filling ... 339
Fig. 17.8 Example of correct securing by filling **a** with pallets and **b** air bags ... 340
Fig. 17.9 Example of how to fill the space in the vicinity of the door ... 342
Fig. 17.10 (l) Example of horizontally bracing, poorly placed in part; (r) well placed horizontal bracing elements ... 342
Fig. 17.11 Bracing an over height case on a 40′ flat rack ... 343
Fig. 17.12 Horizontal lashings with longitudinal components—view from the top ... 344
Fig. 17.13 Horizontal lashings with longitudinal components—view from the side ... 344
Fig. 17.14 (l) Horizontal lashing with transverse and vertical components—lateral view and (r) vertical lashing without horizontal components—lateral view ... 344

Fig. 17.15 Vertical lashings with slight horizontal transverse components—lateral view ... 345
Fig. 17.16 Diagonal lashings with horizontal longitudinal and transverse components and vertical components—view from the top ... 345
Fig. 17.17 Diagonal lashings with horizontal longitudinal and transverse components and vertical components—view from the side ... 345
Fig. 17.18 Cross lashings with horizontal and vertical components—lateral view ... 346
Fig. 17.19 Diagonal lashings with small horizontal transverse components ... 346
Fig. 17.20 Cross lashings with better horizontal transverse components ... 347
Fig. 17.21 (l) Ineffective components for securing against tipping; (r) effective components for securing against tipping ... 347
Fig. 17.22 (l) Optimum lashing angle to protect against tipping, but impossible to implement; (r) the real lashing angle is close to the optimum to protect against tipping ... 348
Fig. 17.23 (l) Round turn lashing; (r) round turn lashings are similar to the diabolo toy principle ... 348
Fig. 17.24 Loop lashing used to secure a piece of machinery—lateral view ... 349
Fig. 17.25 Loop lashing used to secure a piece of machinery—view from the top ... 349
Fig. 17.26 Loop lashing used to secure a piece of machinery—side view ... 349
Fig. 17.27 Examples of different loop lashing techniques ... 350
Fig. 17.28 (l) The lashings have been applied—the tipping process begins; (r) an angle less than 30° has been reached—part of the friction forces are still effective ... 350
Fig. 17.29 (l) The round turn lashing fails, the loop lashing remains firm; (r) impact: the round turn lashing fails even more, the loop lashing still remains firm! ... 350
Fig. 17.30 Head loops with spring lashings in the form of a cross lashing ... 351
Fig. 17.31 (l) Figure of eight head loop; (r) head loop with padding ... 352
Fig. 17.32 Tight-fit securing of rolls of paper using a WisaFix lashing cover ... 352
Fig. 17.33 Friction loop made of steel strapping ... 353
Fig. 17.34 (l) and (r) frictional forces act in all directions ... 353
Fig. 17.35 Determining the vertical component of the pre-tensioning force of a lashing ... 355
Fig. 17.36 Potential movement of the load caused by too little securing force applied to a wide load ... 356

Fig. 17.37 (l) Poor strapping; (r) still poor strapping, but slightly improved ... 357
Fig. 17.38 Examples of gravitational forces ... 358
Fig. 18.1 (l) Steel wire rope; (m) lashing point; (r) steel strap ... 362
Fig. 18.2 Inspecting lashing equipment ... 362
Fig. 18.3 Example of tie-down lashing ... 366
Fig. 18.4 Example of direct lashing with chains ... 367
Fig. 18.5 Package secured with four diagonal direct lashings ... 369
Fig. 18.6 Lateral components twice as large as longitudinal components ... 369
Fig. 18.7 Transverse and longitudinal components same size ... 370
Fig. 18.8 Longitudinal components 25% larger than lateral components ... 370
Fig. 18.9 Angle which is too large to be permitted ... 371
Fig. 18.10 As previously, the angle which is too large to be permitted ... 371
Fig. 18.11 Angle used for tie-down lashings in practice ... 372
Fig. 18.12 As previously, angle used for tie-down lashings in practice ... 372
Fig. 18.13 (l) untrimmed wood log; (r) trimmed wood log in direction of the grain ... 373
Fig. 18.14 (l–r) trunk, balks, squared lumber, battens, laths, planks, boards ... 373
Fig. 18.15 Squared lumber, battens and wedges—wood which has been stained indicates that it has been treated to protect it against the sirex wasp ... 374
Fig. 18.16 Applying the rule of thumb to compressive strength parallel to the grain of the wood ... 374
Fig. 18.17 Applying the rule of thumb to compressive strength parallel to the grain ... 375
Fig. 18.18 High outlay—low strength ... 375
Fig. 18.19 Little material—high strength ... 375
Fig. 18.20 Even less material—high strength with four horizontal bracing elements used in pairs ... 376
Fig. 18.21 It may be that one pair of bracing elements may be sufficient ... 376
Fig. 18.22 Heavy drive shaft on trestles ... 376
Fig. 18.23 Securing with diagonal stays ... 377
Fig. 18.24 Always take note of the smallest contact surfaces! ... 377
Fig. 18.25 Example of filling material ... 379
Fig. 18.26 Using car tyres to fill void space ... 379
Fig. 18.27 Examples of various types of air bags ... 380
Fig. 18.28 Gang nails—(l) without and (r) with nail holes ... 380
Fig. 18.29 Different shapes and sizes of shear transfer plates ... 381
Fig. 18.30 Example of chocks ... 382
Fig. 18.31 Structure of softwood on the grain edge ... 388
Fig. 18.32 Material with a μ value of approximately 0.8 ... 389

Fig. 18.33 Material with a μ value of approximately 0.7 389
Fig. 18.34 Material with a μ value of approximately 0.6 389
Fig. 18.35 Material with a μ value of approximately 0.5 389
Fig. 19.1 Thin ropes for securing tarpaulins 396
Fig. 19.2 Vehicle secured with Hercules lashing 396
Fig. 19.3 Different rope structures: cross-sections and plan views 397
Fig. 19.4 Different rope structures: cross-sections and plan views 398
Fig. 19.5 Different rope structures: cross-sections and plan views 399
Fig. 19.6 Cutting wire ropes with shears: (l) incorrect—small lever arm; (r) correct—large lever arm 400
Fig. 19.7 Example of a typical wire cable clamp 401
Fig. 19.8 Single cable with eye 402
Fig. 19.9 Sling or double run—unsuitable construction 402
Fig. 19.10 Sling or double run—ideal construction 402
Fig. 19.11 Example of lashing which is not homogeneous 402
Fig. 19.12 Anchor shackles 403
Fig. 19.13 Lashing turnbuckle 404
Fig. 19.14 Turnbuckle made of higher strength steel for round steel chains with short links 405
Fig. 19.15 Loads ideally suited for securing with steel straps 407
Fig. 19.16 Lashing chain with long links 407
Fig. 19.17 Lashing chain with short links 407
Fig. 19.18 Inserting the hook 408
Fig. 19.19 Shortening the chain 408
Fig. 19.20 Inserting the tensioning lever 408
Fig. 19.21 Tightening and securing the lever (l) the incorrect way (r) the correct way 409
Fig. 19.22 Permissible shortening of chains 409
Fig. 19.23 Correct shortening of a chain—no chain link is trapped 409
Fig. 19.24 Incorrect shortening of a chain—a chain link is trapped 409
Fig. 19.25 Textile lashing belt 410
Fig. 19.26 Various automobile lashings 411
Fig. 19.27 Various locking devices for use with single-use textile lashing belts 412
Fig. 19.28 Different types of packing straps 413
Fig. 19.29 Different types of end fittings 413
Fig. 19.30 Belt protectors 413
Fig. 19.31 Belt tensioner 414
Fig. 19.32 Knotted textile belts 414

List of Tables

Table 1.1 Largest container ships built between 2017 and 2022 17
Table 1.2 Ships on order (as of August 2023) 21
Table 1.3 Container loading records (as of August 2023) 22
Table 4.1 Standard topics to be included in training programmes for personnel responsible for the packing and securing of cargoes in containers 75
Table 4.2 Example calculation of permitted stack pressure in containers 82
Table 4.3 Crush pressures for different pallets 84
Table 4.4 Drum contact area versus force [N] versus area [m^2] versus pressure [kPa] 85
Table 5.1 Summary of ship motions 93
Table 5.2 Extract from the container packing guidelines 95
Table 5.3 Examples of accelerations that may arise during road transport 101
Table 5.4 Examples of accelerations that may arise during rail transport 105
Table 7.1 Container dimensions as stipulated in DIN/ISO 668 or DIN 15,190, Part 1 128
Table 7.2 Container dimensions as stipulated in DIN/ISO 668 or DIN 15,190, Part 1 129
Table 7.3 Container widths 129
Table 7.4 Non-standard container heights 130
Table 7.5 Suitable and unacceptable container sizes for international use 130
Table 7.6 Container sizes used in the US 130
Table 7.7 Maximum mass for a 20′ container according to the previous standard 130
Table 7.8 Current internal container dimensions 131
Table 7.9 ISO standard maximum gross weights for inshore containers 132

Table 7.10 ISO standard maximum gross weights for inshore containers 132
Table 7.11 Minimum dimensions for loading areas and volumes for standard ISO containers 132
Table 10.1 Container identification including country, size and type codes 194
Table 10.2 Container identification including country, size and type codes 199
Table 10.3 Container identification including country, size and type codes 199
Table 10.4 Container code G 200
Table 10.5 Container code V 200
Table 10.6 Container code S 201
Table 10.7 Container code R 201
Table 10.8 Container code H 202
Table 10.9 Container code U 202
Table 10.10 Container code P 203
Table 10.11 Container code T 203
Table 10.12 Container code A 204
Table 15.1 Misleading versus clear formulation 310
Table 17.1 Influence of the lashing angle on the vertical component 355
Table 17.2 Securing forces depending on pre-tensioning force, lashing angle and friction 355
Table 18.1 Determining the maximum securing load from the breaking load 363
Table 18.2 Common securing materials and accepted breaking loads 364
Table 18.3 Calculation for the longitudinal, transverse and vertical securing forces using the lashing angles α and β 367
Table 18.4 Additional frictional forces produced by the vertical component 368
Table 18.5 Additional frictional forces produced by securing forces 368
Table 18.6 Additional frictional forces produced by securing forces 369
Table 18.7 Guideline values for the MSL at different estimated angles 372
Table 18.8 Sliding friction coefficients for selected materials 390
Table 18.9 Sliding friction coefficients for selected materials 390
Table 18.10 Sliding friction coefficients for selected materials 390
Table 18.11 Sliding friction coefficients for selected materials 391
Table 18.12 Sliding friction coefficients in accordance with the VDI Guidelines 392
Table 19.1 Calculating the diameter of a polypropylene strap—5 mm 396
Table 19.2 Determining the diameter of polypropylene-covered Hercules—12 mm 397
Table 19.3 Determining the diameter of a steel wire rope—16 mm 398

Table 19.4 Determining the diameter of a steel wire rope—16 mm 399
Table 19.5 Calculating the breaking load and MSL for steel strapping 406
Table 19.6 Calculating the breaking load and MSL for blue strapping ... 406
Table 19.7 Diameters of chain steel 410

Part I
Development of Containerisation

Chapter 1
The Development of Containerisation

In May 2001, the man often considered the "Father of Containerisation", Malcolm P. McLean, passed away aged eighty-seven. McLean used to say that he had the idea of rationalising goods transport by avoiding the constant loading and unloading from one means of transport to another way back at the end of the 1930s at the Port of Hoboken, New Jersey, when still operating as a small-scale haulier. To start with, McLean would load complete trucks onto ships, to transport them as close as possible to their destination. The development of standardised containers and trailers, moved by tractors, made it possible to ship the trailers with only the containers, and in doing so, saving money on both space and costs. Later, the trailers were also left behind with the ships just transporting the containers themselves. Initially, shipowners were more than a little sceptical about McLean's novel idea. This prompted McLean to become a shipowner himself and he appropriately named his newfound company Sea-Land Inc. At the end of the 1990s, McLean sold his company to the Danish shipping conglomerate, Mærsk A/S, however his contribution to the development of containerisation lives on in the name Sealand, now a proud member of the Mærsk groups of companies.

The first ship McLean used to provide his novel idea was the former WWII era T2 tanker, *SS Ideal X* (Fig. 1.1). The *SS Ideal X* had the external pipework removed from the weather deck to allow the placement of large metal boxes, which she did for the first time on 26 April 1956, carrying fifty-eight containers from the Port of Newark, New Jersey, to the Port of Houston, Texas. Despite the *SS Ideal X* having the honour of being the first seagoing vessel to ship cargo in metal boxes, the distinction of being the first vessel to carry *only* containers goes to the converted tanker *SS Maxton* (Fig. 1.2), which following conversion as a T2 class tanker, had the capacity to load as many as 60 containers. Despite the initial success of the *SS Ideal X* and *SS Maxton* to carry and deliver their cargoes safely and on time, it would be another decade before the first container ship would come alongside in Europe. The first container to settle on European soil was set down by the *SS Fairland* at Bremer Überseehafen, Germany, on 6 May 1966. The first containers used by Sealand in Northern Europe

A. A. Olsen and F. Karkori, *Containerized Cargo Handling and Stowage*, Springer Series on Naval Architecture, Marine Engineering, Shipbuilding and Shipping 22,
https://doi.org/10.1007/978-3-031-63329-4_1

Fig. 1.1 Photograph of the *SS Ideal X*

were 35′ ASA containers, which were designed to specific American standards. In other regions, 27′ ASA containers and other ASA dimensions were more often used. Quick to realise the potential, both European and Japanese shipowners recognised the advantages of the container and so invested in this new emergent technology. As applying uniquely American standards to Europe and other countries was costly and beset with technical difficulties, an agreement was eventually reached with American shipowners and hauliers to standardise the size and shape of individual shipping boxes. The result was the establishment of the ISO standards which provided for the lengths of 10-, 20-, 30- and 40-foot metal containers. The width of each container was fixed at 8 ft (2.4 m) and the height at 8 ft (2.4 m) and 8 ft 6 in (2.62 m). For inshore transport within Europe, further agreement was reached on a 2.5 m (8.2 ft) wide inshore container, which is used for primarily for integrated road and rail logistical operations (Fig. 1.3).

Today, the vast majority of containers used by the global shipping industry comply with the ISO standards for combined transport units (CTU), with the twenty-foot (20′) equivalent (TEU) and forty-foot (40′) equivalent (FEU) containers being the most popular. For some years, the ISO standard has come repeatedly under pressure as shippers have sought to increase the size and internal capacity of these standard containers. As stowage factors have increased for most goods, many freight forwarders have demanded ever longer, wider, and higher containers, preferably combined into one unit. Some shipowners have relented to this pressure and offer carriage for containers of dimensions larger than those provided for by the ISO standard. These so-called "jumbo" containers are typically forty-five and forty-eight feet in length, with widths of 8.6 ft (2.6 m) and heights of 9.6 ft (2.9 m). In the United States, even larger containers (e.g., 24 ft (7.43 m) and 49 ft (14.4 m) long, 2.6 m (8.5 ft) wide and 2.9 m (9.5 ft) high) have been adopted for inshore domestic transport. Containers as long as 53-feet have been in use for some time in the US, whilst some American states even allow for 57 ft (17.3 m) containers. In Europe, and indeed across most of the world, the primary limiting factor for determining maximum container dimensions are narrow roads and existing infrastructure. Subsequently,

Fig. 1.2 Photograph of the *SS Maxton*

many countries—and in particular developing countries—are understandably reticent to amending the existing ISO standards. Even so, the huge investments made in containerisation have paid off and container traffic is continuing to grow exponentially year on year. Although it is unlikely growth will continue to be as unbridled as in the past, it will undoubtedly continue until all conventional transport operations have, within the container's limits of capacity and weight, been containerised. By then, it is estimated there will be in excess 8,000 vessels in operation worldwide with a total slot capacity of between nine and ten million standard sized containers. Furthermore, it is estimated there will be a similar quantity of containers ashore being packed or unpacked, awaiting stuffing or unstuffing, or being transferred from one location to another. The majority of these containers will be standardised forty-foot box containers with a smaller quantity of twenty-foot containers also available. Unless the shipping industry develops a new and innovative means of shipping goods en masse, it is more than likely demand for containerised transport will remain significant long into the future.

1.1 Introduction to the Container Vessel

Anyone professionally involved in the business of using containers for maritime transport should, at a minimum, have some knowledge of the kinds of vessels which are used for this purpose. Often, when people who are not shipping professionals read press articles regarding gigantic new container ships, they often find it hard to imagine

Fig. 1.3 Photograph of the first containers to be loaded onboard the *SS Ideal X*

the kind of problems these behemoths of the ocean experience, especially when at sea. It is also very easily forgotten that containers which have been shipped for a majority of their voyage on ultra-modern vessels may have to be loaded onto and carried for the remainder of their voyage onboard less salubrious forms of transportation—be it by road, rail or water. To that end, in this section of the book, we will briefly discuss the various types of vessels which may be called upon to carry containers. In essence, different ships may be distinguished by the basis of various criteria: for example, vessels are often categorised by their type of propulsion; their region of service; their function, tonnage, or measurements; the arrangement of any internal decks or superstructures; whether they are geared or not; and so forth. For shipowners and ship operators, these are fundamental questions. For the majority of freight customers, however, it is very likely the type of propulsion will not be a major concern to them as all modern cargo ships are powered by marine diesel engines. Even so, it is worth briefly explaining the various prefixes that are used in ship naming conventions, as these are often annotated on official shipping documents. To begin with, the vast majority of container vessel names are prefixed with 'MV,' the abbreviation for *Motor Vessel*, or with 'MS,' which is shorthand for *Motor Ship*. In this instance, the words 'vessel' and 'ship' are interchangeable. Some shipowners, typically those who operate large and multi-type fleets, may decide to prefix their container vessel names with an additional 'C' to denote the vessel is a pure container ship.[1] This is particularly helpful for companies such as Mærsk which operate separate fleets of container ships, tankers, car carriers, and many other ship types.

At this point the reader may be asking why this is in any way important, and quite rightly so. The justification for denoting different vessel types comes from the fact that container vessels are designed to carry weight, rather than to operate at speed.[2] This is similar in fact to oil tankers except the latter proceed at a snail's pace in comparison to the most modern container ships.[3] Most cargo ships are powered by huge diesel engines which produce more horsepower than speed. For ships where speed is of paramount importance, different forms of propulsion are frequently used such as marine gas turbines. Vessels fitted with gas turbine engines are given the prefix 'TS' for turbine ship. Very rarely turbine powered container vessels may be assigned the prefix 'CTV' although this is a rarity indeed. Now that we have covered the basic conventions for naming container ships according to their method of propulsion, it is worth discussing the various other categories of merchant vessels which may be called into container service. As we now know, most container ship names are prefixed with MV or MS; this is a relatively steady state for ships involved in liner services. It should be noted however that in many cases containers may be delivered to their primary port of call but are then transported to their end destination using other modes of transportation, including seaborne platforms such as sailing ships and flat top barges.[4] Furthermore, some distinction may be made according to the vessel's area of operation; this is common in North America and Canada where the ships that operate on the Great Lakes and the St. Lawrence Seaway are often referred to as 'lakers.' Moreover, 'coasters,' which are smaller agile vessels, frequent smaller ports along multinational coastlines such as around the Baltic and North Seas. Last of all, a rather peculiar design of vessel may be employed on inshore container routes. These river boats or flat top barges are commonly found on the European waterways such as the rivers Rhine, Seine, and Danube (refer to Fig. 1.4).

For shipping professionals, the prefix alone is often sufficient to provide an indication of the nature and appearance of the ship and how it will accommodate the cargo to be carried on board. A ship's function is also generally reflected in its classification, for example, underwater vehicle, fishing vessel, tugboat, buoy-laying vessel, warship, and so forth. That said, differentiation by tonnage, measurement, arrangement of superstructures, etc. is usually only of interest to the maritime specialist.

[1] Thereafter, the prefix CMV or container motor vessel may be used.

[2] In the sense that cruise ships, for instance, are designed to carry people over long distances and are significantly lighter and more aerodynamic in comparison to most cargo vessels.

[3] Container ships often proceed at a speed over ground of around 20 kn. (37 km/h; 22.9 mph), whereas most oil tankers and ore carriers proceed at a speed over ground of around 12–15 kn. (22–28 km/h; 13.6–17.3 mph).

[4] The latter being quite common throughout the Philippines, Malaysian, and Indonesian archipelagos.

Fig. 1.4 Photograph of the river barge *Aladin*, navigating the Seine, France, undated

Laypeople will be unconcerned with whether goods are being carried on a full scantling, open-deck, shelter-deck or flush-deck vessel. Again, it is worth noting that gross register or net register tonnages are not directly indicative of the size of a vessel, although gross and net tonnage are slightly more useful. To clarify matters further, it is more common than not to refer to a ship's deadweight (dwt/DWT) as a reference to the value of the ship's payload in metric tonnes. Alternatively, the ship's displacement or displacement tonnage is the calculation of its total (gross) weight (i.e., the sum of the weight of the empty, operational ship[5] plus the maximum permitted payload).

Now that we have covered these basic principles, we can begin to look at some of the essential distinguishing features of those merchant ships which carry containerised cargoes.[6] Before we do, however, it is worth differentiating the main types of cargo carrying vessels in accordance with the method of cargo handling. This is useful as some vessels employ more than one means of loading and discharging cargo. There are seven categories of vessels which are of interest to the container handling professional: lift-on/lift-off (LoLo); roll-on/roll-off (RoRo); stow and roll

[5] I.e., the ship's hull and superstructure, propulsion mechanisms, crew, food, fuel and ballast.

[6] It is worth noting that often media articles refer to the "type of freight" in this relation, however this is incorrect. Technically, 'freight' is the term used to describe the *amount of money* the shipowner or carrier receives in compensation for transporting the cargo, rather than describing the nature of the cargo itself.

(StoRo); float-on/float-off (FloFlo); truck-to-truck (T2T); lift and roll (LoRo); and walk-on/walk-off (WoWo). *Lift-on/Lift-off (LoLo).* Vessels which employ a lift-on/lift-off (LoLo) handling method lift the cargo into and out of the ship's hull using lifting gear. These might include onboard derricks, cranes or gantries, or quayside derricks, cranes, or gantries. This is the most common method for loading and discharging containers and is used widely around the world. *Roll-on/roll-off (RoRo) vessels* (sometimes referred to as container RoRo or ConRo) are specially designed ships with a ramp fitted to the stern. Rolling cargo is stowed inside the hull of the vessel, whilst containers are stowed on the weather deck. By having a stern fitted ramp, rolling cargo can be driven directly on and off the vessel either under its own power or under tow by a terminal tractor unit. In the case of the latter, once the cargo is placed onboard the vessel, the terminal tractor unit is disconnected and returned ashore for the next load. When the vessel arrives at the port of discharge, terminal tractor units drive onto the vessel, are coupled to the load, and drive off the vessel with the load attached. The process repeats until all cargo destined for that port is discharged. Throughout the 1980 and 1990s RoRo ships gained mass appeal with shipowners and operators due to their flexibility and cargo carrying capacities. As the 1990s turned into the 2000s, pure container vessels started taking over from RoRo ships as they increased in size and efficiency. As of 2023, the number of RoRo ships remaining in service is significantly lower today than at the beginning of the 1980s when they first emerged.

With *Stow and roll (StoRo)* vessels, the cargo is rolled on or off the ship using one of several methods, including Mafi trailers or forklift trucks. It is uncommon for StoRo vessels to carry containerised cargoes as this is not the most efficient means for loading and unloading containers. More commonly, StoRo vessels are used in the shipment of forestry products such as timber and paper. As the name suggests, *float-on/float-off (FloFlo)* vessels operate by having goods loaded onto floating cargo carriers which are then floated in and out of dock-like holds in the hull of the ship. Alternatively, the carrier vessel semi submerges, moves under the cargo, re-floats, and lifts the cargo into the predetermined stowage space. The cargo is then unloaded by reversing the process. This method may be used for container transport where the floating cargo carrier, such as barges, have already been preloaded with containers. Again, this is not a commonly used method in international container transport as it is time consuming and inherently inefficient. Truck-to-Truck (T2T). With respect to the truck-to-truck method, the cargo is set down using ground conveyors, such as forklift trucks, then loaded onto shipboard lifts. These lifts are raised or lowered to the appropriate loading level. The load is then placed using ground conveyors. Discharging proceeds in the reverse order. Again, this method is not suitable for container traffic and is used in the transhipment of ores, grains, and other loose cargo forms. *Lift and Roll (LoRo)* vessels. With the lift-and-roll method, the cargo is lifted onto the vessel using the onboard loading gear or winch platform, then rolled into place. This method is ideal with a special type of barge carrier and may be suitable for container traffic where the containers have previously been loaded onto or into the barge. *Walk-on/Walk-off* (WoWo) vessels may sound comical but are in fact the most common handling method used in the transportation of caged or boxed livestock. In this instance, the animals are either herded onto or off the vessel using special guide tracks or are held inside livestock boxes. These boxes may then be driven or lifted onto or off the vessel.

With many ships, there is often some overlap between the assorted options for carrying goods, which means that individual vessels cannot be easily assigned to one category or another, as the list might suggest. There have always been many diverse kinds of vessel, but in recent years, new vessel designs have seen the introduction of many more. Since the 1990s there has been an unmistakable trend towards the development of multipurpose ships. This makes it difficult to make general statements about what ships look like and how they are equipped, however, ships always exhibit individual differences. Even sister and standard type vessels are not identical in every respect. In terms of naval architecture, there is usually minimal difficulty finding suitable vessels for cargo transport operations. Organisationally, however, it is significantly harder to find the right ship for the right job. It may be that certain vessels only serve specific ports or that certain cargoes can only be carried in ships of a particular nationality.[7] One major global issue facing the modern merchant navy is the increasing lack of skilled crews onboard ships, and of shore-based stevedores. This seems like a problem which is unlikely to be resolved anytime soon. In fact, the issue is compounded by a perfect storm of increasing technological advancement in some parts of the world, slower adoption of refined techniques and processes in other parts of the world, combined with a general disinterest in seaborne and logistical careers. For our purposes, we need not concern ourselves too much with these problems, for as any ships officer knows, non-seafarers frequently underestimate the hazards of maritime transport and so for the most part inadequately pack and poorly secure cargoes in their containers. As we will see throughout this book, this can have devastating effects on both the cargo, the container and the ship herself.

1.2 Container Ship Architecture and Design

In this section, we will observe a select few from the enormous range of different ships which may be of interest to container logisticians, as well as cargo importers and exporters. General cargo ships are only included to the extent in many sea trading areas they also carry containers in addition to their normal cargoes. For the sake of brevity, the many diverse kinds of vessels will be described only briefly. At the beginning of container traffic, most containers were carried on conventional general cargo vessels. Such vessels were designed to carry dry cargoes. Each hold comprised one or more low 'tween decks' and a lower hold. The ships had a large quantity of light cargo handling gear. Every ship was equipped with appropriate loading gear for heavy-lift cargoes in the central hatch area. Lockers were provided for carrying particularly valuable cargoes. Some ships also had small capacity sweet oil tanks. General cargo ships of this old design are no longer built, but many are still operational

[7] For example, due to cabotage laws.

around the world. The sea keeping ability of most such vessels at sea, low stowage heights, and the like, mean that carriage is safe and loss-free if individual containers are in favourable stowage spaces and are carefully secured (Fig. 1.5).

Modern general cargo ships are built to perform different transport functions; to distinguish them from other multipurpose ships, they are sometimes known as break-bulk ships. In addition to the equipment which makes them suitable for carrying break-bulk cargoes, they also have facilities which allows them to accommodate containers. Such ships are of an open construction, i.e., the hatch area is exceptionally large relative to the deck area. This ensures that the lifting gear can gain direct access to the containers inside the ship's hold(s). Specific construction features may increase the handling efficiency of these ships when in port and help reduce loss ratios. Semi-container vessels are suitable for carrying both general cargoes and containers. In this instance, the hold dimensions, deck loading values, the load-carrying capacity of the loading gear, etc. are tailored to the carriage of standard shipping containers. Such vessels have 'tween decks with flush-closing mechanical hatch covers. All-container ships are designed to carry only containers and are specially equipped for this purpose. If the ports of the region of service are equipped with sufficiently powerful lifting gear, container ships are operated without the use of loading gear. In other regions of service where port facilities are lacking, container ships need loading gear in the form of derricks, cranes, or gantries. All-container ships are designed with the principle of open construction as it must be possible to gain direct access to each

Fig. 1.5 Photograph of the *Grimaldi America*, departing Falmouth, England, 2014

Fig. 1.6 Photograph of a top spreader with an ISO container and tank container

container with top spreaders as efficiently as possible. To obtain smooth, square holds, these vessels are often constructed with a double hull. Any holds which are unsuitable for carrying containers are often fitted out as tanks. Because there is no need for them, all-container ships are fitted with 'tween decks' (Fig. 1.6).

Particular attention must be paid to the hydrodynamic design of container ships which operate at high cruising speeds. The tall, heavy deck loads can cause problems with righting capacity. To ensure adequate stability, most all-container ships must carry special solid or liquid ballast and/or be designed broader amidships. The capsize risk of these vessels may be kept within acceptable limits by high values of the roll moment of inertia. Large ballast capacities and high-power pumps are essential, both for trimming and for offsetting longitudinal bending moments. Shipbuilders can tailor characteristics by selecting appropriate ratios between the ship's length, beam, moulded depth, draft, and various other dimensions. The deadweight and hold capacity of container ships may be stated in metric tonnes and cubic metres (Fig. 1.7).

For our purposes, the number of available slots for 20′ or 40′ containers, however, is more meaningful. TEU is the standard industry abbreviation for "Twenty-foot Equivalent Unit." Alternatively, "Forty-foot Equivalent Unit" is abbreviated to FEU. All-container ships are divided into generations depending on their container carrying capacity. Roughly speaking, the generations can be divided as follows:

1st Generation	Up to	1,000	TEU	
2nd Generation	Up to	2,000	TEU	

(continued)

Fig. 1.7 Photograph of the *Mairangi Bay*, Rotterdam, Holland, 1983

(continued)

3rd Generation	Up to	3,000	TEU	
4th Generation	More than	4,000	TEU	Panamax
5th Generation	More than	6,000	TEU	Post-Panamax
6th Generation	More than	8,000	TEU	

Figure 1.8 shows a second-generation container ship. It is a rendition of the *Bremer Vulkan* standard vessel which was built in diverse sizes over several generations: BV 1000, BV 1600, BV 1800, BV 1800 S, BV 1900, and BV 2200, 2200 OH and BV 3800. The variant shown does not have onboard lifting gear fitted. Figure 1.9 shows the *Bremen Express*, a third-generation all-container vessel, which has a total container capacity of 2,950 TEU.

One of the first fourth generation vessels, which have slots for more than 4,000 TEU, was put into service by the US shipping company, American President Lines. This was soon followed by Hapag Lloyd's *Hannover Express* class of container ships, which were put into service in 1992. These class of container ships were unique in that their size was the maximum permissible to enter the Panama Canal at that time. For this reason, they earned the moniker *Panamax* class vessels. Any vessels built larger than this were called *Post-Panamax*, as they were too long or too wide to safely enter the Canal's lock system. Fortunately, the Panama Canal (and Suez Canal) have undergone extensive enlargements to enable even the largest oceangoing vessels to pass through with relative ease.[8] Because of the unique design

[8] Notwithstanding the odd incident, such as that which befell the *Ever Given* in early 2021.

Fig. 1.8 Photograph of the *Bremen Express*, Hamburg, Germany, 2008

Fig. 1.9 Photograph of the *Emma Mærsk*, Aarhus, Denmark 2019

of the container ship, approximately 48.25% of containers can be carried on deck with the remaining 51.75% carried below deck. For the average sized sixth generation container ship, this means it is possible to carry up to eleven rows of containers below deck and up to thirteen rows on deck, with heights of up to ten tiers below deck and five tiers above. The fore body (i.e., forwards of the superstructure) has twenty-nine bays available, whilst the after body (behind the superstructure) can have as many as six. The versatility of the ship is enhanced by the possibility of lowering "flaps" into the guide system so that containers can be stopped at a certain height. This enables ordinary general cargo to be stowed in the holds. The German carrier Hapag Lloyd was the first shipowner to install this system on the *Humboldt Express.* In order, as required by regulations, to be able to transport hazardous cargo containers of certain classes below deck, five holds were specially fitted out for this purpose. This improved segregation also made it possible to increase the carrying capacity for dangerous goods in general relative to other ship types.

1.2.1 All-Container Ships

Suffice it to say, today there is a wide variety of different container ships in service. The most common are sixth generation vessels such as the ten sister ships built by Japan in 1998 for the United Arab Shipping Company (UASC). Each ship has a slot capacity of 3,802 TEU, 2,068 TEU of which, or 54.4% of the containers, can be carried below deck in the hold. In 1998, the first container ships with slot capacities of more than 6,000 TEU came into service. These included the *Karen Mærsk, Regina Mærsk,* and the *Sovereign Mærsk,* which had a stated capacity of 6,600 TEU. With a length of 347 m (1,138 ft), a beam of 42.8 m (140 ft) and a design draught of 14.50 m (47.5 ft), although industry estimates the true capacity was closer to 8,000 TEU in total though the existing IMO regulations at the time prohibited this number of TEU to be loaded. In November 2001 Hapag Lloyd took delivery of its largest container ship to that date, the *Hamburg Express.* With a length of 320 m (1,049 ft) and a moulded beam of 42.8 m (140 ft), as many as seventeen containers could be stowed side by side across a single bay. The vessel's moulded depth was 24.5 m (80.3 ft), meaning containers could be stacked nine high in the hold. When fully laden, the ship had a draught of 14.5 m (47.5 ft). The rated capacity of the *Hamburg Express* was 7,500 TEU or 100,000 metric tonnes, a distinct increase over earlier vessel capabilities. The *Hamburg Express*' propulsion was provided by a 68,640 kW marine diesel engine, corresponding to approximately 93,000 hp, capable of propelling the giant container vessel at a cruising speed of more than 25 kn. In 2003 three further vessels of the *Hamburg Express* class were delivered to Hapag Lloyd. However, by 2006, the *Hamburg Express* had lost the distinction of being the largest container ship afloat with the launch of the mighty *Emma Mærsk.* The first in a class of eight vessels to be built for the A. P. Moller-Mærsk Group, the *Emma Mærsk* has an official carrying capacity of 11,000 TEU (or 14,770 depending on definition). In May 2010 her sister ship, the *Ebba Mærsk,* set a record high with 15,011 TEUs loaded at the Port of Tanger-Med, Morocco. Even this record was not held for long with newer, longer, and wider vessels entering service in 2017. The launch of the *MOL Triumph,* for the Japanese shipping company MOL, commenced a new race in the construction of ever larger container ships. As of August 2023, the largest container ship in operation is the aptly named *Ever Alot*, which boasts a whopping gross tonnage of 236,228 tonnes, a length of 399.9 m (1,312 ft), a beam of 61.5 m (202 ft) and a maximum carrying capacity of 24,004 TEU. Table 1.1 provides a record of the largest container ships to be built from 2017 to 2022, with further vessels planned in the years 2023–2025 (Fig. 1.10).

It remains to be seen whether such vessels can be operated economically. It is an indisputable fact that transport costs per container or "slot costs" will drop in line with improved "economies of scale." However, it should not be forgotten that operating risks increase hand in hand with vessel size. Larger ships will have longer container handling times. To cut operating costs, the larger vessels will have to lengthen their sailing intervals, i.e., they will call into fewer ports when compared to smaller vessels. As a result, more containers will need to be transported in pre-carriage and onward carriage operations to and from the ports which they serve. Many other factors will also have to be considered, such as the rise in automation and impact this will have on crew training and career prospects, ship safety, and shipping economics, but these are issues outside the scope of this book. That said, what does concern us is as container ships continue to expand their capacity, onshore facilities will also need to invest in new infrastructure, quay lengthening programmes, shipping channel dredging, and so forth, to ensure these larger-than-life vessels can be adequately accommodated. It may also be necessary for entire new port facilities to be developed, as has been seen with the Port of Felixstowe in England, the UK's first purpose-built container port, and the DP World London Gateway, which opened in 2013 (Tables 1.2 and 1.3).

1.2.2 Development of the 'Hatch-Less' Container Ship

An interesting development in container ship designed occurred in the early 1990s with the introduction of the hatch-less container ship. These vessels were intended to make cargo handling more efficient. Between 1993 and 1994, the Howaldstwerke-Deutsche Werft AG (HDW) in Germany delivered four hatch-less container ships to the Chilean shipping company Norasia Line Ltd. These 2,780 TEU vessels were equipped with an innovative cargo protection system and marked a further development in what was branded the "ship of the future". Except for holds one and two, which were equipped with pontoon hatch covers to allow for the carriage of hazardous materials, these vessels were designed with no hatch covers. However, as the ships were pressed into service, it became increasingly obvious that the lack of hatches posed many a problem, namely around stability, and especially so when operating in regions subject to heavy rainfall. To counterbalance this design flaw, the company decided to equip holds three to seven with 12 rain protection covers fabricated from lightweight steel. Each rain cover rested on the transverse coaming of the container cell guides and was secured by way of rapid fastenings. The coamings extended in each case up to the highest container which, in some holds, amounted to twelve tiers. Before and after cargo handling, the roofs were removed and replaced

Table 1.1 Largest container ships built between 2017 and 2022

#	Built	Vessel name	Length overall		Beam		Max TEU	GT	Operator	Flag
			m	ft	m	ft				
1	2022	Ever A lot	399.9	1,312	61.5	202.0	24,004	236,228	Panama	Panama
2	2021	Ever Ace	399.9	1,312	61.5	202.0	23,992	236,228	Evergreen	Evergreen
2	2021	Ever Act	399.9	1,312	61.5	202.0	23,992	236,228	Panama	Panama
2	2021	Ever Alp	399.9	1,312	61.5	202.0	23,992	236,228	Evergreen	Evergreen
2	20211	Ever Arm	399.9	1,312	61.5	202.0	23,992	236,228	Panama	Panama
2	2022	Ever Art	399.9	1,312	61.5	202.0	23,992	236,228	Evergreen	Evergreen
3	2020	HMM Algeciras	399.9	1,312	61.0	200.1	23,964	236,228	Panama	Panama
3	2020	HMM Copenhagen	399.9	1,312	61.0	200.1	23,964	236,228	Evergreen	Evergreen
3	2020	HMM Dublin	399.9	1,312	61.0	200.1	23,964	236,228	Panama	Panama
3	2020	HMM Gdansk	399.9	1,312	61.0	200.1	23,964	236,228	Evergreen	Evergreen
3	2020	HMM Hamburg	399.9	1,312	61.0	200.1	23,964	236,228	Panama	Panama
3	2020	HMM Helsinki	399.9	1,312	61.0	200.1	23,964	236,228	Evergreen	Evergreen
3	2020	HMM Le Havre	399.9	1,312	61.0	200.1	23,964	228,283	HMM	Panama
4	2020	HMM Oslo	399.9	1,312	61.5	202.0	23,820	232,311	HMM	Panama
4	2020	HMM Rotterdam	399.9	1,312	61.5	202.0	23,820	232,311	HMM	Panama
4	2020	HMM Southampton	399.9	1,312	61.5	202.0	23,820	232,311	HMM	Panama
4	2020	HMM Stockholm	399.9	1,312	61.5	202.0	23,820	232,311	HMM	Panama
4	2020	HMM St Petersburg	399.9	1,312	61.5	202.0	23,820	232,311	HMM	Panama
5	2019	MSC Gülsün	399.9	1,312	61.5	202.0	23,756	232,618	MSC	Panama
5	2019	MSC Samar	399.9	1,312	61.5	202.0	23,756	232,618	MSC	Panama
5	2019	MSC Leni	399.9	1,312	61.5	202.0	23,756	232,618	MSC	Panama
5	2019	MSC Mia	399.9	1,312	61.5	202.0	23,756	232,618	MSC	Panama
5	2019	MSC Febe	399.9	1,312	61.5	202.0	23,756	232,618	MSC	Panama
5	2020	MSC Ambra	399.8	1,312	61.5	202.0	23,756	232,618	MSC	Panama
6	2019	MSC Mina	399.9	1,312	61.0	200.1	23,964	228,741	MSC	Panama
6	2019	MSC Isabella	399.8	1,312	61.0	200.1	23,964	228,283	MSC	Panama
6	2019	MSC Arina	399.8	1,312	61.0	200.1	23,656	228,741	MSC	Panama

(continued)

Table 1.1 (continued)

#	Built	Vessel name	Length overall		Beam		Max TEU	GT	Operator	Flag
			m	ft	m	ft				
6	2019	MSC Nela	399.8	1,312	61.0	200.1	23,656	228,741	MSC	Panama
6	2019	MSC Sixin	399.8	1,312	61.0	200.1	23,656	228,741	MSC	Panama
6	2021	MSC Apolline	399.8	1,312	61.0	200.1	23,656	228,786	MSC	Liberia
6	2021	MSC Amelia	399.8	1,312	61.0	200.1	23,656	228,786	MSC	Liberia
6	2021	MSC Diletta	399.8	1,312	61.0	200.1	23,656	228,786	MSC	Liberia
6	2021	MSC Michelle	399.8	1,312	61.0	200.1	23,656	228,786	MSC	Liberia
6	2021	MSC Allegra	399.8	1,312	61.0	200.1	23,656	228,786	MSC	Liberia
7	2020	CMA CGM Jacques Saadé	399.9	1,312	61.3	201.0	23,112	236,583	CMA CGM	France
7	2020	CMA CGM Champs Elysées	399.9	1,312	61.3	201.0	23,112	236,583	CMA CGM	France
7	2020	CMA CGM Palais Royal	399.9	1,312	61.3	201.0	23,112	236,583	CMA CGM	France
7	2020	CMA CGM Louvre	399.9	1,312	61.3	201.0	23,112	236,583	CMA CGM	France
7	2021	CMA CGM Rivoli	399.9	1,312	61.3	201.0	23,112	236,583	CMA CGM	France
7	2021	CMA CGM Monmatre	399.9	1,312	61.3	201.0	23,112	236,583	CMA CGM	France
7	2021	CMA CGM Concorde	399.9	1,312	61.3	201.0	23,112	236,583	CMA CGM	France
7	2021	CMA CGM Trocadero	399.9	1,312	61.3	201.0	23,112	236,583	CMA CGM	France
7	2021	CMA CGM Sorbonne	399.9	1,312	61.3	201.0	23,112	236,583	CMA CGM	France
8	2017	OOCL Hong Kong	399.9	1,312	58.8	193.0	21,413	210,890	OOCL	Hong Kong
8	2017	OOCL Germany	399.9	1,312	58.8	193.0	21,413	210,890	OOCL	Hong Kong
8	2017	OOCL Japan	399.9	1,312	58.8	193.0	21,413	210,890	OOCL	Hong Kong
8	2017	OOCL United Kingdom	399.9	1,312	58.8	193.0	21,413	210,890	OOCL	Hong Kong
8	2017	OOCL Scandinavia	399.9	1,312	58.8	193.0	21,413	210,890	OOCL	Hong Kong

(continued)

Table 1.1 (continued)

#	Built	Vessel name	Length overall		Beam		Max TEU	GT	Operator	Flag
			m	ft	m	ft				
8	2018	OOCL Indonesia	399.9	1,312	58.8	193.0	21,413	210,890	OOCL	Hong Kong
9	2018	COSCO Shipping Universe	400.0	1,312.3	58.6	192.0	21,237	215,553	COSCO	Hong Kong
9	2018	COSCO Shipping Nebula	400.0	1,312.3	58.6	192.0	21,237	215,553	COSCO	Hong Kong
9	2019	COSCO Shipping Galaxy	400.0	1,312.3	58.6	192.0	21,237	215,553	COSCO	Hong Kong
9	2019	COSCO Shipping Solar	400.0	1,312.3	58.6	192.0	21,237	215,553	COSCO	Hong Kong
9	2019	COSCO Shipping Star	400.0	1,312.3	58.6	192.0	21,237	215,553	COSCO	Hong Kong
9	2019	COSCO Shipping Planet	400.0	1,312.3	58.6	192.0	21,237	215,553	COSCO	Hong Kong
10	2018	CMA CGM Antoine de Saint Exupéry	400.0	1,312.3	59.0	193.6	20,954	219,277	CGM CGA	France
10	2018	CMA CGM Jean Mermoz	400.0	1,312.3	59.0	193.6	20,954	219,277	CGM CGA	Malta
10	2017	CMA CGM Louis Blériot	399.0	1,309.1	58.6	192.0	20,568	214,286	CGM CGA	Malta
11	2017	Madrid Mærsk	399.0	1,309.1	58.6	192.0	20,568	214,286	Maersk	Denmark
11	2017	Munich Mærsk	399.0	1,309.1	58.6	192.0	20,568	214,286	Maersk	Denmark
11	2017	Moscow Mærsk	399.0	1,309.1	58.6	192.0	20,568	214,286	Maersk	Denmark
11	2017	Milan Mærsk	399.0	1,309.1	58.6	192.0	20,568	214,286	Maersk	Denmark
11	2018	Monaco Mærsk	399.0	1,309.1	58.6	192.0	20,568	214,286	Maersk	Denmark
11	2018	Marseille Mærsk	399.0	1,309.1	58.6	192.0	20,568	214,286	Maersk	Denmark
11	2018	Manchester Mærsk	399.0	1,309.1	58.6	192.0	20,568	214,286	Maersk	Denmark

(continued)

Table 1.1 (continued)

#	Built	Vessel name	Length overall		Beam		Max TEU	GT	Operator	Flag
			m	ft	m	ft				
11	2018	Murcia Mærsk	399.0	1,309.1	58.6	192.0	20,568	214,286	Maersk	Denmark
11	2018	Manila Mærsk	399.0	1,309.1	58.6	192.0	20,568	214,286	Maersk	Denmark
11	2018	Mumbai Mærsk	399.0	1,309.1	58.6	192.0	20,568	214,286	Maersk	Denmark
11	2018	Maastricht Mærsk	399.0	1,309.1	58.6	192.0	20,568	214,286	Maersk	Denmark
12	2017	MOL Truth	399.0	1,309.1	58.0	190.3	20,170	214,286	Maersk	Panama
12	2018	MOL Treasure	399.0	1,309.1	58.0	190.3	20,170	210,678	MOL	Panama
13	2017	MOL Triumph	400.0	1,312.3	58.8	193.0	20,170	210,678	ONE	Marshall Islands
13	2017	MOL Trust	400.0	1,312.3	58.8	193.0	20,170	210,678	ONE	Marshall Islands
13	2017	MOL Tribute	400.0	1,312.3	58.8	193.0	20,170	210,678	ONE	Marshall Islands
13	2017	MOL Tradition	400.0	1,312.3	58.8	193.0	20,170	210,678	ONE	Marshall Islands
14	2019	Ever Glory	400.0	1,312.3	58.8	193.0	20,160	219,775	Evergreen	Liberia
14	2019	Ever Govern	400.0	1,312.3	58.8	193.0	20,160	219,668	Evergreen	Panama
14	2019	Ever Globe	400.0	1,312.3	58.8	193.0	20,160	219,775	Evergreen	Panama
14	2019	Ever Greet	400.0	1,312.3	58.8	193.0	20,160	219,668	Evergreen	Panama
15	2018	Ever Golden	400.0	1,312.3	58.8	193.0	20,160	219,079	Evergreen	Panama
15	2018	Ever Goods	400.0	1,312.3	58.8	193.0	20,160	219,079	Evergreen	Panama
15	2018	Ever Genius	400.0	1,312.3	58.8	193.0	20,160	219,400	Evergreen	Panama
15	2018	Ever Given	400.0	1,312.3	58.8	193.0	20,160	219,079	Evergreen	Panama
15	2018	Ever Gifted	400.0	1,312.3	58.8	193.0	20,160	219,352	Evergreen	Singapore
15	2019	Ever Grade	400.0	1,312.3	58.8	193.0	20,160	219,158	Evergreen	Panama
15	2019	Ever Gentle	400.0	1,312.3	58.8	193.0	20,160	219,612	Evergreen	Liberia

by onshore gear. Moreover, it was found the steel roofs contributed to the aerodynamic quality of the hull shape. The addition of the rain roofs protected not only the stowed containers, but the ship as well, by preventing the build-up of rainwater in the holds. The inclusion of these covers provided a simple answer to a profoundly severe problem. It is worth noting that in the end, this design was not widely adopted and has since been replaced by other architectural designs (Fig. 1.11).

Fig. 1.10 Photograph of the *MOL Triumph*, arriving Port of Rotterdam, Holland, 2017

Table 1.2 Ships on order (as of August 2023)

Delivery	Operator	Number	Max TEU
2023	Mediterranean Shipping Company	4 × Hudong-Zhonghua Shipbuilding 4 × Jiaangnan Shipyard 2 × Yangzjiang Shipbuilding	24,232
2023	Ocean Network Express	6 × Imabari Shipbuilding and Japan Marine United	24,000
2023	Seaspan ULC	2 × (unconfirmed)	24,000
2022	Evergreen	6 × Samsung Heavy Industries	23,992
2022	Evergreen	4 × Jiangnan Shipyard 4 × Hudong-Zhonghua Shipbuilding	24,004
2023	Hapag-Lloyd	12 × Daewoo Shipbuilding and Marine Engineering	23,500>
2023	OOCL	6 × Nantong COSCO KHI Ship Engineering 6 × Dalian COSCO KHI Ship Engineering	23,000

1.2.3 Refrigerated and Reefer Ships

All container ships have some number of separate slots with electrical connections for refrigerated containers. If these connections are present in large numbers and in a certain ratio to the total TEU capacity, this is reflected in the ship's designation. Ship's that are designed to carry only refrigerated cargoes in containers are referred

Table 1.3 Container loading records (as of August 2023)

Date of departure	Vessel	TEUs handled	Port	Destination
14 Aug 21	Ever Ace	21,710	Yantian	Rotterdam
8 Apr 21	CMA CGM Jacques Saadé	21,433	Singapore	Le Havre
12 Oct 20	CMA CGM Jacques Saadé	20,723	Singapore	Le Havre
8 May 20	HMM Algeciras	19,621	Yantian	Rotterdam
28 Jul 19	MSC Gülsün	19,574	Tanjung Pelepas	Algeciras
1 Jun 19	Monaco Mærsk	19,284	Tanjung Pelepas	Rotterdam
11 Feb 19	MOL Tribune	19,190	Singapore	Southampton

Fig. 1.11 Photograph of the *CSCL Indian Ocean*, Port of Felixstowe, England, 2018

to in the industry as reefer container ships. Ships that have refrigerated or climate-controlled holds, but do not carry containers as their main cargo, are referred to as reefer ships, or reefers. These vessels are commonly used in the shipment of fresh fruit and vegetables leading to their nickname 'banana boats' (Fig. 1.12).

Fig. 1.12 Photograph of the reefer, *Dole Ecuador*, Puerto Barrios, Guatemala, 2009

1.2.4 Container Passenger Ships

Container passenger vessels are a relatively recent phenomenon. In some countries, China, Indonesia, and Russia amongst others, these hybrid vessels are increasingly important to coastal and inter-island traffic. Countries with large archipelagos, such as Malaysia and the Philippines, also operate substantial fleets of container passenger vessels as these provide a vital lifeline to offshore communities which would otherwise suffer from poor logistical supplies with the mainland (Fig. 1.13).

1.2.5 Feeder Container Ships

Feeder ships carry containers between major container terminals and other ports which are not served by the major shipping lines. Most Indian ports, for example, are served by feeder ships from Colombo, while the ports of western Denmark are served from Bremerhaven. The term "feeder ship" provides no indication as to the size of the vessel, nor as to whether the vessel is equipped with its own onboard lifting gear. Most feeder ships, are however, small, or medium-sized vessels. The vessels used in feeder service to and from container ports are often customised to operate within specific areas (Fig. 1.14).

Fig. 1.13 Photograph of the passenger RoRo *Xiang Xue Lan*, Singapore, 2008

Fig. 1.14 Photograph of the feeder vessel *Frederik*, Hamburg, Germany, 2019

1.2.6 Ore Bulk Container Carrier

An odd development in ship design that never really gained much traction but has nevertheless persisted is the ore bulk container carrier or 'OBC.' The multipurpose OBC was first developed in the former German Democratic Republic (East Germany) and were highly versatile vessels. Their double hull construction meant that all the internal surfaces of the hold were smooth. Since these vessels were fitted with high-capacity onboard electrohydraulic cranes with a long reach, even poorly equipped ports could be served with ease. In one specific class of the vessel, alternate holds (I, III and V) were reinforced to carry metal ore. Due to their high strength and large hatches without tween decks, these vessels were well suited to carrying rolling mill products and long goods. A half-height, double-walled longitudinal girder fitted below deck functioned as a grain bulkhead. Excluding hazardous materials and cement, all kinds of bulk cargoes could be carried. In some cases, as many as 576 TEU could be stowed in the ship's holds and on deck. Furthermore, thirty-six reefer connections were available allowing these versatile vessels to carry refrigerated and chilled goods as well as standard ambient cargoes.

Open bulk container carriers also have several unusual features which make such vessels highly versatile, and in some respects, even suiting them for container traffic. These bulk carriers are designed like pure container ships, i.e., an exceptionally substantial proportion of the deck area can be opened, with the hatch lengths arranged in a 40 ft grid. Structures delimiting the hold, such as double bottoms, transverse bulkheads and wing bulkheads are designed to withstand the pressure applied by bulk cargoes. Smooth walls and floors are required to allow for the straightforward loading and unloading of loose cargo with grabs and wheel loaders, which is why the container foundations are recessed in the double bottom and the ISO holes are covered with plastic or steel lids. In the vessel shown, containers can be stowed up to seven high in the holds. The stacks are secured at the bottom by twist locks in the tank deck. Container guides providing lateral restraint are folded out from the bulkheads. In each hold, three blocks are formed using single and double stacking cones, with container guides providing guidance and restraint in two planes, thus making it possible to withstand the forces which arise when the vessel rolls. Such stacking aids are provided between the third and fourth and between the fifth and sixth tiers of containers and are sized in line with the maximum admissible container corner loading of 600 kN. A combination of general cargo and/or bulk cargo plus containers may also be stowed in the holds. This is achieved by folding "container brackets" out from the bulkheads onto which a maximum of four tiers of containers, each applying a load of 300 kN, may be loaded over the bulk cargo or general cargo.

As has already been mentioned, roll-on/roll-off is often abbreviated to RoRo, which is a description of how the cargo is handled. The type of cargo being carried is not automatically known as it may comprise any kind of rolling cargo or cargo which has been made 'rollable.' In many cases RoRo ships also carry other cargoes without this being clear from the vessel's designation. There is, however, one feature common to all RoRo vessels—they can be loaded via bow, stern or side mounted ports. If the ships have several decks, access is provided by way of elevators or ramps. Similarly, ferries are equipped to carry both passengers and rolling cargo, which may comprise automobiles, trucks, chassis, trailers, or railway carriages. In many cases, cargo is carried on "cassettes" which are packed and unpacked in port and transferred on and off the vessel with special terminal tractors. Containers are only loaded or unloaded indirectly using roll trailers or similar. Stern, bow and or side ports help ensure rapid cargo handling. With many types of ferries, a one-way system is used, creating a "first in first out" traffic flow. In ferry transport, a distinction is drawn between accompanied vehicles, where the driver drives on board, travels with the vehicle, and drives off again, and unaccompanied vehicles, where the vehicle is driven onto the vessel, after which the driver dismounts and leaves the vessel. The vehicle is then picked up by another driver at the ship's destination. Only auxiliary loading gear is usually present. Many passenger/railway ferries and ferries capable of carrying both railway vehicles and trailers, chassis or other road vehicles are in service in short-sea RoRo traffic in the North Sea, Baltic Sea, and the Mediterranean Sea. These vessels have bow thrusters to ensure rapid berthing and casting off. Manoeuvring is often facilitated by pitch propellers or a multi screw design. Ferries are often symmetrical in structure and may be operated and accessed from either end. Special cargo securing systems are often provided but are usually used only in heavy sea conditions. Most cargo damage and ship sinking incidents are caused by inadequate stowage and securing on the vehicles, as has been shown by various incidents of loss and accidents over the years (Herald of Free Enterprise 1988; Estonia 1994; Sewol 2014; Princess of the Stars 2008) (Figs. 1.15 and 1.16).

Fig. 1.15 Photograph of the *Herald of Free Enterprise,* Port of Zeebrugge, Belgium, undated

Fig. 1.16 Photograph of the *Viking Sally* (later renamed *Estonia*), departing Stockholm, Sweden, 1989

ConRo vessels carry both containers and rolling cargo. RoRo cargoes are loaded below deck, while containers are primarily stowed on deck. On some vessels, for example, ships of the ACL or Polish Ocean Line fleets, special container shafts are provided in the forward third of the ship to accommodate containers below deck. In regions of service with high cargo volumes, container handling is carried out by onshore gantry cranes. The ships rarely have their own loading gear. The *Finnsailor* is one such ship which carries rolling or rollable cargo below deck and containers, plant, general cargo, etc. on deck. RoRo-LoLo carriers are ships which are capable of loading and stowing cargoes both via ramps using roll-on roll-off methods and with on board lifting gear using the lift-on lift-off method. Such vessels accordingly also have upper deck hatches, and, in some cases, cargo holds divided by transverse bulkheads. Figures 1.17 and 1.18 are examples of RoRo-LoLo container vessels. In the first example, the vessel has a RoRo component in the form of a stern door. On the port side it has an internal vehicle ramp to provide access to the two 'tween decks. The fitted-on board cranes allow containers to be handled in ports without shoreside lifting gear. Barge carriers are specialised container transport vessels.

Barge carriers, though largely capable of operating independently of ports, depend on calm areas of water or "floating areas." This fact explains the great strategic significance of the system as barges can be set down and picked up at any coast. Barge transport operations are only worthwhile between economic areas which have identical volumes of goods passing in each direction, and where inland waterways lead inland so that effective use can be made of the advantages of the floating containers.

Fig. 1.17 Photograph of the *Repubblica Argentina*, departing Casablanca, Morocco, 2022

Fig. 1.18 Photograph of the *Ever Ace*, Port of Hamburg, Germany, 2022

The prohibitive cost of building these types of ships and the capital expenditure for three sets of barges mean that such systems are not economically viable everywhere. There are various systems used onboard barge carriers:

(1) Lighters Aboard SHip LASH
(2) SEEBEE
(3) BACO liners, and
(4) Condocks.

LASH stands for Lighters Aboard SHip. In the LASH system, barges are carried athwartships in holds and on deck. The barges are picked up and set down at the stern by the LASH onboard gantry crane using spreaders. Distinguishing external features of these carriers are the far forward location of the superstructures or deck houses, the location of the exhaust stacks to the aft and sides, the stern outriggers, and the large gantry crane with a lifting capacity of more than five hundred metric tonnes. The most famous LASH was the *München*, which sank in the North Atlantic on or about 12 December 1978. SEABEE carriers are capable of embarking barges at the stern using winch-driven lift platforms with a load-carrying capacity of more than 2,000 metric tonnes. Barges are loaded, usually in pairs, by being floated into the dock-like after body of the ship over the lowered lifting platforms, then lifted to deck height, from where they are rolled into the ship on very flat rail-mounted trolleys. This type of transfer is thus known as "lift and roll." Lykes Line's SEABEE ships can stow thirty-eight barges on three decks. Special fittings allow the upper deck to be loaded with containers instead of barges. SEABEE ships can carry containers and other cargoes on deck, but these ships do not have on board lifting gear for such cargoes. SEABEE barges measure on average 29.72 m × 10.67 m × 5.18 m (97.5 ft × 35 ft × 16.9 ft). With a tare weight of 171.5 metric tonnes, the unladen draught averages at 0.62 m (2.03 ft). With a full payload of 844 metric tonnes, the maximum draught is 3.22 m (10.5 ft). There may be slight differences depending upon the type and lighter. The barges can accommodate six FEU or twelve TEU. As many as 16 FEU can be located on the barge's hatch covers.

BACO liners (barge/container liners) operate in accordance with the dock principle. In this German-developed system, the barges are floated through bow doors into the carrier, which has been lowered into the water by taking on ballast. Once the bow doors closed, the water is pumped out of the dock and the barges are secured with special fittings. Typical BACO barges measure 24 m × 9.50 m (78.7ft × 31.16 ft) and have a deadweight of 800 metric tonnes, and a draught of 4.10 m (13.45 ft). However, the dimensions of the carrier ships are such that other types of barges and even inland waterway vessels and lighters may also be carried. Several tiers of containers can be carried on deck and be loaded and unloaded with the on-board lifting gear. The term 'condock' is a portmanteau of container and dock ship. A large hold can be loaded with rolling cargo via a stern ramp as well as from above using the LoLo method. These vessels have their own, very high-capacity lifting gear for carrying heavy-lift cargoes. Containers can be carried in three tiers on deck. By flooding their ballast tanks, they can be lowered in the water like floating docks, to take on floating cargoes. The dimensions of those vessels which are in service are such that LASH, BACO and sometimes even SEABEE barges can be accommodated. This class of vessel is more versatile than either LASH, BACO or SEABEE. Its gantry cranes are capable of handling individual items of cargo of up to 1,000 metric tonnes via the LoLo method. Using the RoRo method, heavy items of up to 2,000 metric tonnes can be loaded into the hold via a 20 m × 10 m (65.6 ft × 32.8 ft) ramp, while heavy and bulky items can be rolled on deck via two stern outriggers and self-propelled bogies. Floating cargo can be stowed either in the hold using the FloFlo method or on deck using the piggyback system.

For inland waterways, such as the Rhine and Danube Rivers, coastal motor vessels are employed which have a continuous box shaped cargo hold and an average deadweight of 1,550 metric tonnes. The essential difference between other classes of sea/

river-going vessels resides in the vessel's crane, which is of a collapsible construction and so does not exceed the maximum height for navigating the Rhine and inland canals. The vessel's hatch covers can accommodate 20′ or 40′ containers over several bays. The wheelhouse can be lowered and raised hydraulically, while the masts can be folded down. Most inland waterway motor vessels in service throughout Europe are used to carry dry cargoes. Lengths vary between 38.50 m (126.3 ft) and 110 m (360 ft), with beams averaging between 5 m (16.4 ft) and 11 m (36 ft). Draughts range between 2 m (6.5 ft) and 3.50 m (11.4 ft) fully laden. The deadweight of such vessels is typically between 220 metric tonnes and 3,000 metric tonnes. Many of these vessels are equipped to carry containers. Some single hold motor freighters have been specially tailored to sailing the Rhine on the way to Basel, Switzerland. They have the largest possible hold volume for light bulk cargoes but can also carry heavy-lifts and bulky cargoes. Heavy goods can be rolled on board from land via an external ramp onto the fore body, from where they can be rolled into the hold via an internal ramp with a gradient of five degrees. Low loaders, trailers or other special purpose trucks and crawlers can also be used for this purpose. Loading and unloading operations can be assisted by hydraulic equipment, trim tanks, and a pump system.

1.3 Container Flows

From the perspective of container traffic, it would be ideal for there to be a balance between incoming and outgoing containers in a particular region, not only in terms of container numbers, but also in terms of container type and weight. Unfortunately, such a balance is not achievable. There will always be empty containers that need transporting in one direction or another. From the perspective of the shipping company, general purpose (GP) containers that are usable in any situation present a major advantage. Freight forwarders, on the other hand, tend to prefer the use of specialised containers if they can be transported at identical cost, as the packing and securing of awkward cargo is much easier in special containers when compared to using standard GP containers. As an example, steel sheet in coils can very quickly be loaded onto coil containers and secured using a straightforward process. They are more difficult to pack and secure on flat racks, while they are particularly difficult to pack and secure in box containers. To put this into context, shipowners accepting containers loaded with coils for transport to Kenya would, if the rolls of sheet steel were shipped in coil containers, also must take many empty standard ventilated containers to Kenya to transport coffee beans from Kenya to Europe. Moreover, the coil containers, which would be of no further use in Kenya, would have to be transported empty to some other location where they could be used again. As a result, steel sheet in coils will often be transported to Kenya in "coffee containers" which are less suited for carrying such cargoes. To provide another example, to save on the higher freight costs associated with using specially developed tank containers, special tanks or 'flexitanks' are often loaded into standard GP containers. Whilst this makes integrated transport easier to facilitate, the risk of causing damage to the walls of the GP container from the sloshing of the tank contents makes it less palatable than using properly constructed tank containers. A product of the globalised nature of international commerce, the flow of empty containers from Europe to Asia is symptomatic of the imbalance of trade which flows from West to East. It is most common for containers to be shipped full of goods from factories in China and the like to markets in Europe and the US, with empty containers being sent back. Ideally, shipowners would be happier to transport an equal volume of weight each way as empty boxes generate far less revenue (if any) than those full of goods and products.

In this introductory chapter, we have discussed the main types of cargo ships in merchant service today. As can be seen, the development of the cargo ship has undergone a number of iterations starting with the *Ideal X* in the 1950s, culminating in the behemoths of the 2020s. In Chap. 2, we will begin to look at the positioning and securing of containers onboard the ship.

Chapter 2
Positioning and Securing of Containers

In Chap. 1, we discussed the beginning of containerisation and how container ships have progressively evolved and expanded to ever larger proportions. Compared to the first container ships of the 1950 and 1960s, today's behemoths truly are masters of the sea. In this chapter, we will discuss some of the fundamental issues around the positioning and securing of containers onboard ships. It is surely no surprise to the reader that this is an extraordinarily complex and specialised undertaking which requires extensive training and experience. Therefore, this chapter is intended for the benefit of an audience other than the professional mariner.

2.1 General Stowage on Ships

On most ships which are specially designed for container traffic, the containers are carried in a longitudinal position, facing in an aft to bow direction (refer to Fig. 2.1). This stowage method is sensible with respect to the interplay of stresses that exert themselves in heavy seas. It also accounts for the loading capacity and weight of individual containers. Stresses in heavy seas are greater athwartships than at the fore and aft therefore the loading capacity of container side walls are designed to be higher than that of the end walls. However, on many ships, containers may be stowed in athwartships bays or transported athwartships for various reasons. This must be taken into consideration when packing and securing cargo in the container prior to despatch (see Fig. 2.2). As is shown in Fig. 2.2, this stowage method rarely sensible with respect to the stresses exerted by heavy seas and the loading capacity of the container. As stated earlier, stresses in heavy seas are greater athwartships than at the fore and aft but the loading capacity of container end walls is lower than that of the side walls. When atypical stowage methods, i.e., where some of the containers are stowed athwartships and others fore and aft, are used, they require greater effort during the packing and securing stage. The two figures (Figs. 2.1 and 2.2) show how

A. A. Olsen and F. Karkori, *Containerized Cargo Handling and Stowage*, Springer Series on Naval Architecture, Marine Engineering, Shipbuilding and Shipping 22,
https://doi.org/10.1007/978-3-031-63329-4_2

Fig. 2.1 Containers stowed lengthwise from fore to aft

Fig. 2.2 Containers stowed athwartship (i.e., athwartships stowage)

important it is to establish the carrier and their way of transporting containers, either in order to rule out certain modes of transport or to be able to match the best cargo securing techniques to the specific mode of transportation. If, however, the method of transportation is not known, then packing and securing must be geared to account for greater than average dynamic stresses.

2.2 General Securing Information

When securing containers onboard, the stresses that result from the ship's movements and wind pressure must be adequately accounted for. Forces resulting from breaking-wave impacts can only be considered to a certain degree. All the containers onboard must therefore be secured against slippage and toppling, with greater care being taken to ensure that the load-carrying parts of the containers are not loaded beyond admissible limits. Except in the case of individually carried containers, securing is achieved by stacking the containers in vertical guide rails or by stowing them in stacks or blocks. The containers are then connected to one another. These are then fixed to static parts of the vessel.

2.2.1 Securing in Vessel Holds by Cell Guides Alone

Every class of all-container type ship are provided with cell guides (see Fig. 2.3) with vertical guide rails serving as securing means for hold-based cargoes. In this instance, the greatest stress the containers are exposed to stems from stack pressure. Since the containers are not connected vertically, lateral stress is transmitted by each individual container to the cell guides. When positioned in such cell guides, individual containers are not able to shift. If the corner posts of one of the containers at the bottom of a stack collapses under the weight of excessive pressure, the containers stowed above it will suffer only slight damage. The risk of damage to containers in adjacent stacks is also kept within tight limits. The containers are guided by the rails of the cell guides during loading and unloading. By referring to Fig. 2.4, we can clearly see how the upper ends of the guide rails each take the form of insertion guides.

2.2.2 Securing Containers in Vessel Holds by Cell Guides and Pins

Feeder ships, multipurpose freighters, and container ships in certain regions must be particularly flexible in their equipment in order to carry containers of varying dimensions. To this end, convertible stowage frames have been developed in which 20', 24½', 30', 40', 45', 48' and 49' containers may be stowed securely. Most of these frames are manufactured as panels which are brought into their required position by cranes. At the base they have fixed cones, which engage in pockets welded into the hold top area. At the sides, the frames are secured by pins, which engage in bushes. These are then let into the wing bulkheads. Such frames are often man-accessible, so that the containers can be locked in place by means of pins. If it is necessary to carry containers 2,500 mm wide, the frames are arranged according to this dimension.

Fig. 2.3 Cell guides in an all-container vessel

Fig. 2.4 Guide rails of two adjacent slots

To secure standard general purpose (GP) containers of normal width, closure rails are then fitted on both sides of the guide rails by means of screw connections. If necessary, these adapters can be removed. Removable container guides have also been developed and constructed for multipurpose freighters, reefer vessels and such like. These guides allow containers to be carried in regular or insulated holds without any risk of causing damage to the holds themselves. If other cargoes are carried, the stowage guides can be removed using the ship's own or shore-based loading or lifting gear. Once removed, the equipment is securely deposited in special holders on deck.

2.2.3 Securing Containers in the Vessel Hold by Conventional Securing and Stacked Stowage

On older, conventional general cargo vessels and multipurpose freighters, stacked stowage methods are used in the hold, combined with various securing methods (see Fig. 2.5). The lower containers stand on foundations capable of withstanding the stack pressures which arise. Dovetail foundations, into which sliding cones fit, are provided to prevent slippage. The containers are then connected by single or double stacking cones or twist locks. The entire stack or container block is then lashed using lashing wires or rods and turnbuckles. This system entails a lot of lashing work and material and, moreover, is less secure than securing the containers in cell guides.

2.2.4 Securing Containers in the Vessel Hold by Block Stowage and Stabilisation

Securing containers in the vessel hold using block stowage and stabilisation is less frequently used today than in the past but may still be found on some older conbulkers and other multipurpose freighters. On these vessels, containers are interconnected horizontally and vertically using single, double and even quadruple stacking cones. The top tiers are connected by means of bridge fittings. To the sides, the containers are supported at their corner castings with "pressure/tension elements" (refer to Figs. 2.6 and 2.7 for examples of fastening containers together). This type of container securing method has two marked disadvantages: the first occurs if an individual container disintegrates. In this situation, it is not just one container stack which is adversely affected, but the entire container block. The second drawback concerns the dimensional tolerances and wear and tear that is exerted on the stacking cones. In this instance the entire block can shift in heavy seas. This can lead to the intermediate stacking cones breaking, which can in turn cause the entire block to collapse with devasting effect (Fig. 2.8).

(a)

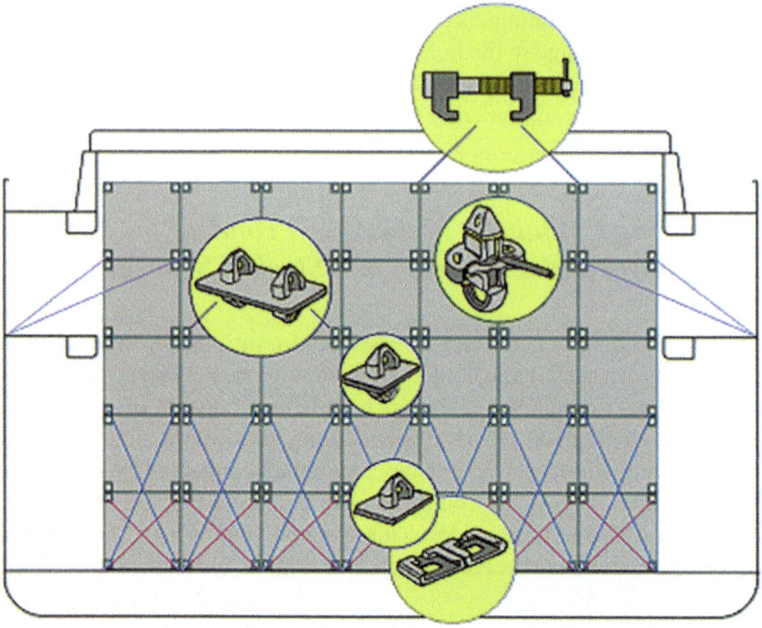

(b)

Fig. 2.5 **a** Example of stacked stowage with conventional securing. **b** Example of stacked stowage with conventional securing

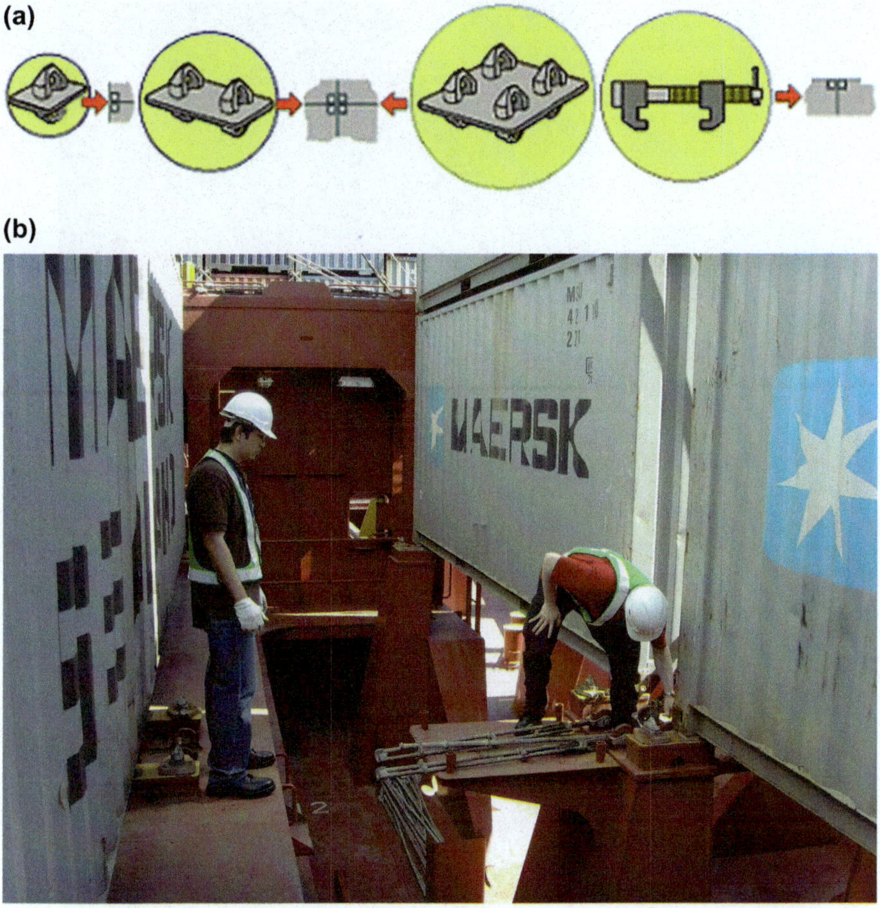

Fig. 2.6 **a** Fastening containers together. **b** Fastening containers together

2.2.5 Securing Containers on Deck Using Container Guides

On some ships, containers may also be secured on deck using cell guides or lashing frames. Some years ago, the American shipping company, Atlantic Container Lines, pioneered the use of cell guides on deck. Similarly, Polish Ocean Lines also developed a combined system. On most other ships, cell guides are pushed hydraulically over the hatch cover as soon as loading below deck is complete and the hatches are covered.

Fig. 2.7 Fastening containers together

2.2.6 Securing Containers on Deck Using Block Stowage Securing

Securing containers on deck using block stowage securing was developed in the early days of container shipping but has since seen its use reduce in part for economic reasons. Refer to Fig. 2.9a, b for examples of block stowage securing on deck. In

Fig. 2.8 **a** Example of block stowage method with stabilisation. **b** Photograph of block stowage method with lower stabilisation. **c** Photograph of block stowage method without upper stabilisation

this instance, the containers in the bottom layer are positioned in socket elements or on fixed cones. Double stacking cones are used between the layers and the corner castings of adjoining containers are connected at the top by bridge fittings. The containers are held together over the entire width of the ship or hatch cover by cross lashings. A distinct disadvantage of this method is reduced flexibility when loading and unloading, since adjoining containers must be moved if access to a specific container is required. Numerous variants, which are not listed in any greater detail here, are available for attaching the lashings. Sometimes the lashings from different stacks can cross one another. Refer to Fig. 2.9a, b for an example of crisscross lashings on different container stacks. Today, this securing method is used almost exclusively on exceptionally large container ships only. In these instances, instructions for lashing procedures must be prominently displayed in English and in the language of the deck crew.[1]

2.2.7 Securing Containers on Deck Using Stacked Stowage Securing

Securing containers on deck using stacked stowage securing methods is the one used most frequently on container ships today. Cargo handling flexibility is its key advantage. The containers are stacked one on top of the other and then connected with twist locks. The containers are then lashed vertically using steel lashing rods. By using this system, no stack is connected to any other. The container lashings do not cross over the lashings from other stacks, except for the "wind lashings" on the outer extremes of the ship. The one major concern is that a container stack of this kind can topple if it is not adequately secured. To avoid this, close attention and supervision from the ship's officers is needed to ensure the lashings are tight and positioned correctly. When searching for potential carriers, it is important for forwarders to avoid "sloppy" carriers wherever possible. This applies quite generally, and not only to the operators of aging ships. Now that we have discussed the general principles of the positioning and securing of containers onboard ship, we can turn our attention to the methods used for planning the stowage of containers (Fig. 2.10).

[1] This avoids miscommunication between the ship's officers, crew and shore-based personnel.

(a)

(b)

(c)

Fig. 2.9 **a** Example of block stowage securing on deck. **b** Photograph of crisscross lashing on different container stacks. **c** Photograph of crisscross lashing on different container stacks

Fig. 2.10 Securing of a 3-tier stack on board a semi-container ship

2.3 Container Stowage Plans

In the first part of this chapter, we examined some of the key issues around container stowage onboard container ships. The positioning of the container onboard is critical both to the safety of the vessel and to the security of the load itself. As we said earlier, it is widespread practice for containers to be stowed facing aft to fore as this helps alleviate many of the stresses that are exerted on the ship's structure, particularly when the vessel is encountering heavy seas. However, knowing how to stow containers onboard the ship is only one part of the equation. The second part of the loading process—as far as the cargo officer is concerned—is determining where best to stow the containers on the ship. This is where the container stowage plan comes into play. Working stowage plans are drawn up to assist the ship's officers during their advance planning. Master stowage plans definitively document the positioning of each container onboard once the finalised cargo manifests are received. The cargo manifest (in theory) states what is inside each container, the mass or volume of the contents of the container, and any crucial details the carrier may need to know to ensure its safe carriage. Once the working stowage plan has been developed and signed off by the ship's master, the cargo officer can begin to plot the location of each incoming and outcoming container in relation to its size, weight, and contents. This is typically done using a bay-row-tier arrangement.

The bay-row-tier arrangement follows a system of numerical coordinates which relate to length, width, and height dimensions. The stowage space of the container onboard the ship is unambiguously stated in numbers and is (always) recorded in the carrier's shipping documents. This means it is then also possible to establish, later, where the container was located during its passage. This is important to know should the customer later lodge a claim against the carrier for loss, damage, or spoilage. According to this principle, bays are defined as the individual container blocks positioned in the transverse plane; rows are lengthwise; and tiers are vertical. In theory, a vessel with thirty-eight bays could number each bay continuously from 1 to 38. However, that would only work if 20' containers are carried. If the ship only carries 40' containers, then it would be more appropriate to number the bays from 1 to 19. Since, however, ships are expected to carry both 20' and 40' containers, neither of these conventions would work in practice. Subsequently, a different numbering system is used which allows for the stowage of both 20' and 40' containers on the same vessel. In this system, the bay spaces for 20' containers are numbered from fore to aft with odd numbers, i.e., 01, 03, 05, 07, 09, etc. The bay spaces for 40' containers are numbered with even numbers, i.e., 02, 04, 06, 08, 10, etc. To illustrate how this works in practice, we need to imagine we are standing at the bow (or stern) of the ship. The rows of containers are numbered with even numbers leading from the centre leftward and odd numbers rightward from the centre.

Where the ship is designed with an odd number of rows, the middle row is always numbered 00.

On close inspection, the photographs in Figs. 2.11, 2.12 and 2.13 shows the left-hand row 16, which is intended for containers only on deck. Rows 14, 12, 10, 08, 06 may be filled both on deck and in the holds. Rows 04, 02, 00, 01 and 03 are likewise designed to be occupied in the hold and on deck (Fig. 2.14).

In this example, rows 05, 07, 09, 11 and 13 are empty in this bay. Row 15 is designed only for on deck occupation and is also empty (Fig. 2.15).

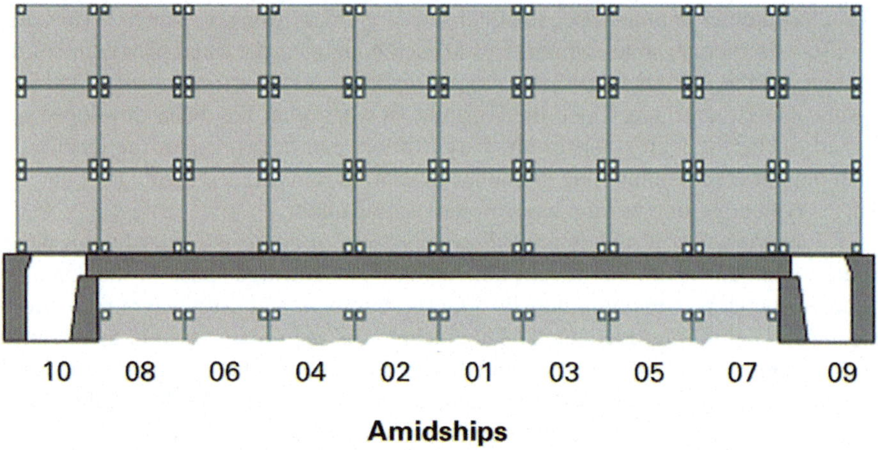

Fig. 2.11 Row numbering where there is an even number of rows

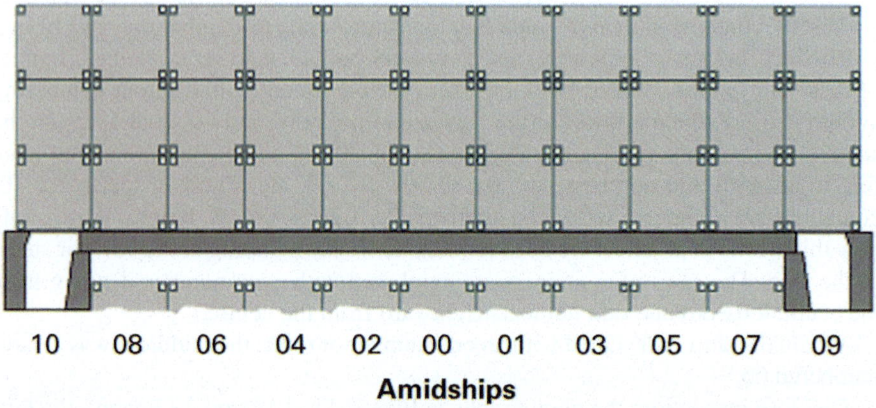

Fig. 2.12 Row numbering where there is an odd number of rows

Fig. 2.13 Numbering of the port rows of a ship

Fig. 2.14 Numbering of the starboard rows of a ship

Fig. 2.15 Row numbers of the aft bay of a ship

The container tiers are allocated even numbers, starting from the bottom. The conventional method is to start with 02 in the hold and then count upwards to 04, 06, 08, 10, etc. In the case of deck cargoes, it is conventional to start numbering with 80 or 82 although there may be slight differences between ships and carriers.

On the ship shown in Fig. 2.16, the containers standing directly on the main deck are numbered 80 and those standing on the hatches are numbered 82. The number is incremented by two for each higher layer. These bay, row and tier numbers are then clearly annotated in the bay plans.

Fig. 2.16 Numbering of horizontal container layers, or tiers

2.4 Bay Plan

The bay plan (as shown in Fig. 2.17) is an important document which contains vital information about each individual container. This information typically includes the container identification number, its alpha prefix, the port of destination or discharge, and any other important details which might be relevant to the present or onward carrier.

In accordance with the bay-row-tier system, and referring to Fig. 2.18, we can clearly see the coloured containers are given the following stowage space numbers:

- 20’ container in the red-coloured slot: 53 12 12
- 40’ container in the blue-coloured slot: 54 07 88

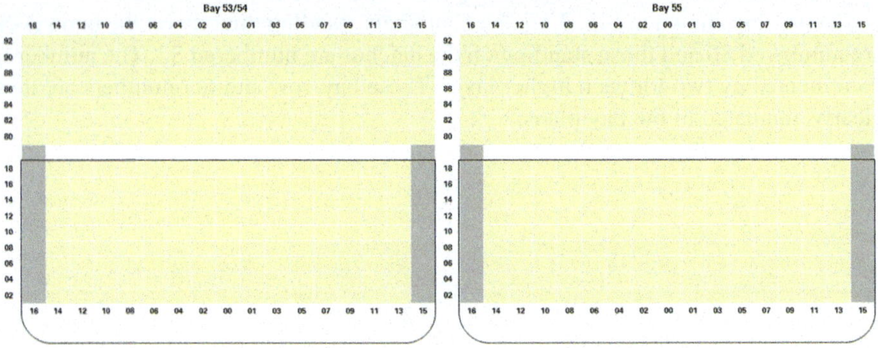

Fig. 2.17 Bay plan

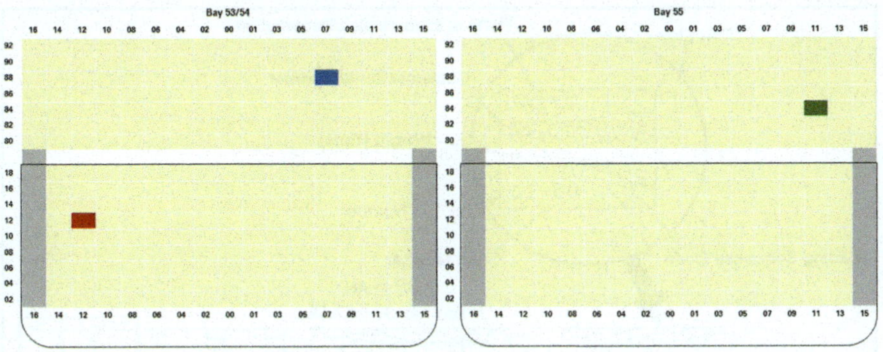

Fig. 2.18 Colour-labelled containers in a bay plan

- 20’ container in the green-coloured slot:55 11 84

The system illustrated above is the most widely used. However, other numbering systems can and do exist such as stating the coordinates in a different order, for example, row-bay-tier and similar combinations. On RoRo ships, where it is more common to carry rolling cargoes in the holds, slots are typically organised according to the lanes which run lengthwise (Figs. 2.19, 2.20 and 2.21).

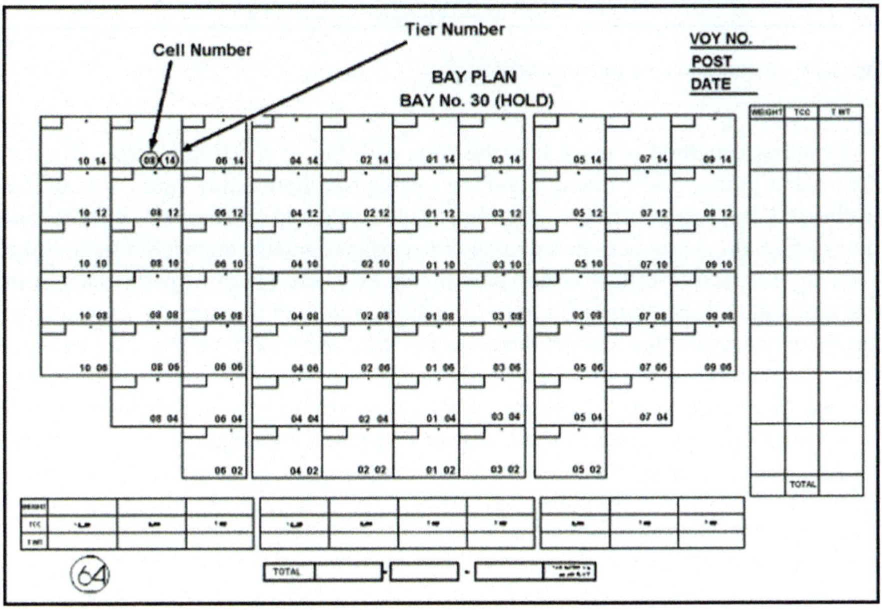

Fig. 2.19 Typical bay plan (Blank)

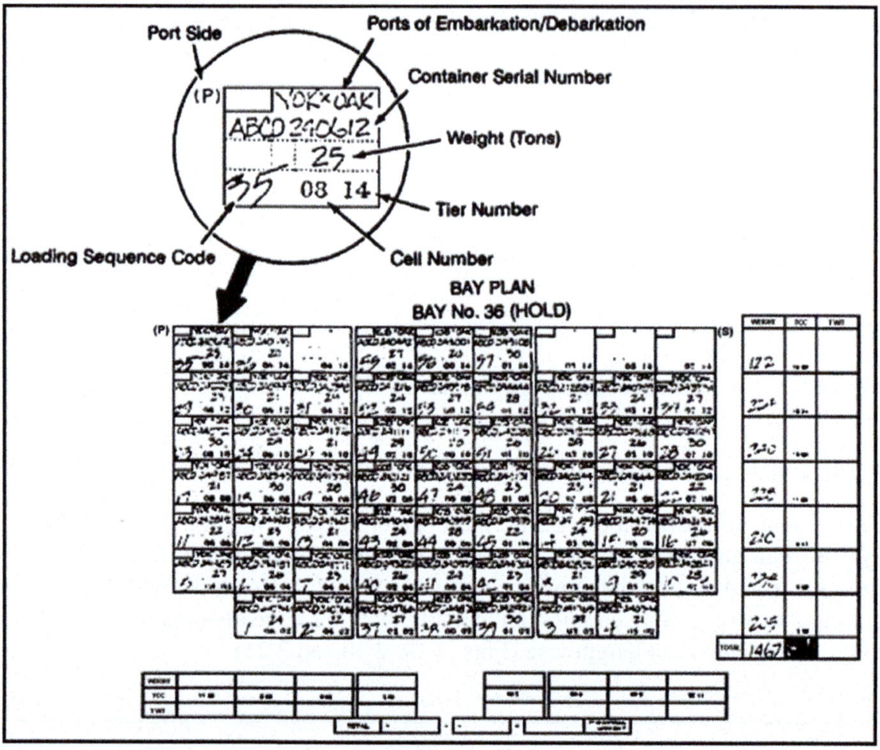

Fig. 2.20 Identification of container codes

Whichever method is used, it is the responsibility of the ship's cargo officer to ensure that proper and efficient planning takes place before any cargo is loaded or discharged. It is recognised therefore that individual cargo officers will often develop their own working practices based on their experience and the nuances of their vessel. Even so, it is best to follow established practices where possible as this aids in the standardisation of containerisation so long as the safety of the ship, her crew and her cargo are accounted for. Unfortunately, even with the best planning, accidents can and do happen at sea. For this reason, it is paramount both the shipper and the carrier have adequate insurance to cover any losses sustained to both the cargo and vessel. For this reason, this complex area is discussed in the next chapter.

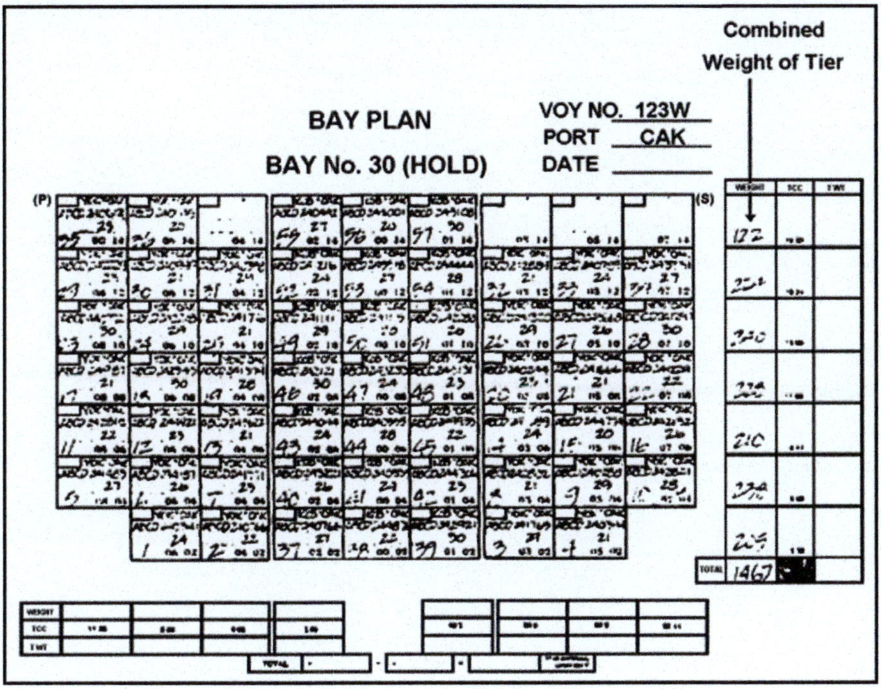

Fig. 2.21 Typical bay plan (Completed)

Chapter 3
Responsibility for Container Packing and Cargo Insurance

In Chap. 2, we briefly discussed the methods for which containers may be loaded and stowed onboard ships in preparation for transit. We also discussed the role and function of the bay plan, and how this can help determine the position of an individual container in event the container, and/or its contents are damaged during transit. In this chapter, we will look at the responsibility for container packing, and the importance of having adequate insurance coverage should the worst happen. The storage, cargo handling, and the transport of goods are covered by both national and international laws and regulations, which set out the minimum requirements for the treatment of goods, vehicle use, and so forth. Most official regulatory guidelines are supplemented by private agreements and terms and conditions of conducting business. This usually includes provisions for the careful packaging and marking of goods for dispatch, which should be accompanied by proper packing, stowing, securing, and unpacking. When carrying hazardous materials, additional provisions must be considered with respect to packaging, labelling, documentation, and other regulatory considerations. In an ever-globalising world, goods are increasingly subject to intermodal and integrated logistical supply chains which involve road, rail, water and air transport. Needless to say, this manhandling can cause stress to the cargo and outer packaging. Sadly, packaging, labelling, stowage and securing deficiencies can only be established during transit by conducting random spot checks. To prevent unnecessary damage caused by minor carelessness, expert packing and securing is essential for care-free carriage.

This book therefore pays particular attention to avoidable negligence, which may result in loss and/or (could) form the basis of claims for compensation. It has as its central theme issues relating to transport legislation, some of which will be explained and commented on later in more detail. The fundamental rules to which reference will frequently be made are contained in the latest cargo transport unit (CTU)

A. A. Olsen and F. Karkori, *Containerized Cargo Handling and Stowage*, Springer Series on Naval Architecture, Marine Engineering, Shipbuilding and Shipping 22,
https://doi.org/10.1007/978-3-031-63329-4_3

packing guidelines (in this book we will refer to the CTU as a 'container'.)[1] The preamble to the container packing guidelines points out that the use of containers, swap-bodies, vehicles or other containers reduces the physical hazards to which cargoes are exposed. However, it also emphasises that improper or careless packing of cargoes into or onto containers or improper cargo securing may cause accidents resulting in personal injury during the handling or transportation of containers. It is also pointed out that considerable damage may additionally be caused to the cargo or the container. It is worth stressing, therefore, that the individual responsible for packing and securing the cargo in the container is often the last individual able to look inside the container before it is sealed for transit. In this respect, it is quite common for a container to go an entire journey from sender to recipient without being opened. The preamble to the container packing guidelines states that …

> Drivers of road vehicles and other road users if the [container] is transported by road; railway employees and others, if the [container] is carried by rail; crews of inland waterway vessels, if the [container] is carried on inland waterways; handling staff at inland terminals when the [container] is transferred from one means of transport to another; dock workers when the [container] is loaded and discharged; ships' crews, who often accompany the [container] under the most severe transport conditions [and] those who unpack the [container] […] all persons, such as the above and passengers, may be at risk from a poorly packed container, swap-body or vehicle, particularly one which is carrying dangerous cargoes.

The preamble to the container packing guidelines makes it abundantly clear that hazards exist simply in the normal carriage of containerised cargoes. By adding human error, laziness or carelessness, these hazards are magnified—especially when the cargo carried is of a hazardous nature. To avoid hazards for occurring, it is critical every effort is made to ensure cargoes are correctly packaged, labelled, stowed and secured inside the container before the container doors are locked and sealed.

3.1 Intermodal Carriage of Containers

3.1.1 Carriage by Road

If containers have been closed, locked and secured with seals, it is impossible for road vehicle drivers to check how well (or not) the container has been packed and secured before it is loaded onto the trailer chassis. Even if the container is freely accessible, it is still only feasible to inspect the cargo located within the immediate door area of the container. Improperly packed containers can significantly reduce the safety of road vehicles through the shifting weight of the contents of the container. This can cause the vehicle to overturn. Containers transported on trailer chassis and flatbeds constitute a particular hazard where they are poorly packed, and the cargo is inadequately secured. It is the shipper's duty to ensure the proper packing and

[1] The abbreviation 'CTU' has the same meaning as the expression "intermodal transport unit" or "ITU".

securing of the contents within the container before the container departs the shipper's premises. Poorly packed, inadequately secured container loads may also put third-parties at risk in the event shifting items of cargo smash through the container walls or cause the vehicle carrying the container to jackknife or overturn. Under normal circumstances, a container chassis driver cannot be held liable for losses where the vehicle is roadworthy arising from improper packing of containers where the latter is sealed and loaded in a recognised safe manner.

In the majority of cases, the haulier's contract of carriage applies only to the transport of the container and not the contents. Their duty to load the container in an appropriate manner is fulfilled where the maximum total weights and axle loads are complied with, appropriate locking of the container is ensured, and hitching is performed correctly. If the container is sealed on the arrival of the driver, then the driver has no way of influencing, supervising, or checking the quality of the packing and securing of the load. Losses caused by such deficiencies cannot therefore lie within the driver/haulier's area of responsibility. Instead, it lies explicitly with the shipper themselves. If the driver comes to notice obvious deficiencies when taking delivery of the container, or on a later occasion—for instance due to the driving dynamics of their vehicle—the driver must respond accordingly. It is advisable in the first instance to contact the shipper for instructions. In the case of unsealed vehicles, an internal visual inspection may be performed. Any gross errors may then be discovered and reported. In practice such inspections are rarely performed, but it is worth noting that in such situations the driver may be held partially responsible if the means to inspect the container were available, but an inspection was not made. To that end, it simply cannot be stressed enough that the shipper must always ensure that cargoes are loaded and secured in such a manner as to be safe for transport. The responsibility of the shipper is not limited to the load itself, but also to any damage or injuries caused to third-parties or property arising from their dereliction of duty. This responsibility is widely enshrined in law, as the following example from the German Traffic Code shows:

> The load, together with clamping chains, gear and other loading equipment, must be stowed so as to be safe during transport and be secured especially so as to prevent it from falling off and from making avoidable noise.[2]

It is worth noting the context of this paragraph is worded very generally and does not prescribe any particular target, individual or group. It must therefore be adhered to by everyone responsible for loading road vehicles (Figs. 3.1, 3.2, 3.3 and 3.4).

3.1.2 *Carriage by Railway*

For containers destined for railway carriages, switchmen and other railway employees are not responsible for checking that the container has been properly

---

[2] Section 22(1). https://bmdv.bund.de/blaetterkatalog/catalogs/327056/pdf/complete.pdf.

Fig. 3.1 Photograph of empty containers at a container depot

packed. Their inspection duties may extend so far as to the proper securing of the container onto the freight wagon, but not so far as to the condition inside the secured container. Similar to lorry drivers, where deficiencies are detected in open containers, railway personnel have a general responsibility to intervene. Poorly packed and inadequately secured cargoes put not only the switchmen but also other railway users and third-parties at significant risk. This is because poorly secured cargoes are frequently known to smash through the end walls of containers during wagon and engine switching routines and on occasion even through the side walls during normal rail transport (Fig. 3.5).

3.1.3 Carriage on Inland Waterways

It goes without saying that the crews of inland waterway vessels cannot be expected to inspect each container that is delivered to them for secure packing. This means cargo handling crews at inland terminals are often put at risk by poorly packed, inadequately secured containerised cargoes, especially when they are transferred from one means of transport to another.

Fig. 3.2 Photograph of a container transfer using a stack loader

Fig. 3.3 Photograph of a container loaded on a skeleton chassis

Fig. 3.4 Photograph of a container loaded on a flatbed trailer

Fig. 3.5 Photograph of containers loaded on a railway carriage

3.1.4 Carriage by Sea

As far as sea transport is concerned, the ship's crew do not have any opportunity, on a day-to-day basis, of influencing how containers are packed or how cargoes are secured within them. In the early days of container shipping, it was common practice for the cargo officer to conduct at least occasional and rudimental spot checks when in port. Today, due to higher cargo volumes, shorter lay times alongside and an overall reduction in crewing levels, this is possible in only very occasional situations. The long-established practice of having containers checked and inspected by experts employed by the shipping companies or loss adjusters instructed by the carriers has also mostly been abandoned for cost saving reasons. These days, containers are only occasionally inspected by experts before loading onto the ship. In most cases, the only inspections to be carried out are those by officials such as customs officers, border control agents, the coastguard and/or a limited number of other competent authorities. Shipment inspections are slightly more common on hazardous cargoes but even then, it is the exception rather than the rule. This situation is accurately reflected in this extract from the container packing guidelines, which clearly sets out the responsibilities for those whose job it is to pack cargoes inside container units:

> From a legal standpoint, an inspection by the ship's command is unnecessary, since the people who pack a container are responsible for ensuring that all regulations are observed. In the case of containers carrying hazardous materials, such compliance must even be confirmed by a signature on the container packing certificate. Accordingly, in the case of FCL/FCL [full container load] and FCL/LCL [less than full container load] containers, the shipper is usually responsible. The bill of lading documents this in the "shipper's load, stowage and count" clause. In the case of LCL/FCL and LCL/LCL containers, those responsible are those instructed to pack. With regard to legal liability, they bear the consequences of any deficiencies or omissions, as far as these are not caused by the customer, for example because the latter fails to pay for necessary work to be carried out or for securing or has only very specifically defined work carried out. If people or the environment are harmed by deficient container packing, those responsible may be prosecuted. The potential for risk is particularly great when packing hazardous materials.

Even so, historically, the results of container inspections have been truly shocking. Inspections of around 70% of containers, swap-bodies, semi-trailers and other "cargo transport units" between 1980 and 2004 have revealed some very obvious and avoidable deficiencies. Theoretically, it is only possible for the crew of a ship to inspect cargo securing where "open" containers are used such as flat racks, open top and side platforms, etc., and only then when the cargo is not additionally covered by tarpaulin or otherwise hidden from view.

3.1.5 Transfer of Containers at Inland Terminals

Cargo handling crews at inland terminals are often put at risk by poorly packed, inadequately secured containers. By the time the container reaches the inland terminal, it is often on its way to its final destination. Depending on the longevity of the journey, and the means of transportation used, the container may have been subjected a vast array of stresses and forces which have the potential to dislodge or damage the container contents. Moreover, those who unpack the container at its final destination are the last people in the logistics chain to be put at risk by packaging deficiencies, poor packing and inadequate cargo securing. To avoid unnecessary and avoidable harm occurring to these workers, it is essential best practices are employed at the very start of the logistics chain.

> ***Note***: *it is worth mentioning the existing container packing guidelines does not mention the dangers posed to personnel who inspect containers in the course of their daily duties such as the police, insurance loss adjusters, and cargo inspectors*

3.2 Packing of Hazardous Materials

A practice which presents particular hazards throughout the logistics chain is the carriage of hazardous materials (HAZMAT). Where HAZMAT is loaded into containers, they should always be located within the immediate door area. This allows easy access in the event of spillage. Unfortunately, it is common practice to place or "hide" HAZMAT under other items of cargo where they are neither easily identified nor reached in the event of an emergency. Without at this point going into individual paragraphs, articles or points of regulations, laws or guidelines, it should be noted that with this type of transport too it is the shipper's duty to ensure proper packing and securing. Poorly packed, inadequately secured container loads may also put innocent third-parties at risk if shifting items of cargo smash through the container walls or cause a vehicle carrying the container to overturn (Figs. 3.6 and 3.7).

In this chapter thus far, we have discussed some of the key points arising from the poor packing of cargoes within containers. When cargoes are insecurely packed, this can lead to lateral movement, which in turn, can destabilise the container. Where the container is stacked onboard a ship, or driven on an open chassis, these lateral movements can result in quite devastating consequences. For this reason, it is of absolute paramount importance that every effort is taken to ensure cargoes are loaded and stowed as securely as possible before the container is locked in transit. It is entirely possible the container may not be opened from the time it leaves the shipper's address to the point of delivery at the receiver's address. This could be hours, days, weeks or even months with the container being manhandled dozens of times in the intervening period. The importance of good housekeeping and securing practices simply cannot be overstated. That said, in this next section, we will examine the causes of damage and loss to cargoes during transit.

Fig. 3.6 Example of an inadequately secured container load

3.3 Container and Cargo Insurance

Approximately three sixths of all the containers used worldwide belong to leasing companies; another two sixths are owned by shipping companies themselves; and the final sixth to other owners. In principle, cargo insurance only covers claims relating to material damage sustained to goods. Exceptionally, the proprietor's interest in the container as packaging may also be insured if a special agreement is reached or through inclusion there of in the damage liability invoice value. Interest in a leased container is not in principle insured. The purpose of packaging is to facilitate the transport of goods and to protect the goods from potential damage. If the above aim has been fulfilled, it is most common for only the packaging to be damaged, and not the goods themselves. To that end, the following may be inferred from the general principles of marine insurance law. If the packaging does not have any separate, intrinsic value, then no independent loss may occur. Packing paper, which is discarded after the arrival of the cargo and suffers wear during transport, cannot therefore form the basis of a claim for damages. If the goods have reached their destination, the only relevant question is whether the goods have suffered loss and not whether the packaging is damaged. This also applies to cases.

Consideration can only be given to any such claim if the goods have depreciated in value solely as a result of the damage to the packaging. This is the case for example

Fig. 3.7 Example of a properly stowed and secured container

with preserved foods and certain branded goods. Another example is cement, which retains its normal commercial value only in its original bags. If cement which is in itself sound is sold in unmarked bags because the original bags have been damaged, it automatically suffers depreciation. In such cases, compensation may be paid despite the damage being limited to the bags. Otherwise, compensation may only be paid for packaging damage if a separate value is assigned to the packaging either in the policy or in the invoice (e.g., barrels or containers). This only applies if the packaging is expected to have a longer service life than the duration of one transport operation. A different matter entirely is the cost of repairing or replacing the packaging, if this has to be performed en route. It may be possible to claim compensation for these costs, though not in the form of packaging losses, but as loss minimisation costs, which are expended to return the goods to a transportable state and to prevent losses on the remaining insured journey. For such a claim to be allowable, the packaging must have been satisfactory at the start of the journey. If the journey is complete or the remainder of the journey is uninsured, the insurer will not meet such costs. This is simply because the insurance contract has expired, which means no more insured losses can be incurred. Because of the uncertainties involved in using cargo insurance to insure a container as packaging, it is better to provide separate hull insurance by way of special conditions for container insurance. As far as the proprietor's interest is concerned, double insurance cover may then be provided by the cargo insurance and the container and hull insurance.

3.4 Container and Hull Insurance

Containers cannot be easily defined as either hull or cargo. To distinguish them more readily for the benefit of insurance, additional conditions may be levied which are known as hull clauses. Nonetheless, the nature of the cargo remains the predominant factor, for which reason the special conditions should only be agreed in conjunction with the stipulated cargo insurance conditions. In general, therefore, containers are not covered as packaging by cargo insurance, but instead by special 'all-risk' insurance by way of special conditions levied as container insurance. However, compensation is only paid for damage to machinery and tarpaulins, if it is an immediate consequence of an accident involving the means of transport or of stranding. The special conditions may also be applied to leased containers, if the lessee bears the risk. However, to clarify the situation, this should be expressly stated in the corresponding "written conditions."

3.5 Insurance Liabilities

3.5.1 Containers Provided by the Shipper

Whether the shipper uses their own container or one borrowed, rented or leased from a third-party, they are using the container as packaging. The transport company is liable for material damage to the cargo and the container under the corresponding carriage conditions, since, when calculating liability, the gross weight is to be understood to mean the total weight including the packaging (Fig. 3.8).

3.5.2 Containers Provided by the Carrier

There are several different conceivable situations to consider, such as whether the carrier uses their own container or one borrowed, rented or leased from a third-party:

- *Permanently mounted container*. If the container is mounted permanently on the chassis of a lorry, i.e., completely fixed thereto, it fulfils the requirements of a fundamental component of the vehicle. This means the transport company is liable for any material damage to the cargo under the corresponding carriage

Fig. 3.8 Generic leased GP container provided by the shipper

conditions. This liability does not include the weight of the container, since the contract of carriage relates only to the cargo and not additionally to the container.

- *Leased containers.* If, on the express wish of the shipper, the carrier makes available to the shipper a freely handleable container which is not connected permanently to the vehicle, for instance because the shipper's intention is for the container to be transferred between several means of transport during carriage, the carrier will demand additional remuneration for handover of the container. This gives rise to a leasing agreement between the carrier and the shipper. The transport company is liable for material damage to the cargo under the corresponding carriage conditions. Material damage to the container is covered by the content of the leasing agreement.
- *Containers as ancillary transportation equipment.* If the carrier makes the container available to the shipper to simplify transport without any special additional agreement of the type mentioned above, the contract of carriage as a rule relates only to the transport of cargo in a vehicle suitable, therefore. The transport company is liable for material damage to the cargo under the corresponding carriage conditions. This liability does not include the weight of the container, since the contract of carriage relates only to the cargo and not additionally to the container (Fig. 3.9).

In the latter part of this chapter, we have discussed some of the pertinent points around container and cargo insurance. Needless to say, this is an extraordinarily complex area, and this book can do it no justice. For further information on maritime insurance, readers are advised to seek out professional publications.

Fig. 3.9 Carrier's container with the carrier's identifying mark clearly legible

Part II
Causes of Damage and Loss During Transit

Chapter 4
Damage and Loss During Transit Caused by Mechanical Stress

There are many companies around the world in which responsible, quality-conscious staff plan and devise optimum transit procedures so that goods can be shipped whilst suffering the very minimum of damage. With such companies, prevention of damage is a priority since it is common knowledge that economic success is to a great extent dependent on customer satisfaction. Considerable effort is taken to avoid shipping damage altogether at best, or at worst, to restrict the extent of such damage. But what about those companies and the people who work for them where the priority is maximising profit over customer satisfaction? It is a sad reality that such companies do exist. Inevitably, their selfish actions or omissions are some of the leading causes of cargo related damage and injuries. In these circumstances it is not the ship's senior management or crew who have fallen short of the standards expected, but those who are responsible for packing the containers in the first instance. In the event of cargo related incidents, it is often quite obvious which containers have been properly packed and stowed and which have not. Analysis of shipping losses frequently focuses on technical issues and takes human shortcomings more or less as read. Such an approach is mistaken, ultimately meaning that the necessary staff training is not provided as it is not regarded as sufficiently important, an attitude that is unfortunately shared by the majority of decision-makers. Insufficient knowledge and inadequate skills on the part of many of those involved in cargo shipping are the root causes of subsequent losses. Errors in the planning and execution of transport operations are for the most part due to human failure.

Poor understanding on the part of many of those in charge about the complex interrelationships involved in the provision of transportation services together with what is in most cases a very low level of training among the personnel involved with the physical handling of cargoes conspire to cost the world's economy billions of dollars every year. Insufficient knowledge and inadequate skills on the part of very many of those involved in cargo shipping are the root causes of subsequent losses. Basic errors are sometimes made as early as during the design and production of

A. A. Olsen and F. Karkori, *Containerized Cargo Handling and Stowage*, Springer Series on Naval Architecture, Marine Engineering, Shipbuilding and Shipping 22,
https://doi.org/10.1007/978-3-031-63329-4_4

the goods which are ultimately to be shipped. While this area will be addressed separately later, some explanatory examples are provided below.

4.1 Examples of Inadequate Securing

Even trained packers would have difficulty packing this cargo unit safely into a container, in fact it would be impossible without incurring additional labour and materials costs (Fig. 4.1).

Given the differing heights of the drums, the cargo unit in Fig. 4.2 is even more difficult to pack into a container (Fig. 4.3).

Designers not only have to take account of a machine's subsequent function, but also of the fact that it must first be transported to where it will be used and installed there without suffering damage. Every machine should have slinging, and lashing points, so that it can be handled and secured without problem.

In the example given in Fig. 4.4, the earth borer has no lashing points. In such cases, securing means are generally attached arbitrarily to components which are not suitable for this purpose. The absence of lashing points is particularly disadvantageous since this machine will be used at many different sites. Every time the machine is moved, more effort will be required for securing, so increasing the shipping risk, which is why attention must be paid to lashing points for transport securing at the design stage (Figs. 4.5 and 4.6).

Fig. 4.1 Non-regulation complaint dangerous cargo unit

Fig. 4.2 Non-regulation compliant dangerous cargo unit

4.2 Loss Prevention Through Training

Referring to the container packing guidelines, the section "Scope" contains the following important statement:

> These guidelines, which are not all-inclusive, are essential to the safe packing of [containers] by those responsible for the packing and securing of the cargo and by those whose task it is to train people to pack such units. Training is essential if safety standards are to be maintained.

In accordance with Sect. 7.2 of the container packing guidelines, training and development is a management responsibility:

> Management should ensure that all personnel involved in the packing of cargo in [containers] or in the supervision thereof are adequately trained and appropriately qualified, commensurate with their responsibilities within their organisation.

Furthermore, Sect. 7.3 states the following requirement for personnel:

> All persons engaged in the transport or packing of cargo in [containers] should receive training on the safe packing of cargo in [containers], commensurate with their responsibilities.

Lastly, Sect. 7.4 offers guidance relating to training; with the various subsections providing as follows:

Fig. 4.3 Machine without securing means (Not fit for shipping)

Fig. 4.4 Earth borer without securing means (Not fit for shipping)

Fig. 4.5 Properly secured container load

> 7.4.1 General awareness/familiarisation training:
>
> All persons should receive training on the safe transport and packing of cargo, commensurate with their duties. The training should be designed to provide an appreciation of the consequences of badly packed and secured cargo in [containers], the legal requirements, the magnitude of forces which may act on cargo during road, rail and sea transport, as well as basic principles of packing and securing of cargoes in [containers].
>
> 7.4.2 Function-specific training:
>
> All persons should receive detailed training concerning specific requirements for the transport and packing of cargo in [containers] which are applicable to the functions that they perform.
>
> 7.4.3 Verification:
>
> The adequacy of the knowledge of any person to be employed in work involving the packing of cargo in [containers] should be verified or appropriate training provided. This should be supplemented by periodic training, as deemed appropriate by the regulatory authority.

In summary, Sect. 7.5 "Recommended course syllabus—overview" states:

> The adequacy of the knowledge of any persons to be employed in work involving the packing of cargoes in [containers] should be verified, in the absence of which appropriate training is considered essential and should be provided. The function-specific training should be commensurate with the duties required to be performed by an individual in the packing and securing of cargo in [containers]. Topics for consideration, to be included in the training as appropriate, are given in Annex 6.

Fig. 4.6 Properly secured container load

Annex 6 of the container packing guidelines presents a list of standard topics to be included in training programmes for personnel responsible for the packing and securing of cargoes in containers. These include (Table 4.1):

Some readers may be aware of the case of the French registered container ship *Sherbro* which lost 88 containers overboard in the English Channel in 1993. As a result, hundreds of thousands of plastic sachets containing the pesticide *Apron*$^{+}$ entered the sea causing serious environmental damage across the coasts of France, Holland and Germany. This incident serves to highlight the importance of comprehensive and regular recurring training sessions to ensure sufficient familiarity with best practices. As far as the *Sherbro* is concerned, an investigation by the French maritime authorities concluded the cargo in the offending container had not been sufficiently secured. As a result, the integrity of the container was undermined from the inside leading to the collapse of the stack (Figs. 4.7, 4.8, 4.9, 4.10 and 4.11).

As the incident involving the *Sherbro* demonstrates, poor packing practices can have wide-reaching consequences that occur long after the container is packed and dispatched from the shipper's premises. In the next part of this chapter, we will look at how mechanical stresses, which are exerted on containers during transit, can lead to structural failure. Often this structural failure is the result of impact stress from poorly packed cargoes rather than from the stresses and forces exerted on the container from the external environment.

Table 4.1 Standard topics to be included in training programmes for personnel responsible for the packing and securing of cargoes in containers

(1) The consequences of badly packed and secured cargo (a) Injuries to persons and damage to the environment (b) Damage to ships and containers (c) Damage to cargo; and (d) Economic consequences
(2) Liabilities (a) Different parties involved in cargo transport (b) Legal responsibility (c) Goodwill responsibility; and (d) Quality assurance
(3) Forces acting on the cargo during transport (a) Road transport (b) Rail transport; and (c) Sea transport
(4) Basic principles for cargo packing and securing (a) Prevention from sliding (b) Prevention from tipping (c) Influence of friction (d) Basic principles for cargo securing; and (e) Dimensions of securing arrangements for combined transportation
(5) Container (CTU) types (a) Containers (b) Flats (c) Swap-bodies (d) Road vehicles; and (e) Rail wagons
(6) Cargo care consciousness and cargo planning (a) Choice of transport (b) Means, choice of container type (c) Check of container prior to packing (d) Cargo distribution in containers
(7) Different methods for cargo packing and securing (a) Lashing (b) Blocking and bracing; and (c) Increasing friction
(8) Equipment for securing and protection of cargo (a) Fixed equipment on containers (b) Reusable cargo securing equipment (c) One-way equipment; and (d) Inspection and rejection of securing equipment

(continued)

Table 4.1 (continued)

(9) Packing and securing of unitised cargo (Bulk) (a) Cases (Boxes) (b) Palletised cargoes (c) Bales and bundles (d) Bags on pallets (e) Flexible IBCs (f) Slabs and panels (g) Barrels (h) Pipes; and (i) Cartons
(10) Packing and securing of non-unitised cargo (Break-bulk) (a) Different types of packaged cargoes loaded together (b) Packing of heavy and light cargoes together (c) Packing of rigid and non-rigid cargoes together (d) Packing of long and short cargoes together (e) Packing of high and low cargoes together; and (f) Packing of liquid and dry cargoes together
(11) Packing and securing of paper products (a) General guidelines for the packing and securing of paper products (b) Vertical rolls (c) Horizontal rolls; and (d) Sheet paper on pallets
(12) Packing and securing of cargo requiring special techniques (a) Steel coils (b) Cable drums (Reels) (c) Wire coils (d) Steel slabs (e) Steel plates (Sheet) (f) Big pipes (g) Stone blocks; and (h) Machines
(13) Packing and securing of dangerous cargoes (a) Regulations for the transport of dangerous cargoes (b) Definitions (c) Packing regulations (d) Packing, separation and securing (e) Labelling and placarding (f) Information transfer when transporting dangerous cargoes; and (g) Liabilities

4.3 Mechanical Stresses in Maritime Transport

Generally speaking, the mechanical stresses experienced during maritime transport are divided into two distinct groups:

- Avoidable shipping stresses, and
- Unavoidable shipping stresses.

Fig. 4.7 Container ship *Seatrain Bennington*, circa 1980

4.3.1 Avoidable Shipping Stresses

Avoidable shipping stresses are caused primarily because of human shortcomings. Incidents of damage often occur because cargo has been inadequately packed, stowed or secured, or because equipment has been used incorrectly. The container packing guidelines attempts to counter these human factors by providing—in Annex 6—a list of topics to be included in training programmes for the packing and securing of cargoes in containers (see previous section and chapter), which requires that packers are instructed in the forces which act on the cargo during transport. This includes

Fig. 4.8 *MSC Nederland* collision, Bayport, Texas, 29 October 2011

Fig. 4.9 Collapsed container

Fig. 4.10 Example of a poorly stacked container

all methods of shipping such as road, rail and sea. This requirement as stated in the container packing guidelines deserves dedicated support as observation of day-to-day practice reveals a frightening lack of knowledge in this area.

4.3.2 Unavoidable Shipping Stresses

Unavoidable shipping stresses are determined by the nature of the transport operation and lies outside the sphere of human influence. Significant numbers of incidents of damage are attributable to the fact that many of those involved in shipping are unaware of shipping stresses or are unable to assess them correctly. This ought not to be the case as information of this kind is available from many sources. As an example, Sect. 4.1 of the container packing guidelines, "General conditions," provides valuable information on how cargoes may be affected during transit, and in particular, during voyages at sea. Information about road, rail and inland waterway transport, and inland cargo handling, are also discussed in a clear and concise format. Most national authorities provide equally important sources of guidance. For road transportation, the *German Vehicular Accident Prevention Regulations* (BGV D 29) sets out the safety parameters for carrying loads by road. Values for the stresses arising from railway transport are often obtainable from the respective national railway

Fig. 4.11 Example of a well stacked container

authorities. In the United Kingdom, this is the Department of Transport (railway division), or in Germany, the Deutsche Bahn AG.

It is not without reason that the container packing guidelines contains the following important statement:

> These guidelines are not intended to conflict with, or to replace or supersede, any existing regulations or recommendations which may concern the carriage of cargo in containers.

In other words, there is no intention for the container packing guidelines to waste time for no good reason; much of the information provided has already been published in other guidelines and specialist publications. The term 'carriage' denotes the movement of cargo using one or more modes of transport. This means in effect the terms 'carriage' and 'shipping' are superordinate terms which cover the storage, handling and transport of goods as well as any associated operations. In terms of mechanical stresses related directly to maritime transport, therefore, we can refer ourselves to the flow diagram below:

It is extremely difficult to put exact values on storage, handling and transport stresses because they are determined by many different parameters. In addition to the stresses shown in Fig. 4.12, other factors of course also play a part, such as the means of transport used, the selected routes, the nature of the roads and railroads or the prevailing weather conditions during sea passage. Protection from shipping damage may be provided with regard to economic viability without consequently

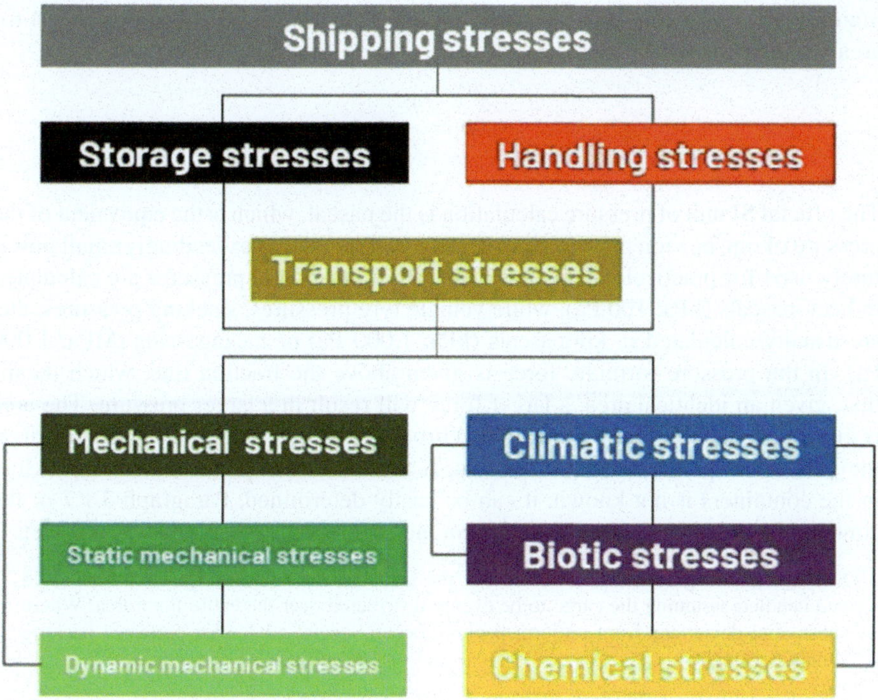

Fig. 4.12 Overview of shipping stresses

being less effective. Effective and economic protective measures can be implemented if enough is known about the stresses to which the cargo is exposed during transport. As experience has shown and measurements have confirmed, the shipping stresses of different modes of transport differ markedly. To that end, the value of good quality training and quality control procedures should be recognised and given due priority in cargo securing.

4.3.3 Static Mechanical Stresses

Static mechanical stresses primarily cause damage through harmful levels of pressure. In the transport sector, cargoes are often thought of as only *light* or *heavy*. This of course is flawed as it neglects the fact these relative terms. It is essential therefore to verify the *actual* weight of the cargo before it is despatched. To calculate the pressure value that is exerted on the load, the following calculation may be used:

$$Pressure = \frac{Force}{Area}$$

Stated again, using conventional abbreviations: (p for pressure, F for force and A for area):

$$p = \frac{F}{A}$$

The official SI unit of pressure calculation is the pascal, which is the equivalent of the pressure of one newton per square metre ($\text{Pa} = \text{N/m}^2$). This exceedingly small unit is rarely used for practical calculations. Instead, atmospheric pressures are calculated in hectopascals (hPa; 100 Pa), while vehicle tyre pressures, stacking pressures, etc. are usually calculated in kilopascals (kPa; 1,000 Pa) or megapascals (MPa; 1,000 Pa). In the pressure formula, force is given above the fraction line, which means that, given an identical area, a larger force will result in a larger pressure. The area is given beneath the fraction line, which broadly translates into the smaller the area, the greater the pressure, and vice versa. If the maximum permitted stack pressure in the containers is not known, it can be easily determined. Paragraph 3.1.7 of the container packing guidelines provides an indirect indication of how this is done:

> Stowage planning should take account of the fact that containers are generally designed and handled assuming the cargo to be evenly distributed over the entire floor area. Where substantial deviations from uniform packing occur, special advice for preferred packing should be sought.

To aid personnel involved in the planning and movement of containers, the permitted payload of a container is calculated as the net mass which is obtained when the tare weight is deducted from the maximum gross mass. The following table (Table 4.2) provides several worked examples for two different 20' and 40' containers:

In typical cargo loading literature, values of 14 kN/m^2 for 20' containers and 10 kN/m^2 for 40' containers are frequently given as the maximum floor loading values. However, as has been shown, this value varies from container to container and can be easily calculated. The maximum stack pressure in a container should be the same as the maximum container floor loading value where the container is packed absolutely evenly. Official pressure units are only very rarely used for calculations in the transport sector. Since it is masses which are handled, pressure is often assumed to be the product of dividing the mass by its area. Instead of the official units (i.e., kg/cm^2), it is more common for kg/m^2 or metric tonnes/m^2 to be used instead. For

Table 4.2 Example calculation of permitted stack pressure in containers

Payload in kilogrammes	Force in newtons	Internal length in metres	Internal width in metres	Pressure in pascals	Pressure in kilo pascals
18,370	180,209.70	5,935	2,370	12,811.77	12.81
21,780	213,663.80	5,895	2,393	15,146.08	15.15
26,700	261,927.00	12,033	2,352	9,254.84	9.25
29,650	290,866.50	12,069	2,350	10,255.45	10.26

English-speaking countries that have retained the imperial system, pounds per square inch (psi) is more often used instead. While, strictly speaking, this is incorrect, such units are so widely used in practice and are more readily understood by the majority of personnel involved, that they have become de facto. The following relations apply when converting such units into the official units and vice versa:

- Hectopascals (hPa) correspond to the old unit millibar (mbar).
- One bar corresponds to 100 kPa or 0.1 MPa.
- Kilogram/square centimetre (kg/cm^2) corresponds to one bar or 100 kPa or 0.1 MPa.
- One kilogram/square centimetre corresponds to 10 metric tonnes/m^2.
- One kilopascal (kPa) = 1,000 N/m^2 and corresponds to 100 kg/m^2 or 0.1 metric tonnes/m^2; and
- One kilogram/square centimetre corresponds to 14 lb/square inch (psi).

For context, a forklift truck tyre pumped up to a pressure of 8.5 bar accordingly has a pressure of 85 kPa, 0.085 MPa or 8.5 kg/cm^2 or 85 metric tonnes/m^2. In the English-speaking world, the tyre is at a pressure of 119 psi.

4.3.4 Stacking Crush Pressure

The 'stacking crush pressure' is defined as the pressure exerted by an item of cargo on underlying items of cargo or components. Crush pressures are significantly higher for some press board pallets and similar types. Unless appropriate measures are taken to distribute the pressure, it is not possible to guarantee that transport will proceed without damage. Extremely high pressures may occur if buckets, pails, barrels and similar cargoes are packed without an intermediate layer. For example, if a barrel has a mass of 183.5 kg, and a total area of the barrel rim of 90 cm^2 is placed on the floor, a pressure of 200 kN is generated. Assuming the barrel rim has a width of 0.5 cm, any offset stowage of another barrel over the top will result in a contact area of 0.00015 m^2, generating a pressure of 12,000 kPa. The only way forwarding personnel can prevent damage due to static mechanical stresses is by taking steps to distribute this pressure, as it is not possible to modify the shipping package themselves or to change the masses involved (Tables 4.3 and 4.4).

However, many warehouses and packing centres do not hold stocks of appropriate materials in the form of walking boards, hardboard, wooden dunnage, etc. Moreover, appropriate intermediate layers are often not used due to ignorance, for reasons of false economy, or other causes. Even so, it is essential to take the nature of the shipping packages into account. A skilled appraisal of the external appearance, such as the base of cases, cargo units and pallets, can provide valuable information as to how to load them. Efficient handling methods using lifting gear or ground conveyors usually require the use of beams or squared lumber to be placed beneath or between the items of cargo. Such use can produce damaging levels of pressure. Even strapping packages may cause damage due to excessive pressure where edge protectors are not

Table 4.3 Crush pressures for different pallets

Pallet structure	Force in newtons	Area of base in m^2	Crush pressure kN/m^2 = kPa
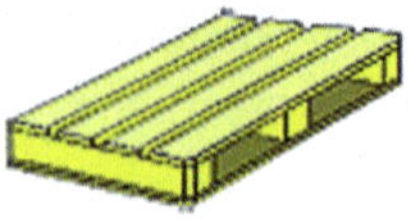	9,810	1,000	9.81
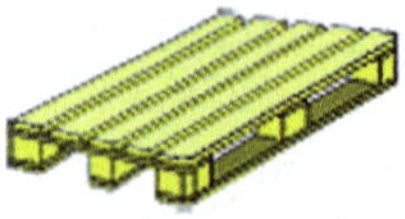	9,810	0.432	22.71
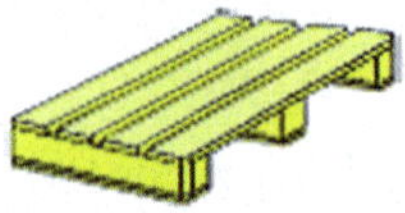	9,810	0.288	34.06
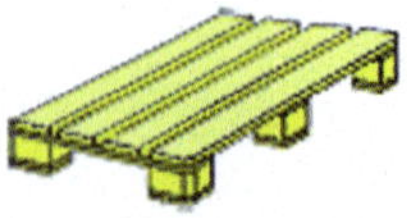	9,810	0.864	113.54
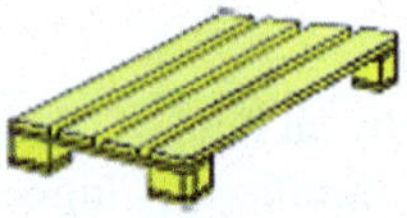	9,810	0.076	170.31

used to distribute the pressure. Extremely high pressures may be generated by narrow or point bearing areas. Negative effects are magnified by multiple tiers or the angles of rest which arise, for example, when stowing pipes, rolls or similar cargoes. When packing goods such as pipes in tiers, this can generate extremely high pressures, as the pipes are only bearing on narrow strips corresponding to the edge lengths of the lumber. The resultant stack pressures are distinctly higher than in either of the two possible cant-line stows. Due to the greater angles of rest or spread, cant-line stow 2 generates greater forces than cant-line stow 1. Further explanations in this respect are discussed in later chapters of this book. The crush pressures bearing down on underlying layers of packages, or the container floor is increased by the ship's motion. It is often forgotten that pressure not only acts downwards from above due to the force of gravity but may also act laterally because of dynamic stresses. In this case, the packages are pressed against one another or against the container walls. This pressure, and in particular slipping and subsequent collision of inadequately secured packages, may result in damage. Very few packers anticipate "home-grown" static mechanical stresses due to inadequate cargo securing, but such stresses are predictable: if cargo stacks can move, critical and dangerously high pressures may arise as tipping occurs due to the resultant small contact area. The reduction in contact

Table 4.4 Drum contact area versus force [N] versus area [m^2] versus pressure [kPa]

Type of packing/Contact areas		Force (N)	Area (m^2)	Pressure (kPa)
	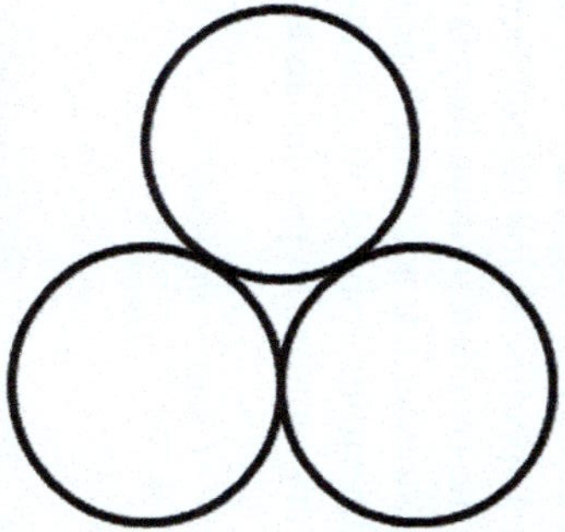	1,800	0.0090	200
	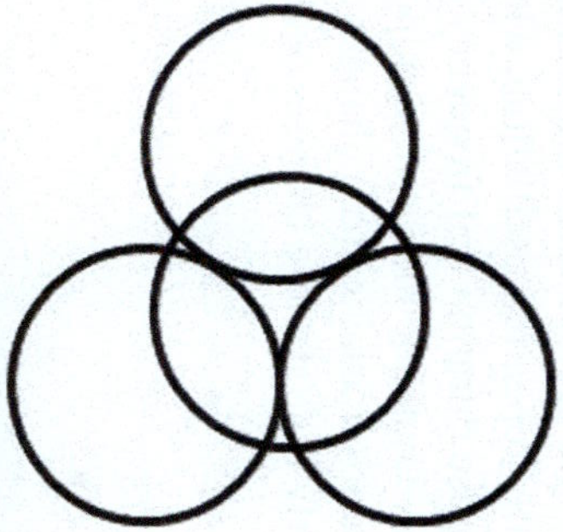	1,800	6×0.5 cm $\times$ 0.5 cm $\times$ 1.5 cm 0.00015 m^2	12,000

area with a simultaneous increase in pressure may cause the edges of the packages to cave in, resulting in damage to the shipping packages and collapse of the stack. In particularly damaging cases, this may result in the loss of whole batches or even of the container. The actual tipping, collision or collapse of packages will be addressed in the dynamic mechanical shipping stresses section.

4.3.5 Dynamic Mechanical Stresses

In relation to dynamic stresses, a primary distinction is drawn between vibration and jolting. In physical terms, the two phenomena are similar, but the distinction is made because they differ in their effects on packages and on means of transport. Vibration comprises periodic oscillations which occur in large numbers, such as vehicle or engine vibration, movement of the ship in heavy seas, and so forth. Jolting comprises occasional events, as may be observed on impact, dropping or tipping. Bumping, kicking or switching impacts are all paint a sufficiently clear picture. The absolute magnitude of both types of stress is measured from their amplitude. For vibration, the second most important parameter is frequency, i.e., the number of periodically repeating oscillations within a specific period (cycles per second). For jolting, pulse duration and frequency are the two determining factors in addition to amplitude. Dynamic mechanical shipping stresses are accordingly primarily caused by acceleration arising from changes in direction or speed. Acceleration values are particularly high if these changes occur very rapidly. To put this another way:

$$Linear\ acceleration = \frac{Change\ in\ speed}{Time}$$

as

$$Speed = \frac{Distance}{Time}$$

Therefore, it follows that linear acceleration may be calculated as follows:

$$Linear\ acceleration = \frac{Distance}{Time \times Time} = \frac{Distance}{Time^2}$$

In other words, if a car accelerates from 0 to 108 km/h in 12 s, its speed has changed by 9 km/h or 2.5 m/s each second. Its acceleration, a, is thus 9 km/h/s or, stated in other terms: 2.5 m s s or 2.5 m/s. If a lorry traveling at a speed of 90 km/h takes five seconds to come to a standstill, it is decelerating at 18 km/h per second or 5 m/s per second. To look at this another way, if we imagine a ship which, while pitching in a heavy sea, suffers a loss of speed from 21 to 9.3 knots within two seconds, she undergoes negative acceleration of –11.7 kn/2 s = –5.85 kn/s = –10.834 km/h/s =

–3 m/s^2. The ship has thus decelerated. Alternatively, acceleration arising from a change in direction may be calculated using the following formula:

$$Radial\ acceleration = \frac{Speed^2}{Curve\ radius}$$

If a road vehicle takes a tight curve with a curve radius of 20 m at a speed of 36 km/h, which corresponds to 10 m/s, the vehicle undergoes radial acceleration of 5 m/s^2. Driving through potholes also causes radial acceleration as there is a change in direction from the horizontal to the vertical and back again. In vehicles with poor shock absorption, this causes considerable acceleration as the curve radii are small and the speeds at which they are driven through are high. In this case, only the suspension and other design parameters protect the vehicle and its load from severe damage. In roll trailers without suspension, rough ground in the terminal may result in elevated levels of acceleration and subsequent damage to the cargo. It is a frequent occurrence for shipping packages to fall off the trailers during such transport operations within cargo handling facilities.

The action of gravity subjects all objects on the Earth to downward acceleration. This is referred to as acceleration due to gravity. The value of this acceleration is constant. Acceleration due to gravity, when rounded to two decimal places, is 9.81 m/s^2 and is stated as '1 g.' For rough calculations, 1 g may be assumed to have a value of 10 m/s^2. In order to appreciate quite what an acceleration of 1 g means, we can imagine a road vehicle accelerating from 0 to 106 km/h in three seconds. The calculation is shown as follows:

$$a = \frac{\Delta v}{\Delta v} \frac{105.948\ \text{km}}{3\ \text{s}} \frac{\cdot 1{,}000\ \text{m}}{\text{h}} \frac{}{\text{km}} \frac{\text{h}}{3{,}600\ \text{s}} = 9.81 \frac{\text{m}}{\text{s}^2}$$

If the magnitude of the acceleration is known, the force it applies to a mass may be calculated using Newton's second law of motion, where:

$$force = mass \cdot acceleration$$

The formulae is conventionally written as:

$$f = m \cdot a \left[N = \text{kg} \cdot \frac{\text{m}}{\text{s}^2}\right]$$

Or

$$\left[kN = \text{t} \cdot \frac{\text{m}}{\text{s}^2}\right]$$

where

$$f = m \cdot a\left(g = \frac{9.81\,\mathrm{m}}{\mathrm{s}^2}\right)$$

On Earth, force and weight are physically identical. As a result of acceleration due to gravity, an object with a mass of 100 kg exerts a force of $F = 100$ kg $\times$ 9.81 m/s^2 $= 981$ N on whatever is supporting it. Stated in other units, this force amounts to 98.1 daN (decanewtons) or 0.981 kN (kilonewtons). Weight-force should accordingly be taken to mean the force with which an object is attracted by the Earth's gravity. This force always acts perpendicularly downwards. In this instance, the formula used is as above, but replacing the general abbreviation F for Force with g:

$$g = m \cdot a\left[N = \mathrm{kN} \cdot \frac{\mathrm{m}}{\mathrm{s}^2}\right]$$

Or

$$\left[kN = \mathrm{t} \cdot \frac{\mathrm{m}}{\mathrm{s}^2}\right]$$

Some publications such as, for example, the VDI Guidelines, mention normal force. The abbreviation F is provided with the index N, giving $FN =$ normal force. Force and acceleration are vector quantities, i.e., quantities in which both magnitude and direction must be stated. Acceleration due to gravity g, however, is always in the same direction as weight. A motorcycle rider traveling around a corner with a radius of 50 m at 72 km/h will experience radial acceleration of 8 m/s^2:

$$ar = \frac{v^2}{t} = 20\frac{\mathrm{m}}{\mathrm{s}} \times \frac{20\,\mathrm{m/s}}{50\,\mathrm{m}} = 8\,\mathrm{m/s}^2$$

In this instance, vector B of acceleration due to gravity of 1 g or 9.81 m/s^2 acts perpendicularly downwards, whilst vector A of radial acceleration of 8 m/s^2 acts horizontally in the direction of the curve and corresponds to 81.5% of acceleration due to gravity (0.815 g). Vector C is the resultant acceleration of 12.658 m/s^2. The direction of the resultant vector may also be described as the apparent vertical. A motorcycle rider must lean into the curve by 39.2°, relative to the vertical, in order to offset this force. Their motorcycle will then be at an angle of 50.8° to the horizontal:

$$\cdot\frac{\tan}{alpha} = \frac{9.81\,\mathrm{m/s}^2}{8\,\mathrm{m/s}^2} = 1.22625$$

The driver will be pressed down towards the ground with approximately 1.3 times their normal force:

$$\frac{12.658\,\mathrm{m/s}^2}{9.81\,\mathrm{m/s}^2}$$

If, instead of being stated in m/s^2, acceleration values are given in fractions or multiples of acceleration due to gravity (g), the forces may be determined with sufficient accuracy directly in decanewtons (daN), if mass is stated in kilograms (kg). If the mass is located within a frame of reference under acceleration, the vector exhibits an inertia force acting on the mass in the opposite direction to the acceleration. An example of this is the automobile driver who, on accelerating, is pressed back into their seat (positive acceleration and negative inertia force) or, on braking, is pulled forward (negative acceleration and positive inertia force).

Friction forces are passive forces, which come into play when objects on a substrate are to be set in motion by forces (static friction) or are to be kept in motion (sliding friction and rolling friction). Static friction also comes into play when the direction of motion of an object (already subject to sliding or rolling friction) is changed. A vehicle cornering is a good example, as all that keeps the vehicle on its desired path is the sufficient static friction of the tyres (in reality, the phenomena at play are more complex than shown here). This is because friction forces counteract incident forces. The following formulae may be used to calculate the size of friction forces:

$$\begin{aligned} \textit{static friction} &= \textit{normal force perpendicular t sliding surface} \\ &\quad \cdot \textit{friction coefficient for static friction} \end{aligned}$$

$$f_0 = n \cdot \mu_0$$

$$\begin{aligned} \textit{sliding friction} &= \textit{normal force perpendicular to sliding surface} \\ &\quad \cdot \textit{friction coefficient for sliding friction} \end{aligned}$$

$$f = n \cdot \mu$$

The normal force perpendicular to the sliding surface may be stated in N, daN, kN or similar units of force. Since the various coefficients of friction are dimensionless, the determined friction force has the same calculation dimension as for the normal force perpendicular to the sliding surface. Since friction is of great significance to cargo securing, more detailed explanations will be provided in another section. Vibration and jolting of the vehicles and equipment used, together with the fundamental and resonant vibration of goods, packaging and cushioning material together largely define the risk of damage to which the goods are exposed during carriage. The magnitude of the pulses, i.e., the duration of forces or velocity of masses, is of vital significance in this connection.

Note: if sensitive goods are to be packaged and carried without suffering damage, it is essential to be aware not only of the frequency spectra which arises during storage, handling and transport, but also of the natural frequencies of the goods being shipped, the shipping packages, cushioning material, etc.

In the high frequency range, extremely high acceleration values of several dozen to a few hundred g have been measured. However, due to the mass inertia of the shipping packages, such short duration impacts rarely cause damage due to inadequate cargo securing. However, the shipper must provide appropriate packaging for sensitive goods in order to mitigate such vibration. In the low frequency range, acceleration values encountered during transport and handling are of the order of some fractions of one g to several g. The longer period of action may result in shifting of the cargo and consequent mechanical damage. These low "normal values" applying to cargo securing on or in vehicles are stated by various organisations. Those involved with shipping should therefore be aware of the typical g values. The effects of dynamic mechanical shipping stresses are frequently underestimated, despite the "normal values" being widely available or easily obtainable. For instance, the container packing guidelines contains various comments on this very issue.

Note: the following comments may be made regarding the tables shown in Sect. 1.9 of the container packing guidelines: the wording/notation is incorrect in that forces are stated in the column headings, while the level of acceleration caused by such forces is stated in the associated fields. If forces are the intended meaning, "g" as a measure of acceleration due to gravity of 9.81 m/s^2 ought to be replaced by "g" as a unit for the normal force. If, however, the values in the table are intended to indicate the level of acceleration, the word "forces" in the column headings ought in each case to be replaced by "acceleration." The same applies to the wording of the footnotes to the tables.

In this chapter we have discussed some of the leading causes for damage and loss resulting from the poor packing and securing of cargoes. It simply cannot be stressed enough how important it is for packers to ensure goods are stowed and lashed appropriately to avoid the risk of damaging the contents of the container, and more importantly, to avoid causing structural damage to the container itself.

Chapter 5
Mechanical Stress During Transport

In the previous chapter, we discussed some of the main causes and consequences for damage and loss to cargo during transit, when caused by mechanical stress. In this chapter, we will look at the various forms of mechanical stress that can cause damage and loss during cargo handling. Before we do, however, it is probably worth noting that Paragraph 1.1 of Sect. 5.1 "General conditions" of the container packing guidelines clearly states that:

> Voyages are made in a variety of weather conditions which are likely to exert a combination of forces upon the ship and its cargo over a prolonged period. Such forces may arise from pitching, rolling, heaving, surging, yawing or swaying or a combination of any two or more.

Moreover, Paragraph 1.2 continues:

> Packing and securing of cargo into/onto a [container] should be carried out with this in mind. It should never be assumed that the weather will be calm, and the sea smooth or that securing methods used for land transport will always be adequate at sea.

Being at the very beginning of the container packing guidelines, this guidance is readily available and as such, there is no proximal excuse for ignorance. With that in mind, we can begin to discuss the mechanical stresses commonly exerted during container transportation.

5.1 Mechanical Stresses During Maritime Transport

The acceleration values which may be anticipated in maritime transport depend on the shape of the surface or sub-surface vessel, its beam, the position of the centre of gravity and centre of buoyancy, and similar parameters which determine the behaviour of ships at sea. All kinds of ship movement may be divided into three types of linear motion and three types of rotational motion. It can, in general, be

A. A. Olsen and F. Karkori, *Containerized Cargo Handling and Stowage*, Springer Series on Naval Architecture, Marine Engineering, Shipbuilding and Shipping 22,
https://doi.org/10.1007/978-3-031-63329-4_5

stated that the outwardly directed centrifugal accelerations brought about by any rotational motion are not significant. This accordingly applies to yawing, pitching and rolling. Yawing involves the rotation of the ship around its vertical axis. This occurs due to the impossibility of steering a ship on an absolutely straight course. Depending on sea conditions and rudder deflection, the ship will swing around its projected course. Yawing is not usually a cause of shipping damage. Alternatively, heaving involves the upward and downward acceleration of the ship along her vertical axis. Only in absolute calm are upward and downward motions at equilibrium. This results in the ship floating at rest. Buoyancy varies as the ship proceeds through wave crests and troughs. If the wave troughs predominate, buoyancy falls and the ship "sinks." If the wave crests predominate, the ship "rises". Such constant oscillation has a marked effect on the containers and their contents. With surging and swaying, the sea's motion accelerates and decelerates the ship forward and backward and side to side. Depending on the lie of the vessel, these movements may occur on all axes and not merely, for example, on the horizontal or vertical axes.

If a vessel's forebody is on one side of a wave crest and the afterbody on the other side, the hull may be subjected to considerable torsional forces. During pitching, the ship is lifted at the bow and lowered at the stern and vice versa. Pitching angles vary with the length of the vessel. On short vessels, the pitching angles are usually within the range of between 5 and 8°, sometimes more. With exceedingly long vessels (such as the *MOL Triumph*), the pitching angles are significantly less, averaging around <5°. To put this into perspective, if a container ship with a length of 300 m length experiences a pitching angle of ~3°, a container stowed in the bay closest to the bow or stern—at a distance of approximately 140 m from the pitching axis—will cover a distance of 29 m within a pitching cycle! The container will rise 7.33 m upwards from the horizontal plane before descending 14.66 m on the downwards plane, before rising 7.33 m again to the central plane. The process then starts again. During the upward motion, stack pressures rise, while they fall during the downward motion. Rolling involves the side-to-side movement of the vessel. The rolling period is defined as the time taken for a full rolling oscillation to occur from the horizontal to the left, back to horizontal then to the right, and back to horizontal. In vessels with a high righting capacity, i.e., stiff ships, rolling periods of ten seconds and below are quite common. The rolling angle is measured relative to the horizontal plane. Even in moderate sea conditions, large vessels can roll as much as 10°. In severe sea conditions, rolling angles of >30° are not unusual. Even the largest container ships can experience such rolling periods. To help counteract this effect, ships are predominantly fitted with stabilisers and other anti-heeling systems to help to dampen the ship's movements (Table 5.1).

Not all systems, however, are usable or effective in severe weather. On rare occasions, rolling angles may reach 45°> and above. It is easy to imagine what that means for inadequately secured container cargoes. Rolling and pitching of a vessel generates upward and downward acceleration forces which are directed tangentially to the direction of rotation, the values of which increase with distance from the rolling or pitching axis and are inversely proportional to the square of the rolling or pitching periods. At an identical distance from the axis, if the rolling or pitching period is

Table 5.1 Summary of ship motions

Linear motions	Rotational motions
Surging is motion along the longitudinal axis	Rolling is motion along the longitudinal axis
Swaying is motion along the transverse axis	Pitching is along the transverse axis
Heaving is motion along the vertical axis	Yawing is motion along the vertical axis

halved, acceleration forces are quadrupled, while if the rolling or pitching period is doubled, acceleration forces are quartered. Rolling or pitching angles generate down slope forces. Steeper tilting, as occurs during rolling, promotes cargo slippage. As already mentioned, the outwardly acting centrifugal accelerations generated by rotational motion are of no significance in rolling and pitching.

Overall, containers and packages may be exposed to such accelerations for extended periods when at sea. Moreover, the oscillations may be superimposed one on the other resulting in their intensification. It must be emphasised that it is not the "hazards of the sea" which causes damage to cargoes, but rather the inadequate securing of goods inside their container. While such damage has indeed occurred in association with the rolling motion of the ship, the root cause is most commonly the "home-grown" forces arising from poor packing and securing (Fig. 5.1).

Slamming refers to the hydrodynamic impact which a ship encounters due to the upward and downward motion of the hull. This is joined by entry into wave crests and the consequent abrupt immersion of the ship into the sea. Oftentimes vibration from the hull is transferred from the ship's frame to the cargo. This means goods are exposed to stresses from the extremely low frequency oscillations generated by sea conditions and by higher frequency machinery and propeller vibrations. Such risks can be mitigated through proper packing and stowage (Fig. 5.2).

The absolute acceleration values encountered onboard ship are not excessively high. In favourable stowage spaces, they may even be lower than those encountered on land or during air transport. In many cases, the values stated in Table 5.2 are excessive. That said, the frequency with which the motion occurs must definitely be borne in mind. At a rolling period of 10 s, a ship moves side to side 8,640 times daily. Over several days' of heavy seas, the cargo will thus be exposed to alternating loads at a rate of tens of thousands of times. In relation to Table 5.2, Sect. 1.7 of the container packing guidelines provides examples of accelerations which could arise during transport operations though it includes the caveat "… national legislation or recommendations may require the use of other values" (Fig. 5.3).

The values stated in the footnotes (a), (b) and (c) to Table 5.2 of the container packing guidelines describe the principles of acceleration in the vertical plane. Such accelerations are particularly high during pitching and rolling movements and, in exposed positions during inclement weather, can easily reach 1 g. The container packing guidelines provides a maximum safe acceleration of 0.8 g. Vertical acceleration reduces friction forces and increases stack pressure. The annex to the *Code of Safe Practice for Cargo Stowage and Securing* (the CSS Code) contains an assembly of tables for determining acceleration forces as a function of the stowage space

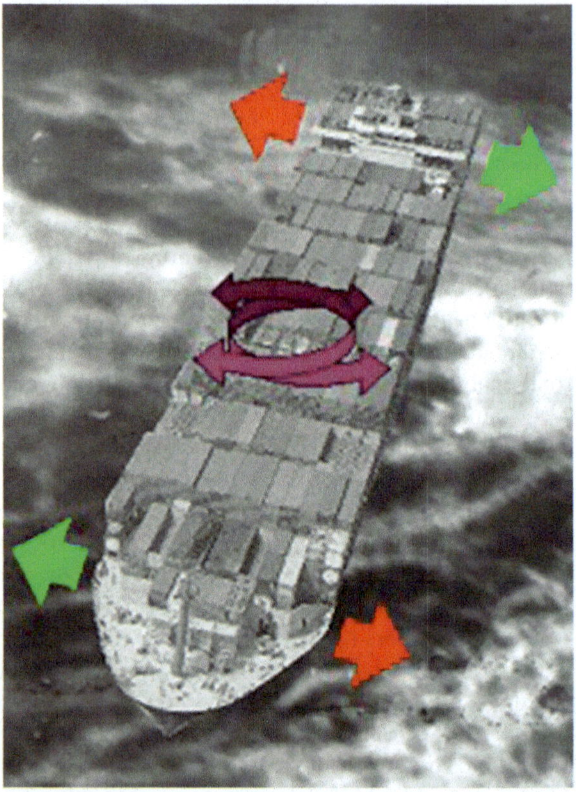

Fig. 5.1 Demonstration of a ship motions (Yawing is motion around the ship's vertical axis)

onboard, the ship's length, and speed. However, these tables are not suitable for use when packing containers and securing cargoes in containers. This is because where containers, road vehicles, railway wagons, or similar, and road trailers, roll trailers and semi-trailers are loaded inland for maritime transport, their ultimate stowage space onboard is an unknown quantity. The least favourable conditions should thus always be considered. As a rule of thumb, loads of 1 g in the vertical direction and 0.8 g in the horizontal direction should be anticipated for worldwide transport. Shipowners should not accept any attempt by the shipper to specify a particular container slot in advance. Even notes on the bill of lading requiring loading below deck are inadvisable. All shipping packages must accordingly be constructed so as to withstand 0.8 times the weight of all adjacently stowed cargo and twice the mass of the cargo loaded on top. If this is not the case, appropriate protective measures must be taken. In these situations, the use of additional rigid receptacles, frames, false decks and similar measures may be helpful.

Modern cargo handling procedures and the ships developed for this purpose have made maritime transport faster and cheaper and, in particular, have reduced cargo handling stresses in port. In order to ensure great flexibility in terms of loading and unloading, modern ships, in particular RoRo ships and ferries have poorer seakeeping

(a)

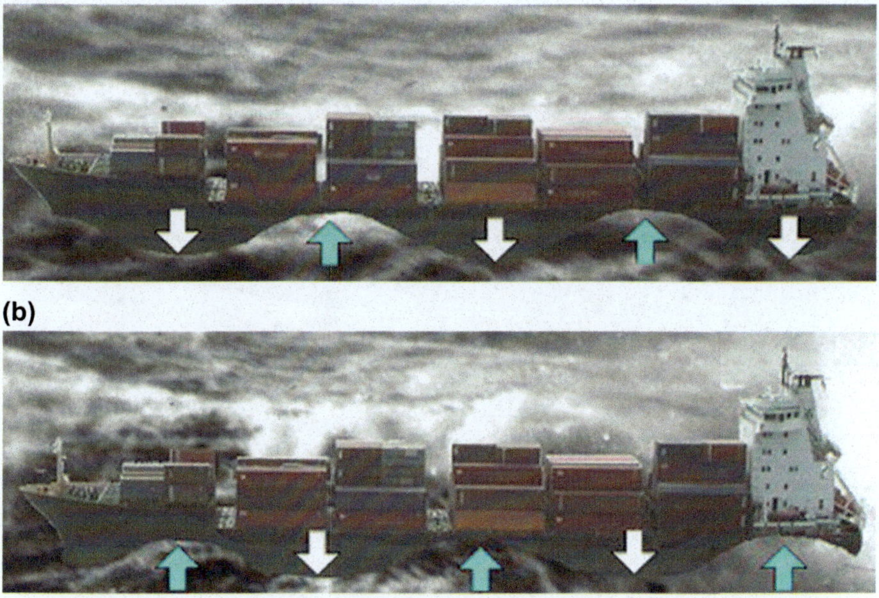

(b)

Fig. 5.2 Demonstration of ship motions (Heaving is motion along the ship's vertical axis)

Table 5.2 Extract from the container packing guidelines

Mode of transport: ocean going vessel	Forward acting forces	Backward acting forces	Sideways acting forces
Baltic sea	0.3 g (b)	0.3 g (b)	0.5 g (b)
North sea	0.3 g (c)	0.3 g (c)	0.7 g (c)
Unrestricted	0.4 g (d)	0.4 g (d)	0.8 g (d)
1 g = 9.81 m/s^2 The above values should be combined with static gravity force of 1.0 g acting downwards and a dynamic variation of:			
	(b) $= \pm 0.5$ g	(c) $= \pm 0.7$ g	(d) $= \pm 0.8$ g

ability than conventional general cargo or heavy-lift vessels. For reasons of stability during loading, they require a high righting capacity. As "stiff" vessels, they initially oppose heeling movements with an extremely high righting moment. The high roll moment of inertia of these vessels entails shorter rolling periods and high transverse acceleration forces. Due to the particular nature of RoRo shipping, the ship's officers are not able to influence the stability behaviour of their vessel by adjusting the weight distribution. This means the risk of accidents is particularly high because, given the large free surfaces in the ship, overturning cargo and the possible consequent ingress of water may result in an abrupt capsize. No doubt most readers will recall at least

(a)

(b)

Fig. 5.3 Rolling is the movement of a ship around its longitudinal axis, the rolling angle in this case being 10°

one major accident that has occurred in the past 20 years.[1] Inadequate cargo securing in transport receptacles such as containers, swap-bodies, etc. may consequently have a significant impact on ship safety.

'Home-grown' acceleration forces in maritime transport are some of the most common causes for cargo damage onboard ships. Because container packers often do not have the appropriate knowledge and skills, they underestimate the effect of gaps in the stow. The consequent motion has a devastating effect on the cargo. 'Home-grown' acceleration can be readily identified onboard if the stowage spaces have been subjected to similar forces, but only some of the goods have suffered damage. It is even clearer when goods stowed in an exposed location remain undamaged, while other goods suffer damage despite being exposed to lower acceleration forces. If we refer to Fig. 5.4, we can clearly see the cases on the platform at the top left were exposed to higher acceleration forces than those cargoes in the containers, which were stowed beneath or further inwards. Although the cases were only secured with a single belt each, they have only shifted slightly, while the containers and their contents have been completely destroyed. In the lower container (1), the poorly secured machine has been set in motion and has forced the container doors open. The tank on the platform (2) stowed above is secured with only two belts and is thus also inadequately secured. Nevertheless, it has withstood the acceleration forces and has slipped by a small margin. This is a clear indication that the acceleration forces were still relatively slight. If we look at Figs. 5.5 and 5.6, these figures clearly show the results of 'home-grown' acceleration forces. It should be noted that all the containers have been exposed to stresses from the inside out, i.e., they have bulges rather than dents. The fibre structure of the plywood walls of the container clearly reveals that the forces were acting from the inside out. It can be deduced the container was destroyed by gaps in the stow. These gaps resulted in extremely high acceleration forces and shocks.

Again, the annexes of the CSS Code provide tools to assist in calculating wind pressure and the effects of sea spray on above deck cargoes. The details provided in the CSS Code may be of assistance in dimensioning cargo securing on open containers such as flat racks, platforms, etc., but they are otherwise of interest only to the ship's officers, and not to container packers working inland. As an approximate guide wind pressure may be estimated for the need to fasten tarpaulins, etc., at 100 daN/m^2. Closed containers are spray-tight provided they have no technical defects. Cargoes stowed on deck may be exposed to breaking-wave impact. Even for experts, the magnitude of these forces and their effects are difficult to estimate. Additional securing measures cannot prevent such effects but reduce them to a limited extent. While securing can never withstand breaking-wave impact, cargoes on open containers should, as a precaution, be secured against floating away. In conventional shipping, damage prevention is the responsibility of the ship's officers. Responsible

[1] For example, the *MV Tricolor*. The *MV Tricolor* was a 50,000 tonne Norwegian-flagged vehicle carrier built in 1987, which has the unfortunate distinction of having been involved in three English Channel collisions within a fortnight. Ultimately the vessel capsized and sank in the English Channel some 17 nautical miles (20 mi) north of the French coast on 14 December 2022.

Fig. 5.4 Demonstration of a rolling angle of 45°

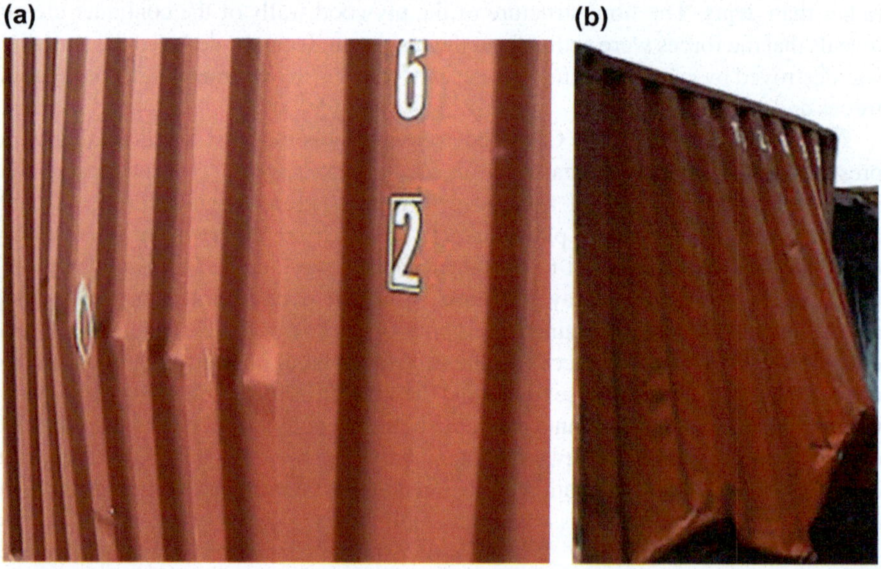

Fig. 5.5 Photograph of bulges in containers as a result of home-grown acceleration forces

Fig. 5.6 Photograph showing a container "ripped apart" as a result of home-grown acceleration forces

ship's officers will accordingly use any means available to them to minimise the effects of rough seas and breaking-wave impact to be as small as possible. Cargo officers will stow cargoes which are particularly sensitive or require particularly extensive securing in locations which are subjected to less acceleration. In the container trade, no consideration can be given to specific requirements with regard to stowage space for particular containers. Moreover, the central stowage planning officers, which prepare the initial plans, and the ship's officers, have no knowledge as to what is loaded in the containers. The only exception are dangerous cargoes, which must be declared in advance. In this case, the contents are known, and the containers receive special stowage consideration.

5.2 Mechanical Stresses During Inland Waterway Transport

With respect to the acceleration forces encountered, an inland waterway vessel is regarded by the container packing guidelines as an insignificant risk means of transport, an opinion which is borne out in practice. Subsequently, Sect. 1.6 states:

> Inland river and water transport is generally smooth. It will not normally exert any forces higher than those of road transport on the cargo and the [container]. Diesel engines of inland river and waterway vessels may create some low-frequency vibrations which under normal conditions should not give reason for any concern.

Higher frequency vibrations which could result in cargo damage are not a cargo securing but instead a packaging issue. Harmful vibration must be avoided in this case by means of appropriate cushioning materials. One error which is frequently encountered is providing excessive cushioning travel. In very general terms, it can be stated that containers may be exposed to vastly different stresses during maritime transport than they are in road, rail or inland waterway transport. Unless the voyage proceeds very calmly in mild weather, the containers and their cargoes will be exposed to oscillation and vibration which is primarily caused by rolling and pitching. It is exclusively during rolling, due to the tilt and or rolling angles which arise, that shipping packages are pressed against the container walls and are squashed against the walls or the surfaces of adjacent shipping packages. The same occurs in "open" containers if parts of the cargo are pressed against the lashings or bracing. The oscillations of rolling and pitching alternately increase and reduce stack pressure. These changes peak at the moment the motion is reversed. Assuming a vertical acceleration of 1 g, a package can thus alternate between "twice its weight" and "weightless". Appropriate deductions or additions may be made for other acceleration values. When containers are incorrectly packed or the cargoes inadequately secured, the packages may shift, be dented, squashed, jumbled up, etc. This then concludes this chapter on the mechanical stresses cargoes experience during transit at sea. In the next section, we will turn our attention to the mechanical stresses exerted on cargoes during land transport. Again, this is not an aspect that the ship's officers have any control or influence over, but it is certainly advantageous to understand how road transport can impact on the integrity of internal container loads before they are loaded on board ship for onward transit.

5.3 Mechanical Stress in Land Transport

Section 1.4 of the container packing guidelines is devoted to the mechanical stresses which occur during road transportation. Accordingly:

> Road transport operations may generate short-term longitudinal forces upon the cargo and the [container]. They may also cause vibrations that may vary considerably due to different suspension systems, different road surface conditions and different driving habits.

The container packing guidelines also contain several explanatory sketches. In summary, anticipated stresses are determined by the structural parameters of the goods and of the vehicle or vehicle combination, the position of the goods on the loading area, how fully laden the vehicle is, the condition of the roadway, the driver's habits, and so forth. Special transport operations entail obtaining comprehensive information about the circumstances to be considered. Stowing and securing measures may then be tailored to the particular transport situation. If precise data is unobtainable, the values stated in the container packing guidelines, or the assumed loads specified by the relevant national organisations, should be considered. In relation to Table 5.3, it is stated in Sect. 1.7 of the container packing guidelines that examples of accelerations are given which could arise during transport operations however.

> ... national legislation or recommendations may require the use of other values.

It is unsatisfactory that the tables state no values for vertical acceleration, although any road user will know just how much road vehicles can vibrate up and down. The Swiss road haulage association *Les Routiers Suisses*, for instance, allows for upward vertical accelerations of 1 g. A forward acceleration of 0.8 g (according to the VDI guidelines) or 1.0 g (according to the container packing guidelines) means that forces of 80% or 100% respectively of the package weight will act on the packages in this direction. Both sets of guidelines assume backward acceleration of 0.5 g, i.e., forces of 50% of the package weight will act on the packages in this direction. Again, both sets of guidelines assume identical sideways acceleration values of 0.5 g. Forces amounting to 50% of the package weight will accordingly also apply sideways. As stated, the guidelines make no mention of vertical acceleration. Most road users will know from their own observation that loads can lift during road transport. Values of just above 1 g vertical are quite readily achieved in the worst circumstances (Figs. 5.7 and 5.8).

In a brochure published by the former German federal railroad company, Deutsche Bundesbahn, it is stated:

Table 5.3 Examples of accelerations that may arise during road transport

Mode of transport: road	Forward acting forces (g)	Backward acting forces (g)	Sideways acting forces (g)
VDI guidelines	0.8	0.5	0.5
CTU packing guidelines	1.0	0.5	0.5
Swiss road transportation regulations	1.0	0.5	0.5
British road safety regulations	1.2	0.5	0.8

Fig. 5.7 Acceleration in longitudinal direction according to VDI guidelines

Fig. 5.8 Acceleration in transverse direction according to VDI guidelines

> Considerable stresses .... in the vertical direction also occur during road delivery operations. (...) Forces of up to 0.85 times the weight-force of the load must be anticipated in the vertical direction. Measurements, filtered at 32 Hz, carried out by the Institute for Automotive Transport at Hanover University of Technology.

As for every means of transport, the values stated in the guidelines apply only to loads which have been secured in such a way that they cannot shift. If load items are able to move, higher acceleration forces may come into play. In one incident, 8,000 kg of goods in cartons had been loaded into a container. The articulated lorry carrying the container had to brake harder than usual at a set of traffic lights. Although the braking deceleration was less than 0.8 g, we will use this maximum value for the purposes of the calculation. The coefficient of sliding friction of the cartons on the container floor and between the individual cartons will be assumed to be $\mu = 0.3$. The cartons would then have been able to exert a force of 8,000 kg $\times$ 0.5 g = 3,924 daN against the front-end wall of the container. As will be described in greater detail elsewhere, the loading capacity of the container end wall must amount to 0.6 times the payload. In this incident, this would amount to 10,595 daN. The forces which actually occurred were at least 2.7 times higher, indicating a noticeably clear case of 'home-grown' acceleration as a result of gaps in container packing. As a further example, a 1,000 kg package had been loaded onto a road vehicle. The vehicle was going round a curve with a radius of 25 m (82 ft) at a moderate speed of 36 km/h. This generated a sideways acceleration of 4 m/s^2. A force of 392.4 daN was accordingly applied to the load. Assuming a coefficient of sliding friction of $\mu = 0.3$, this force was counteracted by passive friction forces equal to 294.3 daN. The difference of 98.1 daN would never have been enough to smash through the sideboards and cause damage to the vehicle. However, because the package was lying unsecured in the middle of the loading area, it was able to get a "run-up" and develop sufficient destructive force to cause considerable damage.

Subsequently, the provisions governing the loading of road vehicles and the relevant accident prevention regulations specifically state that the load must be packed and secured in such a manner that, under normal traffic conditions, nothing can fall over, drop or roll off the vehicle and that any risk to life must be mitigated. Usual traffic conditions in road transport include evasive manoeuvres, hard braking and unevenness of the roadway which may occur simultaneously and not separately. It should accordingly be assumed, as the basis for proper load securing, that a vehicle on a poor section of road may suddenly need to brake hard and additionally execute an evasive manoeuvre. If the load can pass through such a situation unscathed, it may be deemed secure. It should be borne in mind that the acceleration values which must be assumed for road vehicles according to VDI guidelines and the BGF, the German road vehicle maintenance employers' liability insurance association, correspond to the following tilts in degrees:

- 0.8 g forward acceleration in road transport corresponds to a tilt of 53°.
- 0.5 g backward acceleration in road transport corresponds to a tilt of 30°; and
- 0.5 g sideways acceleration in road transport corresponds to a tilt of 30°.

The loading capacity of cargo transport units is specified in standards and the CSC Code. The relevant sections provide comprehensive information in this respect. However, the values will also be stated here by way of comparison with shipping stresses. With a uniformly applied load the front-end walls of standard swap bodies and containers must be able to withstand forces of 0.4 times the payload. This corresponds to an angle of 23.58°. With a uniformly applied load, the side walls of standard swap-bodies must be able to withstand forces of 0.3 times the payload. This corresponds to an angle of 17.46°. With a uniformly applied load, the side walls of standard containers or of swap bodies with a CSC plate which are used in maritime transport must be able to withstand forces of 0.6 times the payload. This corresponds to an angle of 36.87°.

5.4 Mechanical Stresses During Railway Transportation

Section 1.5 of the container packing guidelines concerns the stresses that occur during railway transport; specifically:

> Rail transport, in addition to subjecting cargo to vibrations (16 Hz), may also lead to shocks as a result of shunting [switching] operations. Many railways have organised their operations in such a way as to avoid shunting [switching] of railway wagons [rail cars] incurring high forces (e.g., by operating dedicated block trains) or by moving [containers] on wagons [cars] with high performance shock absorbers that are normally able to reduce shunting [switching] shock forces. It may be advisable to ensure that such operational features have been established for the rail journey.

Several explanatory sketches are also included. The container packing guidelines defines the terms shunting [switching] and block train as:

> Shunting [switching] means the operation when single railway wagons [cars] or groups of railway wagons [cars] are pushed to run against each other and be coupled together.

Operationally, switching includes quite a wide-range of activities which will not be addressed in greater detail here. In modern switching yards, trains are divided up and reassembled by being pushed (for example with a switching engine) up and over a hump yard, therein automating the hump switching operation. The speed of the railway carriages as they run down the hump is kept within the desired range by special braking or driving systems. Major switching impacts are outdated. In a conventional system, the railway carriages running down the hump are brought to a standstill with drag shoes. If the worker laying the drag shoes makes a mistake, switching impacts involving considerable deceleration may occur. If no hump yard is available, groups of railway carriages are "kicked" by switching engines and the individual carriages run into the appropriate sorting sidings, where they are again conventionally brought to a standstill with drag shoes. In this case too, high deceleration values may occur when the carriages collide. Lower levels of deceleration occur when the carriages are pushed to activate the couplings. The UIC specifies an acceleration value of 4 g for normal switching operations. Containers are not constructed for such stresses and should not be exposed to such acceleration forces. Railway carriages with high-performance buffers, as mentioned by the container packing guidelines, cut the "normal" switching impacts of 4 g in half to 2 g. On the German railway network, railway carriages with high-performance buffers have a lower-case j designation in their name, for example **Lgjs**, **Sgjs**, **Sgjkkms**, etc. In accordance with the German railway operating procedures, a range of different railway carriages must not be switched normally. This is determined by the type of railway vehicle, the type of loading, or certain other features. Railway carriages operated under this scheme are either specially marked "from the outset," for example, by stating "**HUMP YARD SWITCHING AND KICKING PROHIBITED**," "**SWITCH CAREFULLY**," etc., or they may be provided with special tickets bearing a yellow triangle standing on its apex, a red circle ticket or the standard caution tickets. Railway carriages operating under the caution ticket scheme are divided into three classes.

The container packing guidelines defines block trains as follows:

> Block train means a number of permanently coupled railway wagons [rail carriages or cars], normally running directly between two selected terminals or entities without shunting [switching].

Such block trains are known as "point-to-point" services or "rail carriages not switched in transit." A customer of the European UIC railway system can assume that any railway carriages loaded with containers or other containers will be transported in block trains or handled under the caution ticket scheme. Accordingly, the only values which need to be considered are from the container packing guidelines table (refer to Table 5.4) which states the acceleration forces applying in railroad transport are those for combined transport in Europe. In respect to Table 5.4, it is stated in Sect. 1.7 of the container packing guidelines that examples of accelerations are given which could arise during transport operations:

> ... however, national legislation or recommendations may require the use of other values ...

The statement in the table "the above values should be combined with static gravity force of 1.0 g acting downwards and a dynamic variation of: (a) = ±0.3 g" means that the container packing guidelines have adopted the UIC railway values for acceleration. In accordance with these values vertical acceleration is calculated at 0.3 g. The same values are also specified by the VDI guidelines for intermodal transport. In road-railway intermodal transport, the following values are thus obtained: 1 g in the longitudinal direction, 0.5 g sideways and 0.3 g vertically. From the perspective of the cargo, a longitudinal acceleration of 1 g is the same as if the container had

Table 5.4 Examples of accelerations that may arise during rail transport

Mode of transport: rail	Forward acting forces (g)	Backward acting forces (g)	Sideways acting forces (g)
Rail cars subject to shunting [Switching][a]	0.8	0.5	0.5
Combined transport[b]	1.0	0.5	0.5
$1\ g = 9.81\ m/s^2$	1.0	0.5	0.5

The above values should be combined with static gravity force of 1.0 g acting downwards and a dynamic variation of: (a) = ±0.3 g

[a]The use of specifically equipped rolling stock is advisable (e.g., high-performance shock absorbers, instructions for shunting [switching] restrictions)

[b]"Combined transport" means "wagons [cars] with containers, swap-bodies, semi-trailers and trucks, and also block trains (UIC and RIV)"

been stood up vertically, first in one direction and then the other. Sideways acceleration values in road-railway intermodal transport are identical for both modes of transportation. They correspond to a tilt of 30°. The effects of acceleration of ±0.3 g are similar to each package in the container alternately becoming 30% heavier and then 30% lighter than its true weight (Figs. 5.9 and 5.10).

Fig. 5.9 Acceleration in longitudinal direction according to VDI guidelines (rail)

Fig. 5.10 Acceleration in longitudinal direction according to VDI guidelines (rail)

> ***Note:*** *with respect to the carriage of cargo by railway companies on continents other than Europe, information should be obtained locally about the prevailing conditions in each case. In some situations, stresses may be higher than those permitted in Europe. For example, in winter, many containers destined for receivers close to the Great Lakes are unloaded in Halifax and carried by railway to Detroit as the sea route* via *the St. Lawrence Seaway is impassable due to ice. Similar situations may occur in the northern and eastern extremes of Russia.*

So far, we have covered the main mechanical stresses that occur during transportation (by sea, by road and by rail). In the next chapter, we will look at mechanical stresses that occur during cargo handling.

5.5 Mechanical Stress During Cargo Handling

The definition provided in the container packing guidelines for cargo handling reads:

> [Cargo] handling includes the operation of loading or unloading/discharging of a ship, railway wagon [railway carriage], vehicle or other means of transport.

Although most people universally understand the meaning of "ship," the container packing guidelines provides the following definition:

> Ship means a seagoing or non-seagoing watercraft, including those used on inland waters.

With these two points clarified, the stresses occurring during cargo handling can be divided into two classes:

(1) stresses which act on an empty or packed container due to the use of suitable equipment when handling the whole container; and
(2) those stresses which arise during packing and unpacking of the containers by manual methods and or by using mechanical aids and equipment [such as forklift trucks].

The container packing guidelines define a forklift truck as follows:

> Forklift truck means a truck equipped with devices such as arms, forks, clamps, hooks, etc. to handle any kind of cargo, including cargo that is unitised, overpacked or packed in containers.

In addition to the general distinctions that may be drawn between handling the cargo during the loading and unloading of a container and the handling of the container itself, it is also possible to draw a distinction between avoidable stresses and unavoidable stresses. Packing and unpacking containers or other cargo transport units involves procedures which are no different from those previously used in conventional loading. Indeed, the same risks are encountered and must be considered. Purely manual handling of packages entails exposure to more impact and dropping than is the case when using mechanised handling methods. Properly packaged and palletised goods are at negligible risk when forklift trucks and similar ground conveyors are used. Conversely, the risk is distinctly greater for incorrectly packaged and palletised goods. However, if personnel are untrained, the risks associated with the use of mechanical aids are particularly high. If containers are handled with the equipment specially developed for this purpose, the impact stresses to which the various cargo transport units are exposed are comparatively uniform, irrespective of whether they are being transferred between road or rail vehicles or watercraft. The container packing guidelines (Sect. 1.8) also provides some indications about cargo handling:

> Container movements by terminal tractors may be subject to differing forces as terminal trailers are not equipped with suspension. Additionally, ramps can be very steep, causing badly stowed cargo inside [containers] to be thrown forward or backward.

Terminal tractors (such as tug master units) and terminal chassis are normally only used for moving containers on a flat plane. Various methods may be used in terminals for this purpose. With regard to ramps, the guidelines relate to RoRo tractors in conjunction with roll trailers. Both types of tractors require a hydraulically liftable fifth wheel coupling to allow access to the ramps. The normal chassis or semi-trailers pulled by the tractors have additional suspension. The stated effects on badly packed cargo have already been mentioned several times.

The container packing guidelines (refer to Sect. 1.9) provides some information about the use of lifting gear and ground conveyors:

> Considerable forces may also be exerted on containers and their cargoes during terminal transfer. Especially in seaports, containers are transferred by shore-side gantry cranes that lift and lower containers, applying considerable acceleration forces and creating pressure on the packages in containers. Lift trucks and straddle carriers may take containers, lift them, tip them and move them across the terrain of the terminal.

Even when equipment is expertly operated, stresses of approximately 1 g must be anticipated during cargo handling. Skilled operation assumes that the goods are lifted and set down gently. Jerky lifting and setting down may generate higher g values. Normal setting down impacts cannot always be avoided and additional crush pressures must thus be anticipated. This applies both to setting down the container and when setting the spreader down onto the container. If spreaders without flippers are used carelessly (the flippers centre the spreader onto the container), damage to container roofs must be expected. There is still a risk of damage even when manual spreaders or "over height frames" are used. Especially in the case of onward carriage in countries with poorly managed infrastructure, higher levels of stress must be anticipated during handling and unloading operations. So that no damage is caused during unpacking of the cargo transport units, packing must always be designed in such a way that the containers can be stripped as simply as possible. It must also be expected that containers will not always be handled with due care, and that during onward carriage they may be unloaded under the most rudimentary conditions. "Bumping"

Fig. 5.11 Photograph of the Port of Cape Town container terminal, South Africa

against obstructions is not at all unusual under such circumstances. Packing and securing should be designed to withstand such conditions. Handling stresses during packing and unpacking of containers are the result of carelessness. It is not unusual for personnel to walk on cargo which cannot withstand such weights; packages are dropped during manual handling; poorly packed pallets can come apart during forklift truck operations and individual packages fall out; packages can be punctured with forklift truck forks; packages may be crushed or squashed with the forklift truck; and damage may be caused by the incorrect use of tools and cargo securing materials, etc. The best protection from these and the above-stated risks is superior quality training, adequate staff supervision, and sufficient motivation to work to expected standards.

In this chapter, we have discussed some of the salient points relating to the stresses exerted on cargoes during transit by sea, inland waterway, road, rail and during terminal handling. In the next chapter, we will look at the effect of climatic, biotic and other non-mechanical based stresses on cargo and containers (Figs. 5.11, 5.12, 5.13, 5.14, 5.15, 5.16, 5.17 and 5.18).

Fig. 5.12 Photograph of handling a container with a reach stacker and top spreader, Torshavn, Faroe Islands

Fig. 5.13 Photograph of a skeleton chassis

Fig. 5.14 Photograph of a mobile container gantry crane, Port of Greenock, Scotland

Fig. 5.15 Photograph of an unmanned container shunter, Port of Antwerp, Belgium

Fig. 5.16 Photograph of quayside gantry cranes, Port of Liverpool, England

Fig. 5.17 Photograph of a damaged tank container with hazmat kit

Fig. 5.18 Photograph of a damaged container at the Birzebugga Triq Kalafrana Freeport, Malta, 2008

Chapter 6
Non-mechanical Stress During Transport

In the previous chapter we discussed the most common mechanical stresses which affect containers and their contents during the various modes of transportation. In this chapter, we will look at non-mechanical stresses, i.e., those that result from climatic, biotic, and chemical stresses. These stresses can impact the integrity of the container and its contents as much, if not more so, than the mechanical stresses discussed in the previous chapter. To begin with, we will start by examining climatic stresses.

6.1 Climatic Stresses

An undamaged, tightly closed box container protects its contents from external influences such as ice, snow, rain, salt spray, dust, thermal radiation, ultraviolet light and almost every other conceivable environmental influence. However, box containers cannot provide protection from direct exposure to water as they are only spray-tight and not watertight. To this end, the container packing guidelines states that moisture damage must be avoided wherever possible; hence Paragraph 3.2.7 reads:

> In order to avoid cargo damage from moisture, wet cargoes, moisture-inherent cargoes or cargoes liable to leak should not be packed with cargoes susceptible to damage by moisture. Wet dunnage, pallets or packaging should not be used. In certain cases, damage to equipment and cargo can be prevented by the use of protective material such as plastic films.

The chapter on dunnage indicates how the risks in the first sentence can be combated. The second sentence is especially important although this principle is more often than not ignored. Cargoes, wooden dunnage and other aids are often literally left outside in the rain before being placed in the container. If containers are packed in the open, there is always a risk during wet weather that ground conveyors and personnel will bring moisture into the container with them. Cargoes in a container in which air exchange is prevented may suffer spoilage or be damaged if temperatures and

A. A. Olsen and F. Karkori, *Containerized Cargo Handling and Stowage*, Springer Series on Naval Architecture, Marine Engineering, Shipbuilding and Shipping 22,
https://doi.org/10.1007/978-3-031-63329-4_6

humidities during carriage reach a level which is incompatible with the goods or their packaging. Section 1.3 of the container packing guidelines discusses the negative effects of changes in temperature and humidity on container cargoes; accordingly:

> During longer voyages, climatic conditions (temperature, humidity, etc.) are likely to vary considerably. These may affect the internal conditions in a [container] which may give rise to condensation (sweating) on cargo or internal surfaces.[1] Where cargo is liable to damage from condensation, expert advice should be sought1 (refer to the [container] packing guidelines Annex 1).

If warm air is cooled to below its dew point, the resultant condensation may result in cargo damage. Additional information is provided in the section on "Condensation" in Annex I of the container packing guidelines:

1. Cargoes in transit may be affected by the conditions to which they are subjected. These conditions may include changes in temperature and humidity and particularly cyclic changes that may be encountered). An understanding of condensation phenomena is desirable because condensation may lead to such damage as rust, discolouration, dislodging of labels, collapse of fibreboard packages or mould formation.
2. Solar radiation can produce air temperatures under the inner surfaces of a [container] which are significantly higher than external air temperatures. The combination of these effects can result in a range of day and night cyclic temperature variations in the air adjacent to the inner surfaces of a [container] which is greater than the corresponding range of temperatures just outside.
3. Cargoes closest to the wall or roof will be more affected by external temperature variations than those in the centre of a [container]. If the possible extent of temperature variations or their full significance is not known, advice should be obtained from specialists.
4. Under the circumstances described, condensation may occur either on the surface of the cargo (i.e., cargo sweat) or on the inside surfaces of a [container] (container sweat) either during transport or when the unit is opened for discharge.
5. The main factors leading to condensation inside a [container] are:
   - 5.1 sources of moisture inside the unit which, depending upon ambient temperature conditions, will affect the moisture content of the atmosphere of the unit.
   - 5.2 a difference between the temperature of the atmosphere within the unit and the surface temperature of either the cargo or the inner surfaces of the unit itself; and
   - 5.3 changes in the temperature of the outer surface of the unit which affect the two factors above.
6. Warming the air in a [container] causes it to absorb moisture from packaging or any other source. Cooling the air below its dew point[1] causes condensation.
7. If, after high humidity has been established inside a [container], the outside of the unit is cooled, then the temperature of the unit surface may fall below the dew point of the air inside it. Under these circumstances moisture will form on the inner surfaces of the unit. After forming under the roof, the moisture may drop onto the cargo. Cyclical repetition of cargo or container sweat phenomena can result in a greater degree of damage.
8. Condensation can also occur immediately after a [container] is opened if the air inside the unit is humid and the outside air is relatively cool. Such conditions can produce a

[1] The dew point is the temperature at which air saturated with moisture at the prevailing atmospheric pressure will start to shed moisture by condensation.

fog and even precipitation but, because this phenomenon usually occurs only once, it seldom results in serious damage.

9. The risk of damage and dangerous situations[2] can be minimised if the moisture content of the packaging and securing materials is kept low.

Where necessary, the information provided in Annex 1 "Condensation" may be supplemented by additional information and guidance from other sources.

6.2 Biotic Stresses

Both biotic and chemical stresses are closely interrelated with each other and with climatic stresses. It may be assumed that it is known that biotic and chemical reactions proceed particularly well when exposed to certain temperature and moisture conditions. Biotic stresses are deemed to be present if goods, packaging or parts thereof are damaged by living organisms of whatever kind. Chewing damage may, however, cause considerable losses as may, of course, damage caused by animal excreta. The risk is low in closed containers, provided that no animals have taken up residence before packing, or "sneak in" during packing. Since many rodents are nocturnal creatures, containers which are being packed should be closed during work breaks or at the end of a shift. This is especially important during the hours of darkness. Prior inspection of the container for tightness is a basic requirement. Insects, however, can nonetheless still effect egress into closed containers which, although classed as unventilated, still have openings of small cross-sections. It will be impossible to avoid damage if goods and/or packaging are infested before packing and such infestations are not rectified. In the case of "open" containers, insect infestation can only be prevented by providing sufficiently tight packaging or suitable preservation.

Other frequently observed phenomena include infestation of packed goods or packaging by fungi, including moulds, bacteria and other microorganisms. Heat and high relative humidity encourage the development of mould and the multiplication of bacteria. Slight air movements can also encourage growth. It is therefore advisable to stack cargo securing lumber with spacing battens to provide adequate ventilation during storage. It is equally important to store the lumber under a breathable rainproof cover. In this way, air can pass between the layers of lumber preventing the formation of mould. Whilst the container packing guidelines do not mention biotic stresses specifically, it does indicate that certain countries, wishing to prevent the introduction of pests, place specific conditions on the lumber to be used for packaging and securing. As such, Paragraph 3.3.2 of the container packing guidelines states (Fig. 6.1).

> If a [container] is destined for a country with wood treatment quarantine regulations, care should be taken that all wood in the unit, packaging and cargo complies with the regulations.

[2] For example, when goods of class 4.3 (substances which evolve flammable gases on contact with water) are packed into a container.

Fig. 6.1 Photograph of the Sirex wasp

It is a useful practice to place a copy of the wood treatment certificate in a conspicuous place inside and, where appropriate, outside the unit in a weatherproof pouch.

This primarily relates to the mandatory and usual treatment against the Sirex wasp for cargoes bound for Australia, New Zealand and China. Even where no lumber is packed in the container as cargo or packaging, lumber may nevertheless still be required for cargo securing or as dunnage. In such cases, it is worthwhile having the necessary lumber appropriately treated in advance. Treated lumber should be coloured in order to differentiate it from untreated lumber (as indicated in Fig. 6.2) (Fig. 6.3).

Fig. 6.2 Photograph of lumber with anti-Sirex wasp treatment

Fig. 6.3 Photograph of the *SS Marilu* carrying prefabricated houses for the Queensland Housing Commission from Genoa, Italy, 1952; the shipment was infected with Sirex wasp

6.3 Chemical Stresses

Substances which react together causing damage to the cargo, or its packaging, are of particular interest and relevance to cargo handlers. One of the most familiar chemical reactions is corrosion. Goods must be provided with protection from corrosion by a protective coating, sealed packaging, or by using the volatile corrosion inhibitor (VCI) method. However, this is not in principle the responsibility of packing or cargo securing personnel, but instead of the shipper or their packer. That said, corrosion protection must not be impaired or eliminated as a result of packing or cargo securing operations. Appropriate steps should also be taken to prevent the introduction of corrosion-promoting substances into the container. It must not be forgotten that the salt aerosols present in sea air can have an adverse effect on cargoes in closed containers. Unless special precautions are taken, the risk is greater for open-side and open-top containers. In general, this also applies to all other solid, liquid or gaseous substances which can impair the cargo in some way. To take one example, even a few molecules of a given chemical, e.g., a poison, could "migrate" through one physically tight package into another damaging the goods inside. Preventing chemical effects requires considerable knowledge of chemistry and so is not a responsibility which can be left to loading personnel. Therefore, appropriate instructions are absolutely essential. Simple but nevertheless effective preventive measures against chemical stresses might include:

- Only using containers which are fit for packing, i.e., which are clean and neutral in odour.
- Not ignoring basic stowage rules, for example, never stowing odour-releasing goods with goods which can absorb odours, unless odour contamination can be prevented by specific measures; and
- Protecting any chemically sensitive cargo from extraneous substances which could cause damage.

The container packing guidelines provides useful advice on the fitness of containments for loading. Accordingly, Paragraph 2.2.4 states:

> A [container] should be clean, dry and free of residue and persistent odours from previous cargo.

In addition, Paragraph 3.1.4 mandates:

> Packing should be planned before it is started. [...] The possibility of cross-contamination by odour or dust, as well as physical or chemical compatibility, should be considered.

It is the case that for hazardous materials, compliance with the relevant hazardous materials regulations (IMDG code) is mandatory to prevent the impairment of other goods. Given the handling of hazardous goods is discussed in great detail in the IMDG Code, we will not cover this in any further detail in this book, however readers are strongly advised to consult the IMDG Code where handling of dangerous goods is required. Copies can be sought online and/or from all good bookstores (Fig. 6.4).

Fig. 6.4 Tank container with various HAZMAT labels clearly visible

We have now concluded Part II of the book on mechanical and other container related stresses. It is important to recognise that although the proper packing of containers rests squarely with the shipper or their packer, this does not necessarily signal the cessation of their responsibilities. In the event that cargoes sustain damage during transit, one of the key aspects of any ensuing investigation will be the quality of the packing within the container. Where poor packing is found to be the cause of, or a contributory cause, of cargo damage, it is the shipper and or their packer who will withstand the worst of any claim. This does not absolve the ship's officers or terminal personnel from applying best practice in the handling and stowage of cargo; rather, they contribute to the safe and efficient carriage of the cargo and are instrumental in the cargo reaching its final destination in the same condition as it left the shipper's warehouse (notwithstanding, of course, and follow on transit conditions). In Part III we will begin to examine the design and construction of containers.

Part III
Container Design and Construction

Chapter 7
Container Technical Aspects

This part of the book concerns the technical and cargo securing aspects of container handling. It is not a discussion on the widespread use of containers; nor is it intended to address the economic viability of using containers as a multimodal method of cargo transportation. There is some ambiguity in English around the meaning of the word "container", which derives from the Latin *continere* = to enclose, but for the purposes of this book the word should be taken to mean a shipping container, whether large or small. The American Heritage Dictionary of the English Language defines a container as a "large reusable receptacle that can accommodate smaller cartons or cases in a single shipment, designed for efficient handling of cargo." In common American English parlance, this type of container is often called a "van," although a van is defined in most dictionaries as a motor vehicle, (UK) lorry; (US) truck; or wagon. In the past, the term 'container', when used in the field of goods traffic, was taken to mean standardised transport units of several types, whose dimensions were matched to one another so that they could be combined together as "transport modules." This is still generally true today; however, usage of the term has moved increasingly towards large or bulky containers, which are transported by road, rail, water and (in modified form) by air. This part of the book deals in particular with the packing and securing of cargoes in freight containers. However, the packing and securing instructions are equally applicable to containers as defined in the container packing guidelines. The "*Guidelines for the packing of cargo, other than bulk cargo, into or onto cargo transport units (CTUs) applicable to transport operations by all surface and water modes of transport*", in force since 17 February 1999, defines the following terms accordingly (for the purpose of this guidance the term container has the same meaning as the expression as 'Cargo Transport Unit (CTU)' and 'Intermodal Transport Unit (ITU)'. Accordingly:

> ... cargo transport unit [container] means a freight container, swap-body, vehicle, railway wagon [railroad car] or any other similar unit ...

A. A. Olsen and F. Karkori, *Containerized Cargo Handling and Stowage*, Springer Series on Naval Architecture, Marine Engineering, Shipbuilding and Shipping 22,
https://doi.org/10.1007/978-3-031-63329-4_7

> ... Freight container, or simply container, means an article of transport equipment that is of a permanent character and accordingly strong enough to be suitable for repeated use; it is designed to transport a number of receptacles, packages, unit loads and over packs together from the packing point to its final destination by road, rail, inland waterway and/or sea without intermediate separate handling of each package, unit load or over pack. The term "container" is used throughout these guidelines ...

> ... the "permanent character, sufficient strength and suitability for repeated use" are no use, however, if a container is incorrectly packed and secured.

With that point now clarified, we can complicate matters by describing the several types of containers in use by the global shipping industry. It is understood that the term 'container' is generic in the most common sense of the word. Because of that, there are vast variances in the types of containers that can be used in the transportation of various products and goods. The term 'container' refers to any ITU, swap-body or semi-trailer suitable for intermodal transport. When we talk about TEU and FEU units, we are typically referring to standard ISO 20′ or 40′ containers. These are the most common form of container and make up approximately 70% of all containers in use worldwide.

The Intermediate Bulk Container, (IBC) is a type of rigid, semi rigid or flexible portable packaging that has a capacity of no more than 3.0 m^3 (3,000 L) for solids and liquids. It is designed for mechanical handling and is resistant to the stresses exerted during cargo handling and transportation. Swap body units are a type of container with at least four fittings for accommodating twist locks. According to ISO standard 1161:1983, swap-bodies are not permanently attached to an under frame and wheels or to a chassis and wheels. A swap body need not be stackable but is usually equipped with support legs, designed especially for combined road and railway transportation. 'Vehicle' means a road vehicle or railway carriage or wagon which is permanently attached to an under frame and wheels or to a chassis and wheels, and which is loaded and unloaded as a unit. This term also includes 'trailer,' 'semi-trailer' or similar 'mobile unit' except those used solely for the purposes of loading and unloading.

A distinction is drawn according to the size between large containers, medium containers, and small containers. This book deals in particular with cargo securing in large containers, with the main emphasis being on those which spend at least part of their journey at sea. In principle, the cargo securing rules also apply to other containers. So-called PA containers are medium containers designed especially for rail transport. Small containers are used in groupage traffic and are carried by both road and railway vehicles. Neither of the above-mentioned groups will be dealt with here, however. Depending on container design and the applicable standard, the following distinctions may be drawn: the ISO container (also known as an overseas container or trans container), the ASA container, and the ISO inland container.

To provide consistency, an international standard is necessary for the standardisation of transport and handling operations, for example, to ensure uniform practice in the picking up and setting down of containers using lifting gears and ground conveyors or for the marking of containers. In addition, a series of other national and international publications, some of which have legal force, cover the widest possible range of characteristics of these special containers for rationalised goods transport or provide instructions for their use. We shall not list the individual standards here, as they are of only limited relevance to packing and cargo securing. That said, should a particular standard be of relevance, the reader will of course be referred to it. Technical Committee 104 (TC 104) of the International Organisation for Standardisation (ISO) participated in drawing up many of the ISO standards. Pan-European interests were represented with regard to some standards by the ISO Container subcommittee (SC). The international standards are applied across the European Union as DIN ISO or these days as DIN EN ISO. Up to the United Kingdom's departure from the European Union in 2016, DIN EN ISO also applied to container operations in the United Kingdom. Under the terms of the Withdrawal Arrangement, it is likely DIN EN ISO will continue to apply to the United Kingdom, albeit under the implementation of replacement domestic arrangements. Central to the safe and efficient use of containers is the *International Convention for Safe Containers* which has been ratified by the majority of countries worldwide. The Convention defines the term container as follows:

> Article II
>
> Definitions
>
> For the purpose of the present Convention, unless expressly provided otherwise,
>
> 1. "Container" means an article of transport equipment
>   (a) of a permanent character and accordingly strong enough to be suitable for repeated use.
>   (b) specially designed to facilitate the transport of goods, by one or more modes of transport, without intermediate reloading.
>   (c) designed to be secured and/or readily handled, having corner castings for these purposes.
>   (d) of a size such that the area enclosed by the four outer bottom corners is either:
>   (i) at least 14 sq. m. (150 sq. ft.) or
>   (ii) at least 7 sq. m. (75 sq. ft.) if it is fitted with top corner castings.
>
> The term "container" includes neither vehicles nor packaging; however, containers when carried on chassis are included.

According to the industry publications DIN ISO 668 Series 1 ISO containers—Classification, dimensions and ratings and DIN ISO 830 Containers—Vocabulary, containers are defined as follows:

> Container: a transport container which:
>
> - *is of a permanent character and accordingly is strong enough to be suitable for repeated use*
> - *is specially designed to facilitate the transport of goods, by one or more modes of transport, without intermediate reloading*
> - *is suitable for mechanical handling*
> - *is designed to be readily packed and unpacked, and*

- *has a capacity of at least 1* m^3.

Vehicles and packaging are not containers.

In principle, the only difference between the *International Convention for Safe Containers* and the DIN EN ISO standards is that the former provides for a container surface area of at least 14 m^2 or 7 m^2, as a function of the corner castings, while the latter deems receptacles with volumes of 1 m^3 or more to be containers. We will not at this point go into the specific definitions of the term container given in hazardous materials legislation, but if the need arises, the relevant references will be cited at the appropriate point.

7.1 Container Dimensions and Weights

The lengths of containers are still based on a modular system, for which the starting point was, and indeed still is, 40′ (12,192 mm) in length. The other nominal sizes have arisen by division, with 3 inches being taken off each time to allow the containers to be combined together in practice. Containers under an **A, B, C, D** designation are 8′ high. If the container is 8′6″ high, then the letters are doubled to provide an amended **AA, BB, CC, DD** designation (Table 7.1).

At present, the following lengths are stated and defined in the metric and or imperial customary systems. They relate not only to freight receptacles such as containers but also to swap bodies (Table 7.2).

Current (typical) container widths are (Table 7.3):

In addition, containers of the following heights may also be used (Table 7.4):

Containers 7.43 m (24.3 ft) long are in principle half-size 49′ containers. With a width of 2.50 m (8.2 ft) and a height of 2.60 m (19.6 ft), such containers fulfil the technical requirements and traffic laws of many countries (Table 7.5).

Table 7.1 Container dimensions as stipulated in DIN/ISO 668 or DIN 15,190, Part 1

Design	Length			Height			Width			MGW	
	mm	ft	in	mm	ft	in	mm	ft	in	kg	lbs
1A	12,192	40		2,438	8		2,438	8		30,480	67,200
1AA	12,192	40		2,591	8	6	2,438	8		30,480	67,200
1B	9,125	29	11$^3/_4$	2,438	8		2,438	8		25,400	56,000
1BB	9,125	29	11$^3/_4$	2,591	8	6	2,438	8		25,400	56,000
1C	6,058	19	11$^3/_4$	2,438	8		2,438	8		20,320	44,800
1CC	6,058	19	11$^3/_4$	2,591	8	6	2,438	8		20,320	44,800
1D	2,991	9	$^{93}/_4$	2,438	8		2,438	8		10,160	22,400
1E	1,968	6	$^{93}/_4$	2,438	8		2,438	8		7,110	15,700
1F	1,460	4	$^{93}/_4$	2,438	8		2,438	8		5,080	11,200

Table 7.2 Container dimensions as stipulated in DIN/ISO 668 or DIN 15,190, Part 1

Length code (Character)	Container length		
	mm	ft	in
1	2,991	10	
2	6,058	20	
3	9,125	30	
4	12,192	40	
A	7,150		
B	7,315	24	
C	7,430		
D	7,450	24	4
E	7,820		
F	8,100		
G	12,500	41	
H	13,106	43	
K	13,600		
L	13,716	45	
M	14,630	48	
N	14,935	49	
P	16,154		
USA only		53	
Permitted in some US States		57	

Table 7.3 Container widths

Container width		
mm	ft	in
2,438	8′	
2,500		
2,550		
	8′	6″

The above-mentioned container is identical to the CEN standard C-type swap-body, which is 7.45 m (24.4 ft) long. However, attempts to introduce such a container with widths of over 2.55 m (8.3 ft) and heights of over 2.90 m (9.5 ft) will continue to meet with resistance (Table 7.6).

The appropriate specialist committees have devoted time to discussing the introduction of 49′/14.9 m long containers with a width of 8′6″/2.60 m. However, such a container is unacceptable to most countries in Africa, Asia, Australasia and Europe. The same applies to a half-length container 24′4½″/ 7.43 m long, because of the 2.60 m width (Table 7.7 and Fig. 7.1).

Table 7.4 Non-standard container heights

Height code (Example container type)	Container height		
	mm	ft	in
9 [20′ platform]	337	1	1¼
9 [40′ platform]	610	2	
9	1,219	4	
8 [Half height]	1,293	4	6
0	2,438	8	
2 [Most containers]	2,591	8	6
4	2,743	9	
5 [High-cube]	2,895	9	6
6	>2,895	>9	6

Table 7.5 Suitable and unacceptable container sizes for international use

Containers considered acceptable for international use		Containers not considered acceptable for international use	
Length	7.43 m	Length	7.45 m
Width	2.50 m	Width	2.55 m
Height	2.60 m	Height	2.90 m

Table 7.6 Container sizes used in the US

Length	=	49′	=	14,935.2 mm	(14.90 m)
Width	=	8′6″	=	2 m 590.8 mm	(2.60 m)

Table 7.7 Maximum mass for a 20′ container according to the previous standard

Container size	Maximum gross weight		Regulatory text
	kgs	lbs	
20′	20,320	44,800	[To previous ISO standard]
20′	24,000	52,910	[To ISO 688]
20′	30,480	67,200	[Optional]
30′	25,400		
40′	30,480	67,200	
40′	32,000	70,550	[Optional]

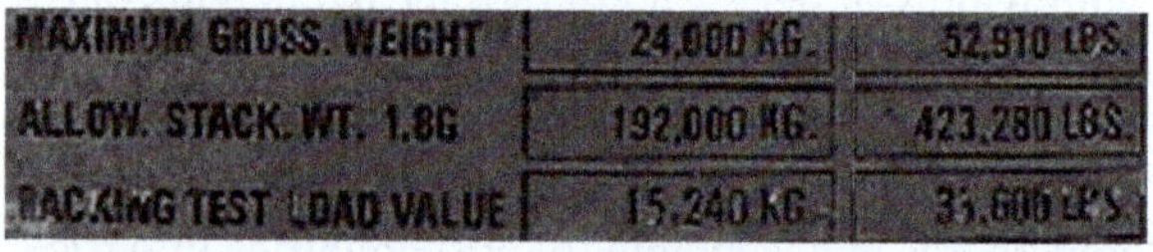

Fig. 7.1 Maximum mass for a 20′ container according to DIN ISO 688

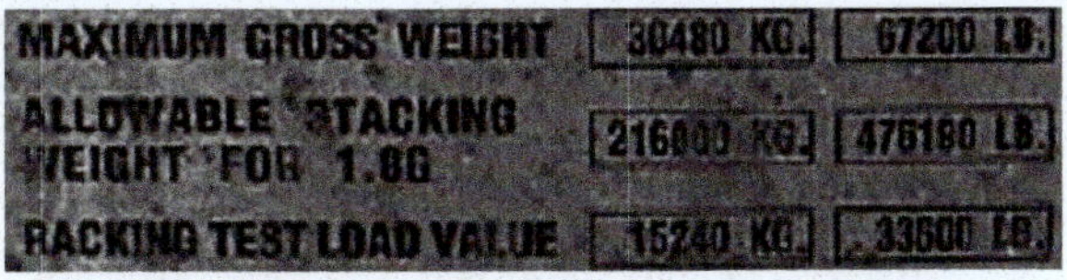

Fig. 7.2 Maximum mass for a 40′ container according to the previous standard or optionally for a more recent 20′

The significant feature of this container is that the term used is the more correct "mass," the unit of mass being correctly stated as the kilogramme (Fig. 7.2).

The term used in this example is weight, as is more usual, but the unit used is that of mass. The correct unit would be the Newton, with the decanewton or kilonewton [daN, kN] also being possible. The external dimensions of the containers are specified in ISO 668, with the maximum allowable dimensional tolerances being ±10 mm. The internal dimensions are stated as minimum values. The current internal container dimensions are dependent on the structural material used and the container type selected (Table 7.8).

Many of the following statements fall equally within the scope of the operational markings covered, where additional information may be found. If real data relating to containers to be ordered and subsequently packed are required for the purpose of forward planning, it is best to refer to the brochures published by shipping companies or other container owners. When packing a container, it is important to cast an eye over the plates affixed to it. Pallet-compatible containers are 2.5 m wide, for example, and usually carry an appropriate warning (Fig. 7.3):

This warning also counts as an operational marking. More detailed information is also to be found in this chapter. The following two tables give the lengths and gross weights for 20′, 30′ and 40′ ISO containers and 10′–40′ inshore containers (Table 7.9).

The ISO standard maximum gross weights for inshore containers are specified as follows (Table 7.10):

The minimum dimensions for the loading areas and volumes of ISO containers are specified as follows (Table 7.11):

Data relating to loading areas are very seldom found. Volumes, on the other hand, are always stated (Fig. 7.4):

Table 7.8 Current internal container dimensions

Standard container dimensions	Length		Width	Height
	20′	40′		
External: mm	6,058	12,192	2,438	2,591
External: '/"	19′10½″	40′	8′	8′6″
Internal: mm	5,867	11,998	2,330	2,350
Internal: '/"	19′3″	39′4$^3/_8$″	7′7¾″	7′8½″
Door opening: mm	N/A	N/A	2,286	2,261
Door opening: '/"	N/A	N/A	7′ 6″	

Fig. 7.3 Photograph of a warning affixed to a 2.5 m wide container

Table 7.9 ISO standard maximum gross weights for inshore containers

Container length in ft	Length in mm		Maximum gross weight in kg
	External (mm)	Internal (mm)	
20′	6,058	5,867	20,320
30′	9,125	8,931	25,400
40′	12,192	11,998	30,480

Table 7.10 ISO standard maximum gross weights for inshore containers

Container length in ft	Length in mm	Width in mm	Maximum gross weight in kg
10′	2,991	2,500	10,000
20′	6,058	2,500	20,000
30′	9,125	2,500	25,000
40′	12,192	2,500	30,000

Table 7.11 Minimum dimensions for loading areas and volumes for standard ISO containers

Container size	ISO designation	Loading area (m^2)	Volume (m^2)
40′	1 AA	28.0	65.7
40′	1 A	28.0	61.4
20′	1 CC	13.7	32.1
20′	1 C	13.7	30.0

Fig. 7.4 Examples of container volume data

Fig. 7.5 Photograph showing the volume in litres on a tank container

As an approximate guide the values state are 30 m^3 for 20′ containers and approximately 67 m^3 for 40′ containers. For tank containers, preferred practice is to state their volume in litres rather than in cubic metres as shown in Fig. 7.5.

As already described earlier, the CSC Code specifies that the maximum gross weight must be marked on the CSC plate. Other regulations require the maximum gross weight and the tare weight to be indicated. This constitutes the minimum information with which the user of the container must be provided by law (Fig. 7.6).

The maximum payload can be calculated as the difference between the maximum gross weight and the tare weight, but this information is already indicated on most containers. Of interest in this context are the different terms selected by the various container owners. It would be altogether desirable if these terms were to be standardised.

Maximum gross weight – tare weight = permissible payload

Tare weight + permissible payload = maximum gross weight

On platform containers, the data is located on the end walls or frame parts (Fig. 7.7).

Volumes are always to be found together with the weight information. On box containers, all the above-mentioned values are usually marked large and clear on the right door leaf (Fig. 7.8).

Nothing should obscure this important data as demonstrated in Fig. 7.9.

The text of the sticker refers to limited quantities of hazardous materials and has nothing to do with the maximum gross weight. Loading area and loading capacity (volume) depend principally on wall thickness and structural height. In practice, the ISO standard minimum values for weights and volumes are exceeded. Depending on the intended goods transport operation, it may therefore be worth making a comparison of, and choosing, the most favourable container. A comparison of the gross stowage factor with the stowage factor of the respective transport receptacles is always appropriate. Containers often carry information about these aspects. Sometimes, technical data regarding dimensions, weights, volumes, etc. are provided on the CSC plates themselves or on separate plates (Figs. 7.10 and 7.11).

> ***Note:*** *an imperial foot is divided into 12 inches and corresponds to 30.48 cm. An English inch corresponds to 25.4 mm. While inches are conventionally each subdivided into 12 "lines", it is usual to state fractions of an inch, such as ½″, ¼″, 1/8″ or 1/16″. An imperial pound (lb) corresponds to 453.6 g. A cubic foot (cu.ft) correspond to 28.35 L. Metric units may be converted by rule of thumb into imperial customary units and* vice versa, *as follows:*
>
> - *Metre* $\times$ *10* $\div$ *3 = foot.*

Fig. 7.6 Photograph showing the minimum information required on a half-height open-top

- *Foot* $\div$ *10* $\times$ *3 = metre.*
- *Kilogramme* $\times$ *2 + 10% = pound.*
- *Pound* $\div$ *2 – 10% = kilogramme.*
- *Cubic metre* $\times$ *35 = cubic foot.*
- *Cubic foot* $\div$ *35 = cubic metre.*

Fig. 7.7 Photograph showing markings on platform containers

7.2 Container Operational Markings

The operational markings found on almost all containers are, in accordance with DIN EN ISO 6346, intended to simplify the use of the container by providing additional information and warnings. These markings also include both compulsory and optional markings. Examples of compulsory markings are those indicating gross weight and tare weight, which naturally have to agree with those on the CSC Approval Plate. The standard prescribes the form and sequence for marking of gross and tare weights, which are shown below. The example on the left demonstrates standard (i.e., compliant) operational markings, whereas the example on the left demonstrates non-standard (i.e., non-compliant) operational markings (Fig. 7.12).

On the example provided, the minimum compulsory marking requirements are fulfilled insofar as the required technical data is concerned, however the sequencing is completely wrong, with the imperial values being stated first. It is also compulsory to affix certain warning symbols. These include pictograms for air/surface containers, the danger posed by overhead power cables and height markings for containers which are higher than 2.6 m (8′6″) (Fig. 7.13).

Fig. 7.8 Photograph showing weights and volumes on 20′ reefers and 40′ general purpose containers

Fig. 7.9 Photograph showing an inappropriately positioned sticker

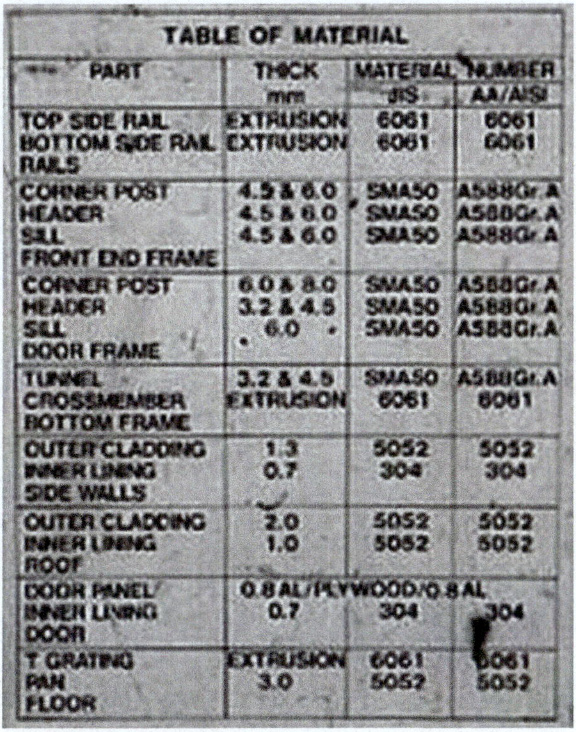

TABLE OF MATERIAL			
PART	THICK mm	MATERIAL NUMBER JIS	AA/AISI
TOP SIDE RAIL	EXTRUSION	6061	6061
BOTTOM SIDE RAIL	EXTRUSION	6061	6061
RAILS			
CORNER POST	4.5 & 6.0	SMA50	A588Gr.A
HEADER	4.5 & 6.0	SMA50	A588Gr.A
SILL	4.5 & 6.0	SMA50	A588Gr.A
FRONT END FRAME			
CORNER POST	6.0 & 8.0	SMA50	A588Gr.A
HEADER	3.2 & 4.5	SMA50	A588Gr.A
SILL	6.0	SMA50	A588Gr.A
DOOR FRAME			
TUNNEL	3.2 & 4.5	SMA50	A588Gr.A
CROSSMEMBER	EXTRUSION	6061	6061
BOTTOM FRAME			
OUTER CLADDING	1.3	5052	5052
INNER LINING	0.7	304	304
SIDE WALLS			
OUTER CLADDING	2.0	5052	5052
INNER LINING	1.0	5052	5052
ROOF			
DOOR PANEL/	0.8AL/PLYWOOD/0.8AL		
INNER LINING	0.7	304	304
DOOR			
T GRATING	EXTRUSION	6061	6061
PAN	3.0	5052	5052
FLOOR			

Fig. 7.10 Example of a CSC plate giving full information

7.2.1 Stackability Markings

The two examples given at Fig. 7.14 show the type of markings which should be applied to containers for air/surface transport, and stackability limitations for both air/surface and sea transportation.

To indicate that these containers have limited stackability, the symbols should be affixed where possible in the upper left-hand corner of the end and side walls of the container, and on the roof. The symbol is almost exclusively shown in black. If the colour of the container makes it difficult to see the symbol clearly, it should be applied to a different background of a more suitable colour, preferably white. According to the standard, the symbols should have the following dimensions: the aircraft should be at least 130 mm (5″), the stackability symbols should be at least 260 mm (11″) high and 280 mm (10″) high for air/surface and sea transport respectively. The capital letters should be at least 80 mm (3″) high. All containers equipped with ladders must be provided with a warning symbol, which indicates the danger posed by overhead power cables, an example of which is provided at Fig. 7.15.

The height of the lightning in the warning symbol must be at least 175 mm (6.875″). The size of the warning symbol, measured between the outer edges of the

Fig. 7.11 Example of a separate plate giving full information

Fig. 7.12 Photograph showing standard operational markings (gross and tare weights)

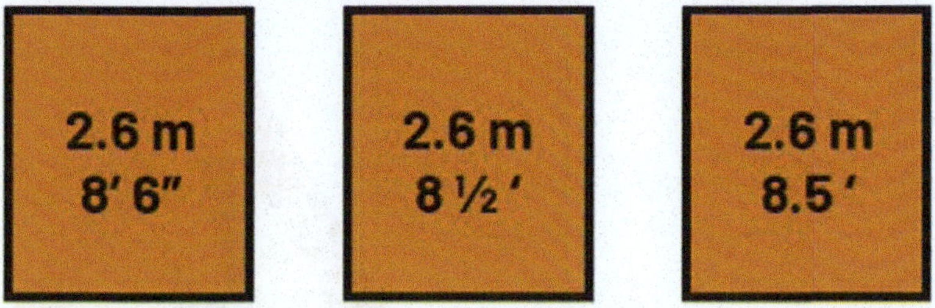

Fig. 7.13 Examples of the different ways of writing 2.6 m and the imperial customary dimension

Fig. 7.14 Examples of the different stackability markings for air transport and air/sea transport

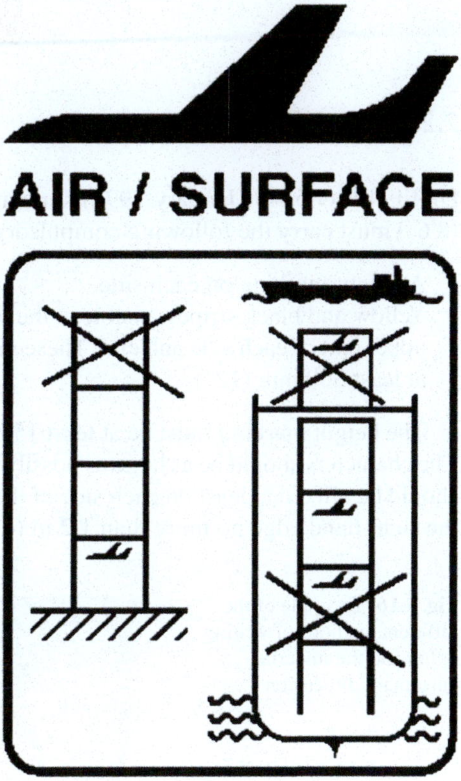

black border, must be no less than 230 mm (9″). The symbol must be affixed in the vicinity of the ladder. It is sensible to affix the lightning symbol to the upper part of the ladder. It is not necessary, however, to affix signs (3) indicating that the container has a height of 2.6 m or 8′6″.

Fig. 7.15 Symbol warning against the danger posed by overhead power cables

7.2.2 Height Markings

DIN EN ISO 6346 (January 1996) stipulates that all containers higher than 2.6 m (8′6″) must carry the following compulsory markings (Figs. 7.16 and 7.17):

- A height marking on each side.
- Yellow and black stripes, visible at the top and side, which must be affixed in the upper part of each side and end. These must start at the corner castings and extend at least 300 mm (12″).

The height marking must be at least 155 mm high and 115 mm wide (6″ × 4.5″). The characters should be as large as possible, so that they are clearly visible. The sign should be affixed at least on each side of the container, in each case in the vicinity of the right-hand edge no more than 1.2 m (4′) from the top edge of the container and

Fig. 7.16 Example of the different methods of writing 2.9 m and the imperial customary dimension

Fig. 7.17 Examples of how yellow and black stripes constitute part of the height marking

no more than 0.6 m (2′) from the right-hand edge (beneath the identification number) (Fig. 7.18).

7.3 Optional Markings

Optional markings, such as mirror images of the height marking at every suitable point (e.g., end wall), may be affixed. It has always been usual to mark high-cube containers with optional markings of various designs. The upper pair of images show a container marked according to the January 1996 edition of DIN EN ISO 6346, while the lower pair show a container marked according to the August 1985 edition of DIN ISO 6346. It is always sensible to emphasise the height of individual containers, even if it doesn't always help. One possible optional operational marking which most operators do affix states the maximum payload or net cargo weight, which is the difference between the maximum gross weight and the tare weight of a container.

Fig. 7.18 Photograph of a container with a full set of operational markings

Although this marking is optional, the standard does stipulate that the weight labelled NET be given first in kilograms (kg) and then in pounds (lb). In general, the various markings are so arranged that the weight data are followed by container volume data.

7.4 Other Markings

Many containers carry instructions about using particular structural elements, which are intended to serve as warnings. Apart from very few exceptions, containers may only be lifted vertically by the upper corner posts. It is not necessary to be able to read the instructions in the picture. The point is simply that there are such instructions describing how a collapsible flat rack is to be erected and collapsed and what points have to be observed. It should go without saying that inexperienced people should not be trusted with such activities. In any case, trained personnel also need to heed such instructions.

7.5 Arrangement of Obligatory and Optional Markings

The relevant standard allows for the areas of yellow and black stripes next to the corner castings to be reduced to a length of 300 mm (12″). It should be noted that the standard may not necessarily conform to national standards; for example, in Germany and various other countries, diagonal black and yellow stripes must always run in the same direction. This symmetrical arrangement of the warning plates, as provided for in the regulations, makes sense. This vehicle with correctly affixed warning is a particularly clear illustration. It would be a good idea if this were taken into consideration when the standards are revised.

The standards allow for optional affixing of size and type codes to the roof and front-end wall. The same applies to affixing a data medium for automatic identification and to marking of the net cargo weight as demonstrated by the respective arrows on the diagrams at Figs. 7.19, 7.20, 7.21, 7.22, 7.23, 7.24, 7.25, 7.26, 7.27, 7.28, 7.29, 7.30, 7.31 and 7.32.

In this chapter, we have covered some of the core technical aspects associated with the design and maintenance of containers in service. Importantly, we have looked at the obligatory markings required to advise of the weight and dimensions of the container, which is necessary for container handlers to make accurate judgements

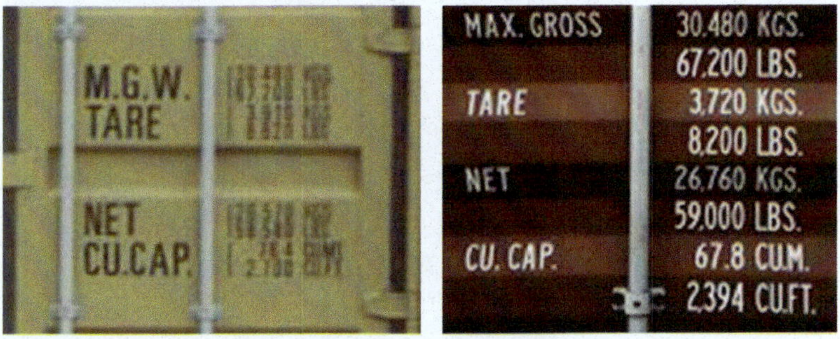

Fig. 7.19 Example of weight and volume markings on 40′ containers

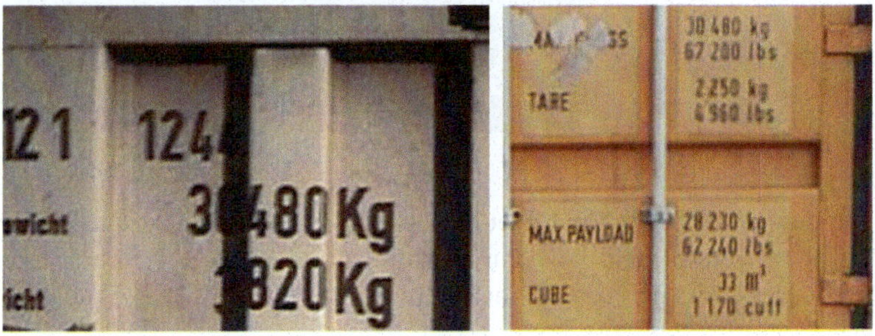

Fig. 7.20 Photograph of partially hidden or covered weight and volume data

Fig. 7.21 Example of how the maximum gross weight may deviate from the standard values

Fig. 7.22 Example of M.G.W. = Maximum Gross Weight ... data

when handling both vacant and occupied containers. In the next chapter, we will turn our attention to container design and how different containers are constructed to different standards according to their anticipated loads.

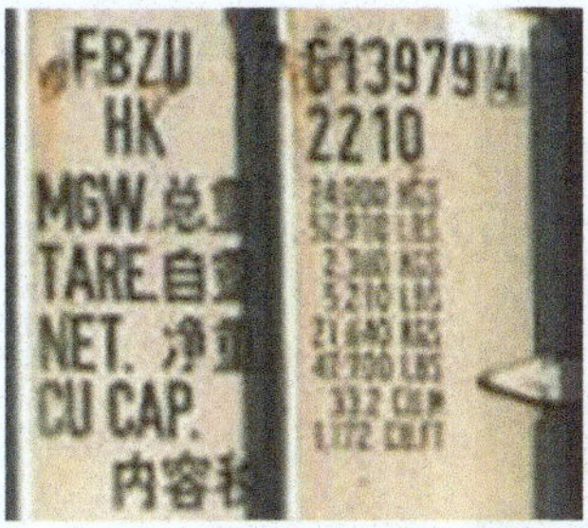

Fig. 7.23 ... and in Chinese characters too!

WARNING

DOOR WEIGHT IS 250KG

USE EXTREME CAUTION WHEN OPENING

Fig. 7.24 Example warning about folding doors on a half-height open-top container

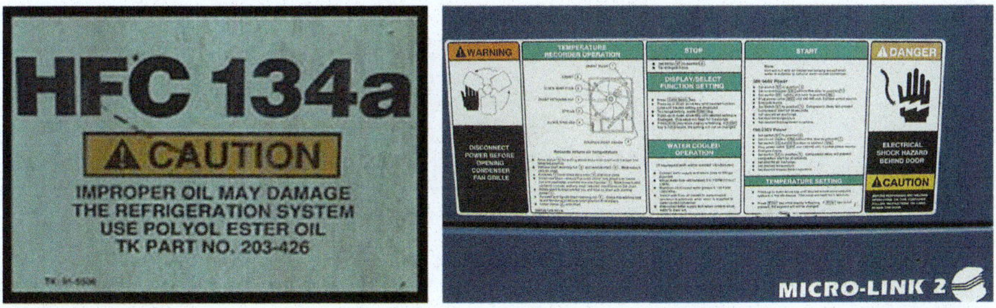

Fig. 7.25 Photograph of operating and safety instructions on a thermal container

Fig. 7.26 Example of a statement of compliance with structural regulations from Bureau Veritas

```
BANGKOK CONTAINER INDUSTRIES
LTD.

HOMOLOGATED BY BUREAU VERITAS

MODEL BCD  3-2      CTTH /
0116/2681

INSPECTED.               1992
[I.I.-T.T.]
```

Fig. 7.27 Example of desirable arrangements of black and yellow warning markings

Fig. 7.28 Photograph of a (UK) van with correctly painted red and white warning

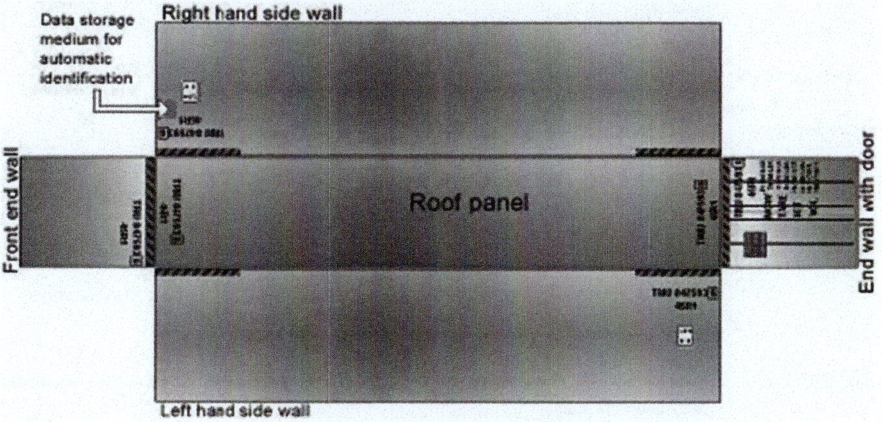

Fig. 7.29 Selective enlargements

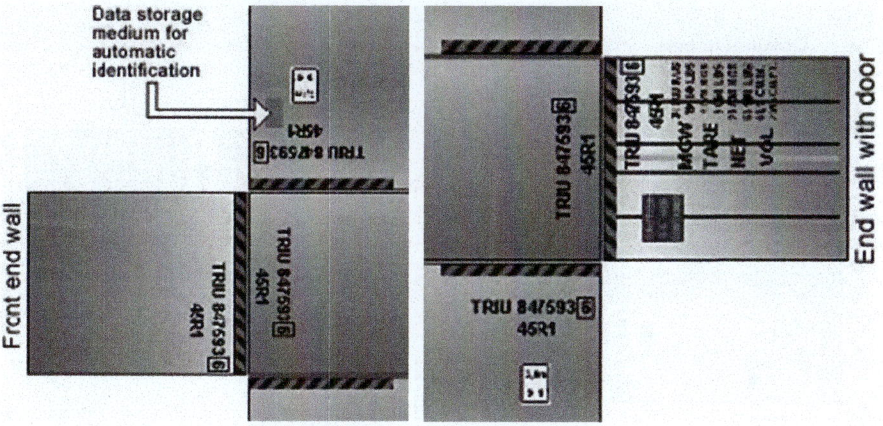

Fig. 7.30 Arrangement of obligatory and optional markings

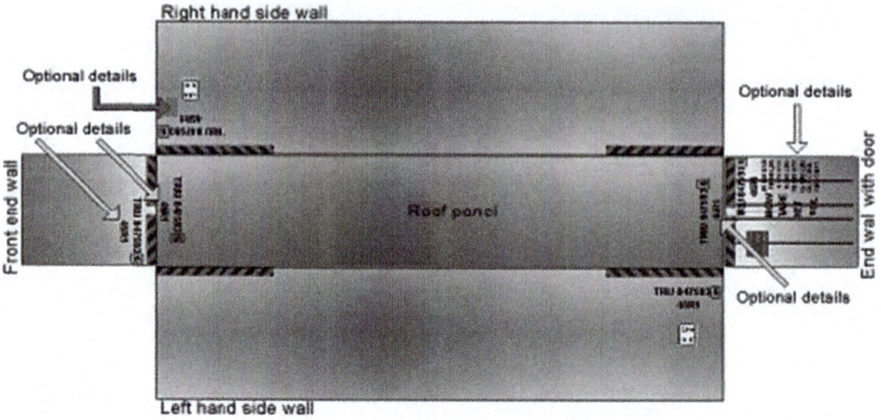

Fig. 7.31 Typical arrangement of acceptable container markings

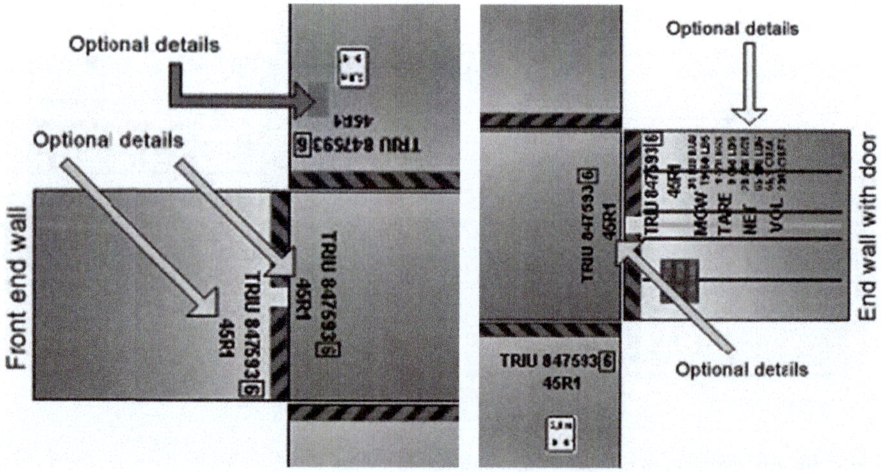

Fig. 7.32 Selective enlargements

Chapter 8
Container Design

In the previous chapter we discussed in some detail the technical aspects of containers and how this information should be presented. In this chapter, we will look more closely at container design as this has a direct influence on how they should be packed, handled and stowed. The container user should be cognisant of the most important structural differences between containers, so that they are in a position to make appropriate preparations for packing and cargo securing and correctly to assess container loading capacities. Problem-free dispatch has been assisted by the standardisation of transport units, such as containers, whilst the standardisation of certain components and some dimensions permit the use of standard handling equipment and means of transport. We will describe fundamental components and designs first of all with reference to standard ISO or box containers. More detailed information is given under the heading *Container types.*

The load-carrying element of all box containers is a steel framework, consisting of four corner posts and two bottom side rails, two top side rails, two bottom cross members, a front top end rail and a door header. Additional bottom cross members are fitted between the bottom side rails, to serve as supports for the floor covering. The side and end walls and the roof are the components of a standard box container which are capable of bearing the smallest load. This naturally depends on the materials used in their construction. Figures 8.1 and 8.2 illustrates the essential design features and construction elements of a standard box container. Not included in the figures are ancillary equipment such the door bar handle and the locking components required for sealing, etc. Where necessary, descriptions of and comments about these components are provided at other points within the book. In the early days of container shipping, the majority of containers were constructed according to American Standards Association standards (ASA), but now the containers used for maritime transport are without exception ISO containers. These differ both in dimension and in the shape of the corner fittings or "corner castings." Most ASA containers, i.e., containers like those used by Sealand, were constructed according to the ASA, have since been adapted to match ISO dimensions. To simplify handling, special universal spreaders

A. A. Olsen and F. Karkori, *Containerized Cargo Handling and Stowage*, Springer Series on Naval Architecture, Marine Engineering, Shipbuilding and Shipping 22,
https://doi.org/10.1007/978-3-031-63329-4_8

Fig. 8.1 Diagram illustrating the essential components of a standard ISO container

Fig. 8.2 Diagram illustrating the part names of the standard ISO container floor

are used, which can handle both types without difficulty. ISO standard 1161 specifies different shapes for top and bottom and mirror images for right and left. The eight corner castings of a container or a corresponding container have to be particularly strong, since they work with the corner posts and the other basic components of the container frame to absorb the forces which lock units or lashings exert on containers when they are stacked on top of one another, during handling and during transport.

DIN/ISO standards specify certain minimum requirements for the loading capacity and stackability of containers. Higher levels of performance may be provided for individual properties, and lower levels may not. It must be possible to stack six ISO containers packed to the maximum weight vertically on top of one another. Maximum offset is set as follows: width wise—24.4 mm (1″) and lengthwise—38 mm (1½″). The actual values of modern containers are generally higher. Many are designed to be stacked in excess of eight or nine high. The maximum stacking load must be marked on the CSC plate (further details are provided in the relevant section of the book). According to safety regulations, stacked containers must, where necessary, be secured against toppling and shifting. On larger container ships, the containers are stowed nine to twelve high in the hold. In such cases, the containers loaded must either be only partly full or be designed to endure greater stackability. The latter is the case with modern containers, but it is possible to use fold-out flaps in the cell guides, which subdivide the stacks. Inland containers are only designed for stacking to three high when fully loaded. Irrespective of the material used to build an ISO container, it is essential for it to be spray-tight. With standard ISO containers, the load-carrying parts are made of steel profiles, this includes as a minimum the entire frame including the bottom cross members and the elements serving as reinforcements, such as bottom side rails in the area of the gooseneck tunnel, etc. Three types of material used for the walls and roof may include:

- Corrugated sheet steel.
- Aluminium sheet in conjunction with stiffening profiles; and
- Plywood with glass fibre-reinforced plastic coatings (plywood + GRP).

These materials are reflected in the container names:

- Steel container.
- Aluminium container; and
- Plywood container.

With respect to sheet steel containers, a wide range of differently profiled corrugated steel sheet may be used for the outer walls. These are protected against corrosion by the application of corrosion resistant paint or similar products. The cost advantages of this type of container have led to its dominance. Of all the containers currently in use, a rough estimate would suggest that 85% are made of sheet steel. Aluminium containers are built either with a pure aluminium skin or with a plywood inner lining. They may be riveted or replete with a smooth or lightly riveted finish. With plywood containers, the outer walls are made of plywood coated with glass fibre-reinforced plastic (GRP). Plywood is a popular material for manufacturing "coffee containers." Container doors are often also made of ply-metal, which consists of a plywood core with sheet metal adhered to it on both sides.

It should be clear from these examples that containers are rarely made from a single material, but instead various material combinations that include steel, aluminium and plywood. Special impregnation against insects or other pests is required for certain regions of service. Most container floors or wooden parts undergo preventive treatment (refer back to the section on Sirex wasp prevention in Chap. 6). ISO containers are provided with coverings of plywood or textured coated board mostly 25 mm thick, or rarely, 30 mm thick. Although wood is expensive, it has substantial advantages over other materials. For instance, wood is strong and resilient, it does not dent, it can be easily replaced during repairs and, when appropriately finished, has an adequate coefficient of friction (although this does not apply to new containers which have a mirror-bright finish). Planking is preferred for flat racks and other similar platform containers. 20' platforms or half-height open-top containers often have a floor of steel, e.g., of "tear drop" or otherwise textured sheet. The floors of ISO containers have to be capable of bearing the evenly distributed payload, the emphasis being on "evenly distributed." The following test criteria applies where ground conveyors have access to container floors:

- Axle load 12,040 lbs 5,460 kg
- Wheel load 2,730 kg
- Contact surface per wheel 142 cm^2
- Wheel width 180 mm
- Wheel gauge 760 mm

The wheel contact area of 142 cm^2 corresponds to the size of a standard postcard. Forklift trucks with a load-carrying capacity of two metric tonnes have axle loads of just under five metric tonnes when loaded. Most 2.5 metric tonne forklifts are within the admissible range. However, some electrically operated 2.5 metric tonne forklift trucks reach front axle loads of over 6,000 kg when loaded. It is of course possible for even heavier forklift trucks to drive into containers, provided they are not fully loaded, and the equipment and cargo dimensions allow it. It is essential to note that add-ons reduce the load-carrying capacity of forklifts but increase the front axle load. Goods may only be stacked in ISO containers using equipment with a suitable telescopic mast. Using equipment with twin tyres may reduce the wheel loads, but it does not completely resolve the issue of axle load. It should not therefore be regarded as a license to use heavier equipment.

The strength of ISO containers is laid down in the relevant DIN standards and/or the *International Convention for Safe Containers*: ISO containers must be capable of absorbing the *horizontal forces* arising during regular service at the level of the end frames. Containers must withstand loads in the lengthwise direction which correspond to external acceleration of 2 g acting horizontally on the floor fastening elements. This considers loads which are transmitted via twist locks and other vehicle locking elements to containers. Special railroad container cars with hydraulic shock absorption limit forces to 2 g; examples of these cars are **Lgjs**, **Sgjs** and **Sgjkmmns** carriages and other carriages with a *j* in their designation. In this instance the *j* indicates high-performance (long-stroke) shock absorbers or buffers. In accordance with the CSC, container end walls must be so constructed that forces of 0.4 times the uniformly applied payload can be safely absorbed, or in other words, 40% of the container payload or 0.4 g. Higher or lower values should be prominently marked on the exterior of the container. With respect to the end wall loading capacity, the loading capacity of the side walls must correspond to 0.6 times the uniformly applied payload, i.e., 60% of the payload or 0.6 g. Higher or lower values should again be marked on the exterior of the container. For side wall loading capacity, the values for end and side walls are valid only for large-area loads, therefore any point loading of the walls should be avoided. Because the weight-carrying capacity of many general-purpose containers is not fully utilised, loading is kept below the maximum values in the case of compact and even packing. However, if the rate of utilisation is high and/or uneven, countermeasures must be taken. In the case of container roof panels, an evenly distributed 200 kg load may be applied to a surface area of 600 mm × 300 mm, so meaning that two people may stand next to one another on the container roof. Under no circumstances should container roof panels be covered with cargo.

Some containers are fitted with forklift pockets for handling by ground conveyors. Appropriate regulations relating to the required dimensions may be found in appendix C of ISO 1496/1. The pockets are cavities formed crosswise in the floor structure which allows for the insertion of the forks from the side. The forks must be pushed fully into the pockets. Forks which are too short must under no circumstances be used for lifting, as they may cause damage to the floor. The forklift pockets only allow for the handling of empty containers. Packed containers must not be picked up in this way unless specifically permitted. No marking is provided and there is no explicit instruction given to pick up only empty containers. To rule out errors, appropriate markings should be made a requirement. Standardised regulations and compliance therewith in practice could help in the avoidance of many losses resulting from the incorrect use of these components. Some containers have a recess along the longitudinal sides which allows the containers to be picked up using straddle carrier load suspension devices for transport within cargo handling facilities. Straddle carriers are specially built (low) van carriers with which loads may be lifted but not stacked. Grappler pockets are slots or recesses in the bottom side rails of containers, especially inland containers and swap-bodies. Grapplers slot into them during cargo handling. Such grapplers may also be used with gantry cranes, if no spreaders are used. Grappler pockets also allow direct pick-up of the containers with the tongs of a van carrier. Many containers have recesses in the bottom of the front end. This centrally located recess is known as a gooseneck tunnel. A large number of containers, especially flat racks have them at both ends. The tunnel does not have any effect on loading space, as the inside of the container floor or the flat rack loading area is flat. Rather, this recess serves in centring the container on a gooseneck chassis. Containers with goosenecks can be carried on both normal chassis and gooseneck chassis. Containers without goosenecks can only be carried on normal chassis. Depending on the construction of the chassis, a lower road vehicle overall height may be achieved with gooseneck chassis. In this way, many articulated trucks can see their height reduced by approximately 150 mm. In accordance with the standards, gooseneck tunnels are only provided on 40' containers (Figs. 8.3 and 8.4).

Fig. 8.3 Photograph showing ISO corner castings (Vertical and horizontal)

Fig. 8.4 Photograph showing the securing of containers on deck

8.1 Container Types

It is estimated that broadly fifty percent of containers available internationally belong to shipping companies. Alongside these are a large number of leasing companies, which lease their containers both to shipowners and direct to customers. Containers belonging to forwarders tend to carry dry and liquid bulk goods, for which bulk and tank containers are mostly used. Specific details about the appearance of the containers, their external and internal dimensions, their weights and volumes, temperature control, cargo securing equipment and other distinctive features may in general be found in the brochures published by the respective container owner. Of necessity, we will only deal with the most common types of containers. According to DIN ISO 6346 which came into force in January 1996, a distinction may be drawn between the following types of containers:

- General purpose containers.
- Bulk container.
- Named cargo containers.
- Thermal containers.
- Open-top containers.
- Platform containers.
- Tank containers; and
- Air/surface containers.

Further distinctions are drawn within these groups depending on design and principal characteristics. Information relating to the respective code, the group and type code, can be found in this book under their respective headings. Over the years, expressions have become established which do not always correspond to the standards or which are used in addition to the standard expressions. Some of these are explained here.

8.1.1 General Purpose Containers

General purpose containers are just that, they are used for general purpose cargo. These containers may be of a standard design and construction, a higher-than-normal construction, they may be fitted with open sides, and an open roof, or both, and may or may not be fitted with some means of passive or forced ventilation. *Standard containers.* The term standard container was used for the first containers in their basic form. As these were closed and were primarily suitable for the loading of general cargo, they were/continue to be known as general purpose containers, dry cargo containers, ISO containers or box containers. The initial height of 8' has already been consigned to history. Most box containers today have an external height of 8′ 6″. Unventilated general-purpose containers have openings on either one or both end(s). These two steel containers are of identical construction and as such can be designated as standard containers. Both containers have an external height of 8′6″ and no gooseneck tunnel. This is stated in the code 2210 on the one hand and 22G1 on the other. Both containers have forklift pockets and straddle carrier recesses. Containers equipped with vents also count as unventilated general-purpose containers, although they have small vents in the upper part of the cargo space. Sometimes these containers have individual vents provided all the way along the upper part of the cargo space. Even so, these containers fall into the category of "unventilated general-purpose container." A further subgroup of unventilated general-purpose containers specified in DIN EN ISO 4346 are those with "openings at one or both end(s) plus full openings on one or both sides". This is separate from those containers with "openings at one or both end(s) plus partial openings on one or both sides" (Figs. 8.5, 8.6 and 8.7).

High-cube containers. The expression 'high-cube container' originally covered all containers higher than the standard height of 8′6″. The expression is now used in practice for containers which have an external height of 9′6″. Particular attention needs to be given to height restrictions when these containers are carried by road and rail. It may be necessary to use special chassis or carriages. The yellow and black marking on the top edges of the high-cube container serve as a warning about their height. 40' containers have a larger volume-to-payload ratio than 20' containers, i.e., they are suitable for goods with a higher cargo stowage factor. *Open-side containers.* Open-side containers (OS) have solid end walls and a solid roof panel. The sides may be closed at the bottom by folding down wooden, aluminium profile or steel sheet side walls. These may be divided into sections. Roof bow holders and roof bows are provided for the upper part, which may be covered with tarpaulins. The container

Fig. 8.5 Photograph of a typical 20' container

Fig. 8.6 Photograph of a typical 40' container

Fig. 8.7 Photograph of a typical 45' container

is packed from the side. Open-sided containers also exist which are open only on one side. If bars are positioned over one open side, the containers can be used for transporting livestock. Another variant is the "folding side wall" container, which is a type of inland container. Open-sided containers are also available with doors at the rear of the container. When reading statements in brochures that claim "lateral load securing consists of planks, which are suspended between the removable central support and the corner posts," this should not be read at face value. It is essential to check what forces these structures are rated to absorb (Fig. 8.8).

Open-top/open-sided containers. Open-top/open-sided containers (OTOS) combine the features of open-sided and open-top containers, i.e., the roofs and sides are open and need to be covered with tarpaulins. Since it is American transport that gave rise to the 8' wide standard overseas container, these containers are not suitable for the interchangeable pool type pallets used throughout Europe, which have dimensions of 800 mm × 1,200 mm. To counter this problem, pallet-compatible containers which are 2.50 m (8.2 ft) wide have been constructed. These must not, however, be confused with the pre-existing inland containers which have a 2.50 m (8.2 ft) width (Fig. 8.9).

General purpose containers with distinctive features. General purpose containers with distinctive features are intended to make it possible to transport particular cargoes which could not otherwise be safely transported without damage. Containers

Fig. 8.8 Photograph of a typical high-cube container

Fig. 8.9 Photograph of an **a** open-sided and **b** open-side and open-top container

for hanging garments belong to this group: they are equipped with clothes rails which are attached to special supporting bars. *Ventilated containers.* Passively ventilated containers, also known as ventilated containers, hardly differ from standard containers in external appearance. They are used to transport organic cargoes with high moisture contents such as coffee and cocoa beans. Special equipment is intended to ensure that, as far as is possible, sweat is prevented from forming. In general parlance, these containers are often named after the type of cargo carried within them, hence the widespread use of the expression, *coffee container*. There are two basic variants. The first are containers which offer *natural ventilation*. These containers take advantage of the pressure differences between the internal and external atmosphere to facilitate air exchange. As the warm air rises within the container and exits at the top through the roof ventilation strips, cooler external air enters in through floor ventilation strips. This results in a continuous and uninterrupted outflow of warm air and in flow of cool air. The second type of container employs forced ventilation methods. These are typically mechanically driven and includes fans and air ducts or ventilation flaps to facilitate the necessary air exchange. Container vent slots or air openings are often constructed as a labyrinth to prevent the penetration of spray or precipitation. Often, there are openings in the bottom and top side rails, which form regular air ducts. Occasionally, relatively small, perforated areas are provided at regular intervals in the outer skin. DIN EN ISO 6346 lists ventilated general purpose containers under code V, allocating to them either group code VH or type codes such V0, V2 or V4. Subsequently, a further distinction may be drawn between containers with non-mechanical ventilation at the lower and upper parts of the cargo space, containers with mechanical ventilation installed in the container, and containers with mechanical ventilation located outside the container. There are plenty of other designs of general-purpose container, in addition to those described above.

8.1.2 Dry Bulk Containers

Dry bulk containers or bulk containers may be used to transport loose, free-flowing goods. DIN EN ISO 6346 distinguishes dry bulk containers, under code B, with the designated group codes BU and BK as well as various type codes for non-pressure-resistant dry bulk containers. The latter are most commonly closed or air-tight and dry bulk containers with horizontal or tipping discharge pressure-resistance at test pressures respectively of 150 and 265 kPa. Externally, normal bulk containers are of identical construction to standard containers except for the loading hatches and discharge outlets. The loading hatches or domes are arranged in the roof. To gain access to these, some containers are provided with swivelable ladders. To prevent contact between the cargo and the container walls, "inlets" or liner bags may be introduced into the containers and fixed into place. The unloading hatches are normally at one of the ends, incorporated into the doors. Sometimes, short hoses may also be incorporated, so as to be able to direct the cargo as it is unloaded. Less frequently,

Fig. 8.10 Photograph of a typical dry bulk container

the discharge outlets are arranged at the side. In all the above cases, unpacking is achieved by the force of gravity, which is assisted by tipping the container towards the ground (Fig. 8.10).

8.1.3 Special Dry Bulk Containers

Some special dry bulk containers resemble tank containers. In addition to emptying by gravity, some containers may be emptied by means of compressed air. By fitting "inlets" or liner bags, normal general cargo containers may also be used as bulk containers. Reusable "inlets" are available, as well as single-use "inlets" or disposable liners. DIN EN ISO 6346 specifies these under code S and group code SN "named cargo containers". Examples of these include containers for transporting livestock (type code S0), containers for transporting automobiles (type code S1) and containers for transporting live fish (type code S2).

8.1.4 Thermal Containers

Thermal containers are divided into refrigerated or refrigerated and heated and insulated only types. A distinction is also drawn between those with fixed and removable thermal equipment. Refrigerated or heated containers allow goods to be transported irrespective of ambient temperature. The question of whether the container heats or refrigerates is relative. Nonetheless, the term refrigerated container or reefer has become established in common parlance. The correct term would be temperature-controlled container. Depending on container quality, the entire outer skin may be insulated, thereby reducing the internal dimensions of the container. According to ISO 1496/2, the internal width must amount to 2,200 mm. The container (reefer) floor is made of T-shaped aluminium profiles and is known as T-grating. The floor is strong enough to allow access by forklift trucks. The type of code used distinguishes between thermal containers according to type of refrigeration unit and coefficients of heat transmission (k values). External refrigeration units may be used to control the temperature of insulated containers. DIN EN ISO 6346 lists thermal containers under code R, accordingly:

- Mechanically refrigerated under group code RE and type code R0.
- Mechanically refrigerated and heated under group code RT and type code R1.
- Self-powered mechanically refrigerated under group code RS and type code R2; and
- Self-powered mechanically refrigerated and heated under group code RS and type code R3.

Code H of the same standard covers refrigerated or heated thermal containers with removable equipment, with group code HR applying to:

- Refrigerated and or heated with removable equipment located externally, coefficient of heat transfer $k = 0.4\ W/(m^2 \times k)$, type code H0.
- Refrigerated and or heated with removable equipment located internally, type code H1; and
- Refrigerated and or heated with removable equipment located externally, coefficient of heat transfer $k = 0.7\ W/(m^2 \times k)$, type code H2.

Group code HI covers insulated containers, namely:

- Type code H5 = insulated, coefficient of heat transmission $k = 0.4\ W/(m^2\ K)$; and
- Type code H6 = insulated, coefficient of heat transmission $k = 0.7\ W/(m^2\ K)$.

These containers also resemble standard containers externally, but inside they have a heat-insulating lining of polyurethane foam-based insulating material. Their wall thickness of 50–100 mm reduces stowage space a little when compared to normal containers. The insulation is intended to protect the cargo from rapid temperature variations. If lower temperatures or refrigeration are required, refrigerants may be added. Today, ice is seldom used as it requires special ice boxes; instead, dry ice is more commonly used. Another method is to vapourise liquefied gases. Such containers are used to carry heat-sensitive cargoes. Insulated containers and refrigerated containers may also be used without refrigeration or heating. Since temperature variations in the container are delayed relative to the external temperature, using such containers can be appropriate for certain goods. Refrigeration and heating can be achieved by using compressors directly driven by electrical motors; compressors directly driven by an internal combustion engine; or by externally generating cooling or heating air. The electrical power required for driving refrigerating/heating machines is supplied by onshore or onboard power supply systems or by a separate generator driven by an internal combustion engine. For the latter, diesel engines are the most common for this purpose.

A distinction is drawn between fixed generators and 'clip-on generators,' which are only used for a temporary basis. Fixed refrigeration units (integral units) reduce the useful length and payload of the container. Clip-on units increase the overall length of the containers. These heating/refrigerating machines are individual units which may be stored onshore or onboard and are temporarily attached to the end walls of the container. Each apparatus may be individually adjusted and does not affect the cargo space atmosphere for other containers. Supply air is input at the bottom and return air is extracted at the top. Such containers are also known as "porthole containers," the name arising from the similarity between these openings and the round ships' windows known as portholes. Externally generated cooling or heating air may be supplied via onshore or onboard central refrigeration systems. Containers supplied by this system have two openings in the end wall for the supply and removal of cargo space air subject to artificial temperature control. The term *Conair* container is a brand name often referenced when discussing insulated containers (Figs. 8.11 and 8.12).

Fig. 8.11 Photograph of a refrigerated (reefer) container

Fig. 8.12 Photograph of a thermal container

8.1.5 Open-Top Containers

Open-top containers are suitable for all types of general cargo but especially for heavy, over height cargo. Since their roofs are adjustable, they may also be packed from above. These containers have doors at the rear and their roofs are either adjustable or entirely removable. The roof covering consists either of a tarpaulin supported by roof bows or a solid, fully removable hard-top. Containers with the latter type of roof covering are known as hard-top/open-top containers. So that it is possible not only to introduce cargoes from above but also to load over height cargoes through the doors, some containers have removable or swing-out top members or top door rails. This also facilitates packing using lifting gear at floor level. The roof bows of an open-top container serve both to support the tarpaulin and to stabilise the whole container. Such containers are therefore made of steel, so that the structure is still sufficiently robust despite the absence of a roof. If over height cargoes are carried, the stabilising effect of the roof bows is absent. In some cases, therefore, it would be sensible to investigate whether it would not be equally possible to use flat racks or indeed whether doing so would not be the better solution from the point of view of cargo securing (Fig. 8.13).

Fig. 8.13 Photograph of an open top container

8.1.6 Half-Height/Open-Top Containers

As their name suggests, half-height/open-top containers are open at the top. There are two types in service: completely open or with a tarpaulin and roof bows. The rear door may be folded down and used as a ramp. DIN EN ISO 6346 lists open-top containers, abbreviated as OT, under code U. The group code for all such containers is UT. The following types are distinguished by the type codes indicated:

- U0—Opening(s) at one or both end(s).
- U1—Opening(s) at one or both end(s) plus removable roof in end frame.
- U2—Opening(s) at one or both end(s) plus opening(s) on one or both sides.
- U3—Opening(s) at one or both end(s) plus opening(s) on one or both sides plus removable top members.
- U4—Opening(s) at one or both end(s) plus opening(s) on one side plus full openings on the other side; and
- U5—Full, solid side and end walls (no doors).

With some containers of this design, the rear end wall can be folded out and used as an access ramp for forklift trucks. In these instances, a warning notice must be clearly visible (Fig. 8.14).

Fig. 8.14 Photograph of a half-height/open top container

WARNING

DOOR WEIGHT IS 250KG
USE EXTREME CAUTION
WHEN OPENING

The old, but still valid, number combination 4351 would today be replaced by the code 42U3. Platforms and flat racks are also known in common parlance as 'open containers.' Although such containers may have special equipment, such stanchions, fold-down end walls, lashing equipment, etc., they are often described in specialist literature as containers without additional equipment.

Platforms and flat racks

Platforms consist of reinforced container floors with sheet steel or planking. A large number of lashing points are provided for cargo securing; these may be welded-on or using recessed lashing lugs or rings, or lashing bars attached to the outsides of the side rails. If agreed with the shipping company, several platforms may be combined to form larger loading areas for carrying oversize goods. In these circumstances, acceptance and delivery then requires special attention and coordination of operations between the ship's command, shipping companies and the terminal operator. When empty, platforms can be stacked into piles to save space during transport. Since loaded platforms are not stackable, they are loaded onboard ship as the top layer of a hold or deck stack. If a ship's cell guides are equipped with flaps, any stowage space can be used. 20' platforms are 335 mm (1′1¼″) high, while 40' platforms are 610 mm (2') high. Some 40' platforms are fitted with gooseneck tunnels at both ends.

Flat racks, also known as flats, consist of a container floor and end walls. Flat racks with folding or collapsible end walls or end frames are commonly known as 'collapsible flats' or 'collapsible flat racks.' Flat racks may also be referred to as "open-top open-sided containers." The advantage of collapsible flat racks is that several unloaded flats can be stacked on top of one another to save space during transport. For cargo securing purposes, the flats are equipped with lashing bars or lugs or hooks of appropriate strength. Stanchion pockets and insertable stanchions are often provided. If agreed with the carrier, a number of flat racks may be combined together so as to be able to transport particularly bulky items of cargo. DIN EN ISO 6346 allocates code P to platform containers, with this edition of the standard combining both pure platforms and flat racks under this letter designation. Pure platforms bear group code PL or type code P0. P1 platforms have two complete, fixed end walls, irrespective of how high they are. P2 platforms have fixed posts, either free-standing or with removable top members. Group code PC covers type codes P3 and P4. P3 platforms have folding complete end walls. P4 platforms have folding posts, either free-standing or with removable top members. Group code PS covers type codes P5 to P9. P5 platforms are open at the top and ends (skeletal). In DIN EN 6346, type codes P6 to P9 had not been assigned, but were kept as spares. Here is an example of how one of these codes has been assigned by the German shipping company, Hapag-Lloyd (Figs. 8.15, 8.16, 8.17, 8.18 and 8.19).

8.1.7 Coil Containers

Coil containers are built like flats or flat racks, i.e., they consist of a container floor and flat or frame-like end walls. The container floor has cargo troughs for accommodating coils/rolls of steel sheet. Cargo securing costs are lower, since less complex lashing and blocking is required than on/in normal containers. However, when overall transport costs, including the return transport of empty coil containers, are considered, normal carriage without a special container may be more economical. Coil containers could be included amongst named cargo containers, but they are not covered by the standard (Fig. 8.20).

Fig. 8.15 Photograph of an open-side/open-end platform

Fig. 8.16 Photograph of an open-side/closed-end platform

Fig. 8.17 Photograph showing the detail on a standard P8 platform

Fig. 8.18 Photograph of a closed side flat rack container

8.1.8 Tank Containers

Tank containers are provided for carrying liquids and gases. The characteristics of the substances to be carried determine the material of which the tank is made, while the pressure under which the cargo has to be transported influences its construction. Tank containers always have as their basis a steel frame, into which tanks of various shapes may be inserted. Various fittings and accessories are incorporated, depending on type, method of filling, and method of emptying. Temperature-controlled tank containers require special heating or cooling devices. They may be filled via domes

Fig. 8.19 Photograph of a frame container

or tubes and emptied downwards via floor drains or upwards via riser pipes and pressure generators or in any other suitable manner. The products to be carried may be any types of liquid, liquefied or gaseous substance, ranging from harmless to dangerous. Tank containers for non-dangerous liquids fall under group code TN. Type codes T0, T1 and T2 distinguish between minimum pressures of 45, 150 and 265 kPa. Tank containers intended for liquid foodstuffs must be clearly marked as being intended for this purpose (Fig 8.21):

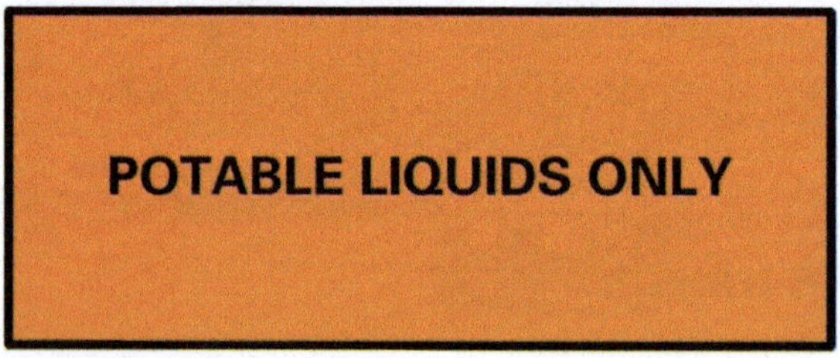

8.1.9 Half-Height Tank Containers

Half-height tank containers or half-height liquid containers are used to carry high-density liquids which cannot be carried in normal tank containers as they cannot be filled to a sufficiently high level and are therefore subject to surging. Tank containers for dangerous liquids come under group code TD. Type codes T3, T4, T5 and T6 distinguish between minimum pressures of 150, 265, 400 and 600 kPa. Such containers are subject, among other things, to the regulations governing the carriage of hazardous materials (i.e., the IMDG code). Tank containers for gases are assigned the group code TG with type codes T7, T8 and T9. T7 and T8 cover minimum

Fig. 8.20 Photograph of a coil palette

Fig. 8.21 Photograph of a typical tank container

pressures of 910 and 2,200 kPa, while T9 does not as yet have a pressure assigned to it. Tank containers must be at least 80% full for safety reasons, so as to prevent dangerous surging of the liquids in transit. As a rule of thumb, they should not be filled more than 95% full, to allow for thermal expansion of the contents. With reference to tank containers, the container packing guidelines states:

> ... they do not cover the filling or emptying of tank containers, portable tanks, or road tank vehicles, or the transport of any bulky cargo in bulk packaging.

For this reason, nothing more will be stated with regard to tank containers at this point in this book. It is worth mentioning, however, that tank containers may be replaced by introducing "flexible tubes" into standard containers, so turning them into temporary tank containers. However, a major problem associated with this is the surging of the fluid inside the container, which may result in damage to the container walls.

Closed containers

Closed containers include all non-ventilatable general purpose, dry bulk and tank containers. These containers allow no or only very slight air exchange. As they are rated gas-tight, these containers are not suited to transporting goods which require an exchange of gases.

Special containers not otherwise discussed

The phrase 'design for ease of opening' is found in container literature in relation to the desire for rapid packing and unpacking. It is often used in relation to dry bulk containers with special openings. Half-height containers are lower than standard containers. The size codes distinguish between those 1,295 mm (4′3″) high and those of a height less than or equal to 1,219 mm (4′). The containers are specially designed for goods with low stowage factors and for high-density cargoes, such as heavy barrels, metal sheet, etc. The following needs to be said with regard to special containers: from a technical point of view, these meet the needs of many forwarders and are desirable with regard to cargo securing. Their special fittings simplify packing and cargo securing, reducing securing costs and shipping risks. However, since they are adapted to one product or group of products to be transported, it is more difficult to use them on all routes, resulting in higher freight rates than for standard containers due to more frequent carriage of empty containers. From the point of view of cargo safety, however, special containers developed for a particular product are preferable to conventional containers. All wooden parts of containers traveling to Australia and New Zealand, including packaging and securing lumber, must be treated, and proof must be provided in the form of appropriate wood treatment certificates. The wooden parts of most containers are impregnated against insect infestation. Australia's and New Zealand's quarantine regulations require proof in the form of a plate. This provides documentary evidence that the wooden parts have been treated against the Sirex wasp.

Cargo securing

Owing to their cargo space being closed on all sides, box containers offer good possibilities for cargo securing by compact loading. For securing individual items, a tight fit may be achieved using bracing, especially in corrugated steel containers. On the other hand, the possibilities for lashing are extremely limited. Useful lashing lugs or rings are provided in all inland containers. Problems may arise if lashing points are over stowed and are no longer available or unreachable. Their strength also presents a problem: in general, their Maximum Securing Load (MSL) or Lashing Capacity (LC) is rated to only 1,000 daN per lashing point. The lashing points for containers are not currently covered by standards and may therefore differ greatly. Not all lashing points may be loaded in all directions. Lashing points must be attached to the frame parts of a container, since these components are adequately capable of absorbing forces. Lugs fitted in the area of the top side rails and sometimes the front top end rails can only be used to a minimal degree for traditional cargo securing by lashing. Lugs are suitable, indeed too strong, for attaching "inlets" or special non-woven fabric for catching sweat. In any case, such measures serve to protect the cargo, and therefore to keep it safe: non-woven fabrics provide protection against moisture or wetness dripping from the container ceiling and "inlets" protect against contact with the container walls. Some containers have lashing bars halfway up the container walls. In corrugated steel containers, they are welded in between the corrugations. The MSL of these bars is not remarkably high, and they can only be used to secure light individual items or small blocks of palletised cargo. The same applies to individual lashing rails fitted into the wooden linings of aluminium containers. Platform containers, i.e., the group comprising platforms and flat racks, have a considerable number of strong securing elements. To achieve a tight fit by application of pressure, stanchions or stakes are often provided, which may be inserted into corresponding stanchion (or stake) pockets. It is regrettably often the case that no details are given about the strength of lashing points. Even if information is provided, it is mostly of no practical use because of the wording chosen. Further explanations relating to this problem are provided in the container packing guidelines in the section on lashing materials, which states accordingly:

> The maximum securing load (MSL) can be calculated using a rule of thumb, according to which the diameter of the lashing points is determined, squared and multiplied by 10; i.e., diameter × diameter × 10. The result of the calculation is in decanewtons (daN) if the diameter is stated in millimetres and kilonewtons (kN) if the diameter is stated in centimetres.

According to the above-stated rule of thumb, the MSL of the lashing point is determined as follows:

$$\text{In decanewtons}: \ 20 \cdot 20 \cdot 10 = 4,000\,\text{daN}$$

$$\text{In kilonewton}: 2 \cdot 2 \cdot 10 = 40\,\text{kN}$$

The rule of thumb can only be applied if the weld seams look good and the weld substrate appears sufficiently strong. If lashing points are welded to excessively thin metal sheet, the rule of thumb must not be used for the calculation. The rule cannot be applied, either, if the lashing points, e.g., rings, are held on with pop rivets, thin screws or the like, despite the rings "looking good." Some platform containers have special securing systems. Some flat racks have fixed belt reels fitted, usually all on one side. The eyelets of the textile belts can be secured on hooks on the other side. Unfortunately, this securing system is unusable, because the "tie-down lashing/over-top lashing/friction loop" method has only a frictional effect, and this only with one-sided, uneven pretension. The belts do not allow a homogeneous tight fit to be achieved. If four of the totals of eight lashing reels were fitted on each side, appropriate securing could be achieved for many types of goods. The containers themselves are secured on ships and on road and railway vehicles by means of the container corner castings. Appropriate securing systems, such as rods on board ship, twist locks or other lock units, etc., may be attached in or to the corner castings, which are also responsible for load transfer if the containers are stacked in ship cell guides.

Occasionally, complaints arise during road-side cargo securing inspections because not all lock units have been firmly tightened. These of course should be. The container packing guidelines provides the following advice on this matter (as per Sect. 6.4):

6.4 Containers on vehicles

6.4.1 [Containers] should be firmly secured on vehicles before the vehicle is moved. The most appropriate points for fixing containers and swap-bodies are the bottom corner castings of the unit. Before and during transport, it should be checked that the unit is properly secured on the vehicle.

6.4.2 For transport on public roads or by rail, containers and swap-bodies should be secured to the road or rail vehicle by all bottom corner castings, in the absence of which, appropriate alternative measures should be taken. The main forces during the transport operation should be transferred to the unit through these bottom corner castings. Some additional forces may be transferred between the road vehicle chassis or rail-car surface and the unit bottom through load transfer areas in the unit bottom and in the vehicle surface. The securing devices on the vehicle may be twist locks, securing cones or securing guides. All such securing devices should be designed in such a way that the "open" or "locked" position of such securing devices is readily apparent.

It is quite common for problems to arise when containers, loaded in particular on road vehicles and attached firmly thereto, have to be carried by combined road/ship/road modes. Only occasionally do the vehicles have the legally prescribed lashing points, and even if they are present, they tend to be positioned unfavourably or to be incorrectly or defectively designed. As of 2023, this is true of approximately 80% of all vehicles which have to be carried by sea. On this matter, Paragraph 2.1.6 of the container packing guidelines states:

> A vehicle should be provided with points for securing it aboard ships (refer to European Standard EN 29367-I (ISO 9367-1): Lashing and securing arrangements on road vehicles for sea transportation on RoRo ships - General requirements - Part 1: Commercial vehicles and combinations of vehicles, semi-trailers excluded, and to EN 29367-2 (ISO 9367-2): Lashing and securing arrangements on road vehicles for sea transportation on RoRo ships - General requirements - Part 2: Semi trailers").

In view of this, it is essential drivers ensure their vehicles are safe and the cargo and container is secured appropriately.

In this chapter, we have examined some of the core design features of the most common containers currently in service. In the next chapter, we will briefly discuss the CSC and structural testing regimes for containers.

Chapter 9
Container Structural Testing Regimes

The technical requirements placed on containers are enshrined in the respective standards in addition to the *International Convention for Safe Containers,* the Convention or CSC Code for short. The aim of the CSC Code is to achieve the highest possible level of safety of human life in the handling, stacking and transporting of containers. The Convention applies to all containers used for international transport, except those containers developed especially for air travel, and which are outside the scope of the CSC Code and therefor beyond the premise of this book. That notwithstanding, article III of the Convention states:

> Each new container shall be approved either in accordance with the provisions for type-testing or for individual testing as contained in annex I.

The CSC Code specifies the precise requirements with which each individual component of the container must comply. Annex II of the Convention provides examples of the appropriate structural safety requirements and tests. Accordingly, the CSC Code states:

> ... a container made from any suitable material which satisfactorily performs the following tests without sustaining any permanent deformation or abnormality which would render it incapable of being used for its designed purpose, shall be considered safe ...

Pursuant to the Convention, every contracting State must ensure that effective procedures are put in place to enforce the regulations provided in annex I of the Convention. This annex sets out regulations for the testing, inspection and approval of containers. The text of the Convention allows all duly authorised organisations to be entrusted with these tasks, with the exception of carrying out container maintenance. In many countries, it is the national classification societies which are entrusted with these tasks, for example Lloyds Register for the UK, Germanischer Lloyd (GL) for Germany, Det Norske Veritas (DNV) for Norway, Bureau Veritas (BV) for France and Belgium, and the American Bureau of Shipping (ABS) for the US. Where the container has been inspected and approved for service, a CSC safety approval plate

A. A. Olsen and F. Karkori, *Containerized Cargo Handling and Stowage*, Springer Series on Naval Architecture, Marine Engineering, Shipbuilding and Shipping 22,
https://doi.org/10.1007/978-3-031-63329-4_9

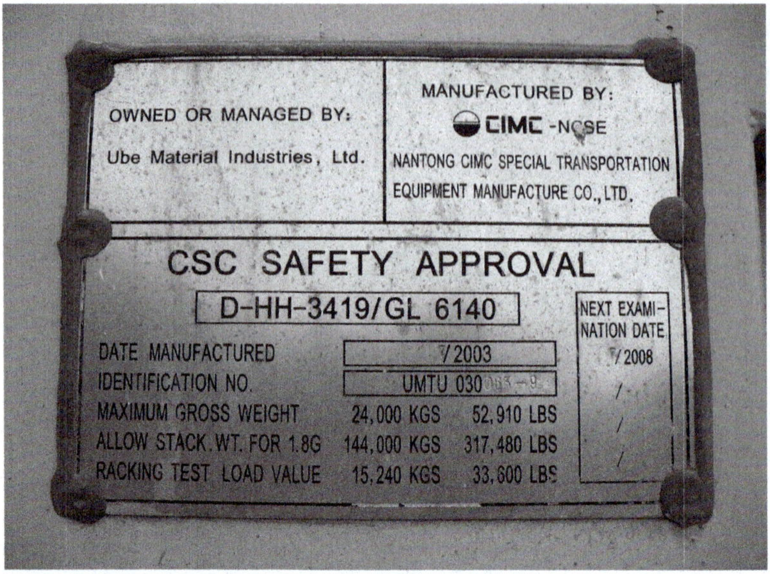

Fig. 9.1 Photograph of a typical CSC plate

must be permanently affixed to the container in a prominent and clearly visible location. This ensures the safety approval plate is easily identifiable. It must be provided in either English or French. The safety approval plate contains vital information about the container including:

- CSC safety approval.
- Country of approval and approval reference.
- Date (month and year) of manufacture.
- Manufacturer's identification number of the container or, in the case of existing containers for which that number is unknown, the number allotted by the approval regime.
- Maximum gross weight (in both kilograms (kgs) and pounds (lbs).
- Allowable stacking weight for 1.8 g (in kilograms (kgs) and pounds (lbs); and
- Transverse racking test load value (Fig. 9.1).

In accordance with the CSC Code, the safety approval plate:

> [...] shall take the form of a permanent, non-corrosive, fire-proof rectangular plate measuring not less than 200 mm by 100 mm. The words "CSC Safety Approval" of a minimum letter height of 8 mm and all other words and numbers of a minimum height of 5 mm shall be stamped into, embossed on or indicated on its surface in any other permanent and legible way.

9.1 Racking Test

The values stated for the racking test must comply with the standards set by the Convention. This is important to note as if the stated units of mass (for example, kilograms) are converted into newtons, a slight difference will result. Importantly, the longitudinal racking test value is not stated on the CSC plate. Even so, the test must be performed from both sides of the container. The ISO standard requires a loading of only 75 kN, whereas Lloyd's Register requires a test load of 100 kN and Germanischer Lloyd applies 125 kN in their load tests. Where important or legally prescribed data is no longer visible or legible on the safety approval plate, a complaint should be filed with the container owner or the container's approving organisation as failure to display said information is a contravention of the Convention. Accordingly:

> ... the country of Approval should be indicated by means of the distinguishing sign used to indicate country of registration of motor vehicles in international road traffic ...

Furthermore, Paragraph 2 (b) states:

> A blank space should be reserved on the Plate for insertion of end and/or side-wall strength values (factors) in accordance with Regulation 1, paragraph 3 and annex II, texts 6 and 7. A blank space should also be reserved on the Plate for first and subsequent maintenance examination dates (month and year) when used.

It is worth noting that some CSC safety approval plates do not carry these details:

> End Wall Strength [should] be indicated on plate only if end walls are designed to withstand a load of less or greater than 0.4 times the maximum permissible payload, i.e., 0.4 P. Side Wall Strength to be indicated on plate only if the side walls are designed to withstand a load of less or greater than 0.6 times the maximum permissible payload, i.e., 0.6 P.

This is based on Paragraph 3 of regulation 1, which provides that:

> "Where the Administration considers that a new container satisfies the requirements of the present Convention in respect of safety and if, for such container, the end and/or side-wall strength value (factor) are designed to be greater or less than that stipulated in annex II such value shall be indicated on the safety approval plate" ... "a blank space should also be reserved on the CSC Plate for first and subsequent maintenance examination dates (month and year) when used."

This is customary practice.

9.2 Container Maintenance and Oversight

Regulation 2 of Annex I of the Convention deals with container maintenance, and reads as follows:

> 1. The owner of the container shall be responsible for maintaining it [the container] in safe condition.

2. The owner of an approved container shall examine the container or have it [the container] examined in accordance with the procedure either prescribed or approved by the Contracting Party concerned, at intervals appropriate to operation conditions. The date (month and year) before which a new container shall undergo its first examination shall be marked on the Safety Approval Plate.
3. The date (month and year) before which the container shall be re-examined shall be clearly marked on the container on or as close as practicable to the Safety Approval Plate and in a manner acceptable to that Contracting Party which prescribed or approved the particular maintenance procedure involved.
4. The interval from the date of manufacture to the date of the first examination shall not exceed five years. Subsequent examination of new containers and re-examination of existing containers shall be at intervals of not more than thirty months. All examinations shall determine whether the container has any defects which could place any person in danger.
5. For the purposes of this Regulation, "the Contracting Party concerned" is the Contracting Party of the territory in which the owner is domiciled or has his head office.

9.3 Approved Continuous Examination Programme (ACEP)

For some years now, owners have in most cases been responsible for examining their own containers. Regulators have provided for this by establishing the "Approved Continuous Examination Programme" (ACEP), in which container owners are invited to participate. ACEP is a recognised repair and maintenance system providing for regular examinations and servicing. To participate in the programme, it is necessary for the container owner to register with the competent authority. This registration is then indicated on the container. Once a member of the ACEP proramme, the container owner assumes responsibility for the necessary inspections. The date for re-examination does not need to be indicated on the CSC plate.

9.4 Approval of New Containers by Design Type

Chapter 2 of the CSC Code sets out the regulations for the approval of new containers by design type, as quoted in detail, below:

Regulation 3—Approval of new containers

To qualify for approval for safety purposes under the present Convention all new containers shall comply with the requirements set out in annex II.

Regulation 4—Design type approval

In the case of containers for which an application for approval has been submitted, the Administration will examine designs and witness testing of a prototype container to ensure that the containers will conform with the requirements set out in annex II. When satisfied, the

Administration shall notify the applicant in writing that the container meets the requirements of the present Convention, and this notification shall entitle the manufacturer to affix the Safety Approval Plate to every container of the design type series.

In addition, regulation 5 of the CSC Code sets out the provisions for approval by design type:

1. Where the containers are to be manufactured by design type series, application made to an Administration for approval by design type shall be accompanied by drawings, a design specification of the type of container to be approved, and such other data as may be required by the Administration.
2. The applicant shall state the identification symbols which will be assigned by the manufacturer to the type of container to which the application for approval relates.
3. The application shall also be accompanied by an assurance from the manufacturer that he shall:
   (a) produce to the Administration such containers of the design type concerned as the Administration may wish to examine.
   (b) advise the Administration of any change in the design or specification and await its approval before affixing the Safety Approval Plate to the container.
   (c) affix the Safety Approval Plate to each container in the design type series and to no others.
   (d) keep a record of containers manufactured to the approved design type. This record shall at least contain the manufacturer's identification numbers, dates of delivery and names and addresses of customers to whom the containers are delivered.
4. Approval may be granted by the Administration to containers manufactured as modifications of an approved design type if the Administration is satisfied that the modifications do not affect the validity of tests conducted in the course of design type approval.
5. The Administration shall not confer on a manufacturer authority to affix Safety Approval Plates on the basis of design type approval unless satisfied that the manufacturer has instituted internal production-control features to ensure that the containers produced will conform to the approved prototype.

Regulation 6 relates to the examination of the container during the production process:

In order to ensure that containers of the same design type series are manufactured to the approved design, the Administration shall examine or test as many units as it considers necessary, at any stage during production of the design type series concerned.

And regulation 7 relates to the notification of the relevant administration:

The manufacturer shall notify the Administration prior to commencement of production of each new series of containers to be manufactured in accordance with an approved design type.

Chapter 3 of the CSC sets out the regulations for the approval of new containers by individual approval. In this regard, Chap. 4 sets out the regulations for the approval of existing containers. In the UK, the "administration" responsible for the type testing of containers is Lloyds Register (LR) whereas in Germany, this responsibility falls to Germanischer Lloyd (GL). Containers are certified if they fulfil the requirements of Annex II with respect to structural safety requirements and tests and of course the other standards that must be complied with. Once the container has been assessed and approved, the test plate is ‘awarded’ and affixed to the left door leaf. Importantly, the Convention, in regulation 1, point 4, makes it clear other competent authorities may set their own requirements in respect of container standards and inspection mechanisms, as stated here:

> The presence of the safety approval plate does not remove the necessity of displaying such labels or other information as may be required by other regulations which may be in force.

The existence of international customs agreements ensures that empty containers do not count as imported merchandise in member States and are thus free of duty provided the containers remain only temporarily within the customs territory. In addition, the container itself is deemed to be a bonded space. The various contracting States each recognise one another’s national approving bodies. Although the member States are bound by the CSC’s structural safety requirements and tests, there is nothing to stop them enforcing additional safety requirements for their own containers. Type testing involves a wide range of test procedures, covering, for example, lifting from top corner castings, lifting from bottom corner castings, lifting using other additional devices on the container such as forklift pockets, grappler arm positions and other methods. The test procedures also extend to stacking and to concentrated loads on roof and floor. Transverse racking is another aspect which is subject to testing. Furthermore, containers must undergo permissible loading capacity tests with regard to longitudinal restraint, the end walls and the side walls. The introduction to structural safety requirements and tests reads as follows:

> In setting the requirements of this annex, it is implicit that in all phases of the operation of containers the forces as a result of motion, location, stacking and weight of the loaded container and external forces will not exceed the design strength of the container. In particular, the following assumptions have been made:
>
> (a) the container will so be restrained that it is not subjected to forces in excess of those for which it has been designed.
>
> (b) the container will have its cargo stowed in accordance with the recommended practices of the trade so that the cargo does not impose upon the container forces in excess of those for which it has been designed.

This is significant in relation to the later chapters of this book in respect to the packing of containers and cargo securing. Only if containers are selected in accordance with the cargo to be carried and properly packed are they in a position to justify the trust placed in them. Testing and certification always relate to a particular type. A valid CSC plate affixed to a particular container tells the user what type of container is before them. Nevertheless, some CSC plates provide further details about this and about other supplementary standards. For example, a CSC plate which displays the character combination ‘1AA’ indicates a 40’ long, 8′6″ high standard container. If combined with a letter C the container is of a closed type. This means the container has been type-tested. If the character combination includes ‘OT,’ this means the container is an open-top. If the container is not subject to ACEP, the CSC plate must therefore carry the date of its next examination, fully in accordance with regulations.

Fig. 9.2 Example of the KOR container class mark

In Fig. 9.2, we can see a CSC plate mark issued by the South Korean Ship Registry, KOR.

In this relatively brief chapter, we have discussed the regulatory and inspection regime for new and existing containers, as provided under the clauses of the *International Convention for Safe Containers.* This regime is critical to the safe and effective movement of containers as it ensures the structural integrity of the container and allows for the early identification of deficiencies. In the next chapter, we will look at the container identification system which utilises container size and type codes to differentiate between the various types and sizes of containers in service at any given time.

Chapter 10
Container Identification Systems and Marking

The current standard, which is responsible for the coding, identification and marking of containers is DIN EN ISO 6346, which came into force in January 1996. Amongst other things, standard DIN EN ISO 6346 specifies that the previous standards with similar content have equal validity, since a number of older versions of containers with different markings naturally remain in service alongside newer containers. This chapter will provide only certain essential explanations with regard to the systems used; for more detailed information, the reader is invited to refer to the corresponding standards and more extensive specialist literature. The foreword to the standard states, among other things, that it includes not only the statutory units, but also corresponding sizes stated in Anglo-American (i.e., imperial) units. In some countries, the use of units nationally and commercially is proscribed by law, for example in Germany, under the *German Units of Measurement Act* of 22 February 1985. Such units may only be quoted to aid business relationships with countries which still use these units. A distinction is drawn between compulsory and optional marking. Compulsory ISO markings must be used on all containers, while optional markings do not. Where optional markings are included, they must be done so on the basis that they improve and promote the uniform application of container marking. However, if a particular style of representation is specified for an optional mark, it must be complied with. The terms "compulsory" and "optional" as used in the standard do not apply to the requirements of any legislative bodies, however.

The following example is a basic version of a horizontal container marking:

SUD	**U**	**307 007**	**9**
Owner code	Product group code	Registration number	Check digit

The container identification system specified in DIN EN ISO 6346 consists solely of the elements shown, which can only be used together: owner code, consisting of three capital letters, product group code, consisting of one of a select group of capital letters (e.g., U, J or Z), a six-digit registration number, and a check digit. The owner code must be unique and registered with the International Container Bureau

A. A. Olsen and F. Karkori, *Containerized Cargo Handling and Stowage*, Springer Series on Naval Architecture, Marine Engineering, Shipbuilding and Shipping 22,
https://doi.org/10.1007/978-3-031-63329-4_10

Fig. 10.1 Example of a typical horizontal container identification mark

(BIC—Bureau International des Containers, currently at 41 Rue Réaumur, 75,003, Paris, France), either directly or through a national registration organisation. A list of current contact details for member organisations is provided at the end of this chapter. As shown in Fig. 10.1, the owner code consists respectively of the letter combinations SUD. The product group code must contain one of the following three capital letters:

U—for all freight containers.

J—for detachable freight container-related equipment.

Z—for trailers and chassis.

The term 'owner code' may also apply to the combination of owner code and product group code, which is also known as an 'alpha prefix.' If the owner code is absent, the container cannot be identified. The registration or serial number consists of six digits. In the example shown, these numbers are 307 007. If the container number consists of fewer than six digits, it is preceded by enough zeros to make a six-digit number sequence. The check digit, which is given as 9 in the example shown, is always a single-digit number. It is usually in a box, to make it stand out from the registration number. The check digit can be used to validate whether the owner code, product group code and registration number have been accurately transmitted. No freight information system (FIS), transport information system (TIS) or similar data processing system will accept a container number if the result of the automatic checking procedure does not show agreement with the check digit. The procedure is deliberately designed to ensure that a number of transmission errors cannot cancel one another out, resulting in the acceptance of incorrect data. The checking procedure is as follows:

An equivalent numerical value is assigned to each letter of the alphabet, beginning with 10 for the letter A (11 and multiples thereof are omitted):

A	**B**	**C**	**D**	**E**	**F**	**G**	**H**	**I**	**J**	**K**	**L**	**M**
10	12	13	14	15	16	17	18	19	20	21	23	24
N	**O**	**P**	**Q**	**R**	**S**	**T**	**U**	**V**	**W**	**X**	**Y**	**Z**
25	26	27	28	29	30	31	32	34	35	36	37	38

The individual digits of the registration number keep their everyday value, i.e., 1 = 1, 2 = 2, etc. The following numerical values are accordingly obtained for the two examples with the alpha prefix "SUDU":

S	**U**	**D**	**U**	**3**	**0**	**7**	**0**	**0**	**7**
30	32	14	32	3	0	7	0	0	7

Each of these numbers is multiplied computationally, as a function of its position, by numerical values of 20–29: the first number by 1, the second by 2, the third by 3 and the 10th by 512. The following calculations are performed for the two examples:

1st #/#	**2nd #/#**	**3rd #/#**	**4th #/#**	**5th #/#**	**6th #/#**	**7th #/#**	**8th #/#**	**9th #/#**	**10th #/#**
2^0	2^1	2^2	2^3	2^4	2^5	2^6	2^7	2^8	2^9
1	2	4	8	16	32	64	128	256	512

If the check digits 9 and 6 respectively are correctly input in each case during data entry, the system accepts the data.

S	U	D	U	3	0	7	0	0	7
30	32	14	32	3	0	7	0	0	7
×	×	×	×	×	×	×	×	×	×
1	2	4	8	**16**	32	64	128	256	512
=	=	=	=	=	=	=	=	=	=

(continued)

(continued)

S	U	D	U	3	0	7	0	0	7
30	64	56	256	**48**	0	448	0	0	3,584
The sum of these numbers equals				**4,486**	And is divided by 11, giving				407.8
The integer multiplied by 11 equals				**4,477**	The difference between the resultant values in the check digit				
				9					

T	E	X	U	4	5	2	1	4	9
31	15	36	32	4	5	2	1	4	9
×	×	×	×	×	×	×	×	×	×
1	2	4	8	**16**	32	64	128	256	512
=	=	=	=	**=**	=	=	=	=	=
31	30	144	256	**64**	160	128	128	1,024	4,608
The sum of these numbers equals				**6,573**	And is divided by 11, giving				597.5
The integer multiplied by 11 equals				**6,567**	The difference between the resultant values in the check digit				
				6					

If someone makes a mistake when transmitting a container number and inputs the alphanumeric sequence TEXU 452,149 into a corresponding system instead of TEXU 452159, the programme will perform the following calculation:

T	E	X	U	4	5	2	1	5	9
31	15	36	32	4	5	2	1	5	9
×	×	×	×	×	×	×	×	×	×
1	2	4	8	16	32	64	128	256	512
=	=	=	=	=	=	=	=	=	=
31	30	144	256	64	160	128	128	1,280	4,608
The sum of these numbers equals				**6,829**	And is divided by 11, giving				597.5
The integer multiplied by 11 equals				**6,820**	The difference between the resultant values in the check digit				
				9					

Since the correct check digit for this container is 6, the system would indicate an error (Figs. 10.2, 10.3, 10.4, 10.5 and 10.6).

The check digit 0 may occur twice, since it arises where the final difference is 0 and where it is 10. To ensure that this does not happen, the standard recommends that registration numbers should not be used which produce a final difference of 10. This is the case, however, with the registration number shown (Fig. 10.7).

Fig. 10.2 Example of a typical vertical container identification mark

Fig. 10.3 Side view markings on containers

T	R	L	U		5	4	3	8	6	2
31	*29*	*23*	*32*		5	4	3	8	6	2
1	2	4	8		16	32	64	128	256	512
31	58	92	256		80	128	192	1,024	1,536	1,024

(continued)

(continued)

4,421	11	401.9	***401***	11	4,411	**10**				

H	**L**	**X**	**U**		**4**	**6**	**9**	**1**	**9**	**2**
18	***23***	***36***	***32***		4	6	9	1	9	2
1	2	4	8		16	32	64	128	256	512
18	46	144	256		64	192	576	128	2,304	1,024
4,752	11	432.0	***432***	11	4,752	0				

10.1 Container Size and Type Codes

In the previous section to this chapter, we briefly discussed the framework for the identification of containers. In this next section, we will look at how containers are identified using their size and type codes. Neither of these components are new to us, as we covered them in the previous chapter. The following figure is a rendition of a typical horizontal marking which additionally includes the country, size and type codes (in accordance with the 1985 version of DIN ISO 6346, use of which is still permitted) (Fig. 10.8).

Appropriate abbreviations are used for each country. In this instance, GB refers to the United Kingdom (i.e., Great Britain). The first digit of the size code indicates the length of the container, with the number **4** standing for a 40′ container. The second digit indicates the container height and whether or not a gooseneck tunnel is present. In our example, the number **3** stands for a height of 8 foot 6 inches (8′6″) with gooseneck tunnel. The first digit of the type code indicates container type. Here, **1** means that the container is a closed container with ventilation openings. The second digit of the type code relates to distinctive features. The **0** means there are openings at the end of the container. In the associated standard, columns are set out vertically:

Marking on containers: size and type code			ISO 6346
1	2	3	4
Length	Height	Type	Features

The corresponding data are laid out underneath. The abbreviation ‘g.n.t.’ is shorthand for gooseneck tunnel. A differently laid out size code summary is shown in Table 10.1. In the type code, the first digit (i.e., the third digit overall) indicates the container type and the second (i.e., the fourth digit overall) indicates distinctive features. However, there is a connection between the respective digits, which the following representation is intended to explain:

Fig. 10.4 Markings on the door of a container

The digit sequence **4364** means: 40′ long container 8′6″ high with gooseneck tunnel and folding, free-standing posts. It should carry the digit sequence **4363**, since the platform container has folding end walls. Included below are examples of container markings in accordance with this older standard. According to the previous annex F "size code designations" annexed to DIN ISO 6346 of August 1985, the two containers are 40′ long (number 4) and 8′6″ high with a gooseneck tunnel (number 3). According to annex G "type code designations," both containers are partially ventilated closed containers, which have passive vents in the upper part of the cargo space, the total cross-sectional area of the vents being less than 25 cm^2/m of container length (number combination 10). However, the coding does not indicate that the green container is 2.50 m wide.

10.1.1 Examples of Container Size and Type Coding Conventions

In the example below, the container is 40′ long (**4**), 8′6″ high and has a gooseneck tunnel (**3**). The type code **00** states that it is a general-purpose container which has opening(s) at one or both end(s).

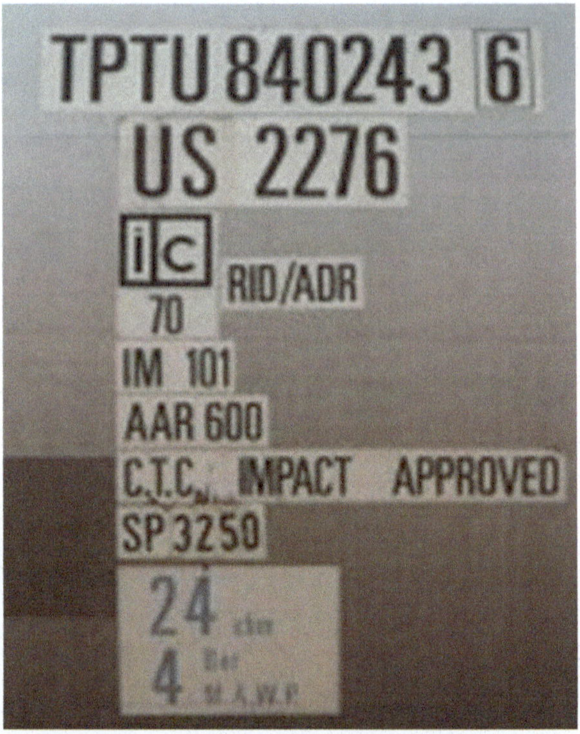

Fig. 10.5 Typical container markings (tank container)

Fig. 10.6 Disadvantage of check digit [0]

H	L	X	U		4	6	9	1	9	2
18	***23***	***36***	***32***		4	6	9	1	9	2
1	2	4	8		16	32	64	128	256	512
18	46	144	256		64	192	576	128	2,304	1,024
4,752	11	432.0	***432***	11	4,752	0				

Fig. 10.7 Correct identification using check digit [0]

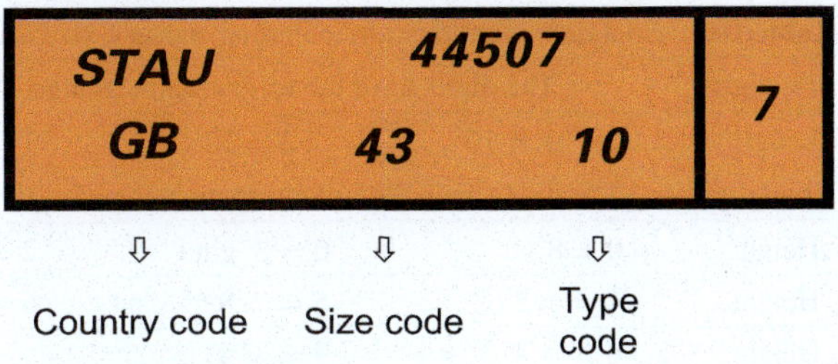

Fig. 10.8 Container identification including country, size and type codes

This container is 20′ long (2) and 8′ 6″ high (2). It is a ventilated standard container (**1**), with narrow, passive vents in the upper part (0).

This container is 40′ long (4), 8′6″ high and has a gooseneck tunnel (**3**). It is a ventilated standard container (**1**), with narrow, passive vents in the upper part (**0**).

Both of these are 20′ containers (**2**), 8′6″ high and constructed as tank containers (**7**) suitable for transporting hazardous materials to a pressure of 6.0 bar.

Table 10.1 Container identification including country, size and type codes

1st digit	The digits have the following meanings			
Length	1 = 10′	2 = 20′	3 = 30′	4 = 40′
2 nd digit	The digits have the following meanings			
Height	0 = 8′	0 = 8′ g.n.t	2 = 8′6”	3 = 8′6″ g.n.t
Height	4 = > 8′6″	5 = > 8′6″ g.n.t	6 = > 4′3″	7 = > 4′3″ g.n.t
Height	8 = > 4′ 3″ < 8′	9 = < 4′		

Marking on container: size and type code			ISO 6346
1	2	3	4
Length	Height	Type	Features
0	Closed general purpose	0	End opening(s)
		1	End & full side opening(s)
		2	End & part side opening(s)
		3	End & roof opening
		4	End & roof side opening(s)
1	Closed vented	0	Smaller passive vents upper part
		1	Larger passive vents upper part
		3,4	Passive vents at upper & lower parts
		6	Mechanical ventilation, located inside
		8	Mechanical ventilation, located outside
2	Thermal insulated heated named cargo	0	Insulated
		1	Insulated
		2	Heated
		5	Named cargo: livestock
		6	Named cargo: cars
3	Thermal refrigerated and heated	0	Refrigerated, expendable refrigerant
		1	Mechanically refrigerated
		2	Refrigerated and heated
		1	Mechanically refrigerated
		2	Refrigerated and heated
4	Thermal refrigerated and or heated removable equipment	0	Refrigerated, expendable refrigerant
		1	Mechanically refrigerated
		2	Refrigerated and heated
5	Open top	0	End opening(s)
		1	& Removable top member in end frame
		2	& Side opening(s)
		3	& Removable top member in end frame
6	Platform based	0	No end walls
		1	Complete fixed end walls

(continued)

Table 10.1 (continued)

Marking on container: size and type code			ISO 6346
1	2	3	4
Length	Height	Type	Features
		2	Fixed free standing posts
		3	Complete folding end walls
		4	Folding free-standing posts
		5	With frame and roof
		6	With frame and open top
		7	Skeletal, with open top and ends
7	Tank container	1	Non-dangerous liquid, 0,45 bar
		2	Non-dangerous liquid, 1.5 bar
		3	Non-dangerous liquid, 2.65 bar
		4	Dangerous liquid, 1.5 bar
		5	Dangerous liquid, 2.65 bar
		6	Dangerous liquid, 4.0 bar
		7	Dangerous liquid, 6.0 bar
		8	Dangerous liquid, 10.5 bar
		3	Dangerous liquid, 22.0 bar
8	Dry bulk	0–9	No specification
9	Air/surface	0–9	No specification

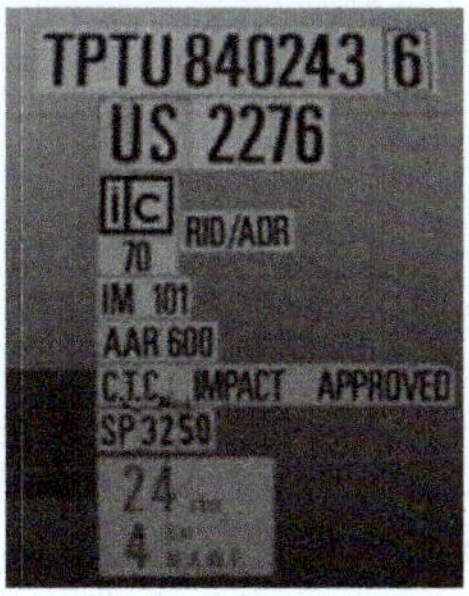

According to the previous annex F "size code designations" in DIN ISO 6346 of August 1985, the two containers are 40′ long (number **4**) and 8′6″ high with a gooseneck tunnel (number **3**). According to annex G "type code designations," both containers are partially ventilated closed containers, which have passive vents in the upper part of the cargo space, the total cross-sectional area of the vents being less than 25 cm^2 per metre of container length (number combination **10**).

However, the coding does not indicate that the red container is 2.50 m wide. The January 1996 edition of the standard bearing the same number includes another "size code", which has eradicated these inadequacies.

Like the above containers, this container is 40′ long, 8½′ high and has a gooseneck tunnel (number sequence **43**). The type code **51** states that it is an open-top container, with openings at one or both end walls and a removable top member in the end frame. The number sequence **43** stands for a length of 40′ a height of 8′6″ plus gooseneck tunnel. In each case the container is a flat rack, but the number sequence **63** is the code for a platform with complete folding end walls:

And the number sequence **64** is the code with a platform with folding, free-standing corner posts.

This flat rack allows verification of the coding, identification and marking according to the previous DIN EN ISO 6346 (1985), showing clearly that it is still valid, since the photo was taken in 2001. The marking beneath the container number reads **RCX 4364** (Fig. 10.9).

In accordance with the previous country code, **RCX** represents China (Taiwan). The size code **43** states that it is a 40′ container 8′6″ high and with a gooseneck tunnel. The number combination **64** states that it is a platform with folding, free-standing corner posts. Similar flat racks exist which are 20′ long and 8′6″ high, of course, but they do not have goosenecks. This is clear from the number sequence **22**. The type code **64** is the same as for the previous example: platform with folding, free-standing corner posts. Both containers have a nominal length of 40′ but are higher than 8′6″ and each have a gooseneck tunnel (size code number sequence **45**). According to type code **10**, both containers are partially ventilated closed containers, which have

Fig. 10.9 Flatbed container

passive vents in the upper part of the cargo space, the total cross-sectional area of the vents being less than 25 cm^2/m of container length. According to the previous 1984 standard:

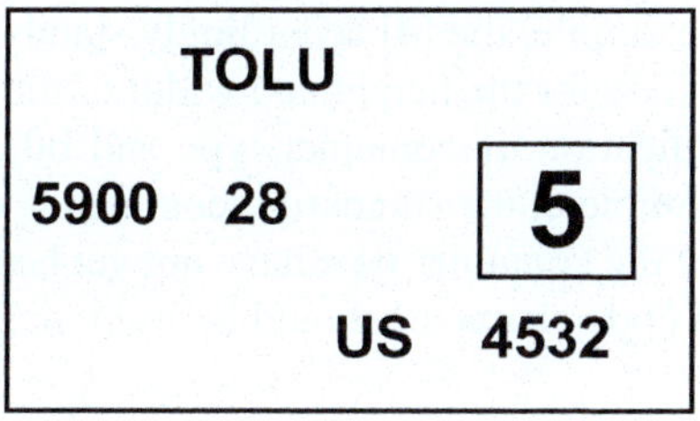

- **4** = 40′ container.
- **5** = > 8′6″ high with gooseneck tunnel.
- **3** = thermal; and
- **2** = refrigerated and heated.

Figure 10.10 shows clearly that the container is a 20′ long refrigerated container. According to the size code, the container height is > 8′6″—this is correct, as the container is 9½ feet high. The number combination **32**, given as the type code, states that the container may be refrigerated or heated.

Fig. 10.10 20′ long refrigerated container

Figure 10.11 shows a variant of the standard horizontal markings with size and type codes:

In this instance, the container is 12,192 mm (40′) long, 2,438 mm (8′) wide and 2,591 mm (8′6″) high. This is a general-purpose container without ventilation and has vents in the upper part of the cargo space. The current standard, used for this marking, is not limited to numbers but also uses letters to make marking more precise. In addition, swap-bodies and the like are also covered by the coding. The superordinate term for both numbers and letters is character. The introductory country code has been discontinued. The first character of the size code indicates the length of the container. In the example, the [**4**] accordingly stands for a 40′ container. The second character is the code for the height and width of the container. The standard type code allows identification of container type and other characteristics. Since this system is not yet complete, the standard recommends use of the group code if special characteristics of the container type have not yet been fixed or are unknown. The highest unassigned code character should be used as a provisional mark when

Fig. 10.11 Size, and type codes additionally stated in accordance with current standard

it is necessary to represent significant characteristics which are not yet included in Tables 10.2 through 10.12 (Tables 10.3, 10.4, 10.5, 10.6, 10.7, 10.8, 10.9, 10.10 and 10.11).

Figure 10.12 shows a variant of vertical container marking, with size and type code (on the left). The identifying details are more or less clearly visible.

In this example (Fig. 10.12), the identifying mark plus size and type codes are in the horizontal arrangement:

NFIU 491076 9
4EG1

Table 10.2 Container identification including country, size and type codes

Character	Container length			Character	Container length		
	mm	ft	in		mm	ft	in
1	2,991	10		D	7,450	24	6
2	6,058	20		E	7,820		
3	9,125	30		F	8,100		
4	12,192	40		G	12,500	41	
5	Spare			H	13,106	43	
6	Spare			K	13,600		
7	Spare			L	13,716	45	
8	Spare			M	14,630	48	
9	Spare			N	14,935	49	
A	7,150			P	16,154		
B	7,315	24		R	Spare		
C	7,430						

Table 10.3 Container identification including country, size and type codes

Container width	Character	Container height			Character for container width	
		mm	ft	in	>2,438 mm >2,500 mm	>2,500 mm
2,438/8′	0	2,438	8			
2,438/8′	2	2,591	8	6	C	L
2,438/8′	4	2,743	9		D	M
2,438/8′	5	2,895	9	6	E	N
2,438/8′	6	>2,895	>9	6	F	P
2,438/8′	8	1,295	4	3		
2,438/8′	9	<1,219	<4			

Table 10.4 Container code G

Container code	Group code	Type	Type code	Principal characteristics
G	GP	Unventilated general purpose	G0	Openings at one or both end(s)
			G1	Vents in upper part of cargo space
			G2	Openings at one or both end(s), plus "full" openings at one or both ends
			G3	Openings at one or both end(s), plus "partial" openings at one or both ends
			G4	Spare
			G5	Spare
			G6	Spare
			G7	Spare
			G8	Spare
			G9	Spare

Table 10.5 Container code V

Container code	Group code	Type	Type code	Principal characteristics
V	VH	General purpose container with ventilation	V0	Non-mechanical ventilation at the lower and upper parts of the cargo space
			V1	Spare
			V2	Mechanical ventilation installed inside the container
			V3	Spare
			V4	Mechanical ventilation installed outside the container
			V5	Spare
			V6	Spare
			V7	Spare
			V8	Spare
			V9	Spare

The container is 12,192 mm or 40′ long (number **4** of first character of the size code). Its height is 2,895 mm or 9′6″ and its width is greater than 2,438 mm or 8′ and is less than or equal to 2,500 mm (letter **E** or second character of the size code). As regards type, the container is a general-purpose container without ventilation, whose main characteristic is that it has vents in the upper part of the cargo space (characters **G1** of the type code). The owner code **NFI** and product group code **U** are also known

Table 10.6 Container code S

Container code	Group code	Type	Type code	Principal characteristics
S	SN	Named cargo containers	S0	Livestock carrier
			S1	Automobile carrier
			S2	Live fish carrier
			S3	Spare
			S4	Spare
			S5	Spare
			S6	Spare
			S7	Spare
			S8	Spare
			S9	Spare

Table 10.7 Container code R

Container code	Group code	Type	Type code	Principal characteristics
R	RE	Thermal containers • Refrigerated • Refrigerated and heated • Self-powered refrigerated and heated	R0	Mechanically refrigerated
			R1	Mechanically refrigerated and heated
	RT		R2	Mechanically refrigerated
	RS		R3	Mechanically refrigerated and heated
			R4	Spare
			R5	Spare
			R6	Spare
			R7	Spare
			R8	Spare
			R9	Spare

as the alpha prefix. The check digit is often written in a box: [**9**]. Below are a number of examples of marking according to DIN EN ISO 6346, January 1996:

According to the two characters **42** of the size code, the container is 12,192 mm or 40′ long, 2,591 mm or 8′6″ high and 2,438 mm or 8′ wide. The type code characters **G0** state that the container is a general-purpose container without ventilation but with openings at one or both end(s).

Table 10.8 Container code H

Container code	Group code	Type	Type code	Principal characteristics
H	HR	Thermal containers • Refrigerated and or heated with removable equipment	H0	Refrigerated and or heated with removable equipment located externally, coefficient of heat transfer K = 0.4 W/m^2 K)
			H1	Refrigerated and or heated with removable equipment located internally
			H2	Refrigerated and or heated with removable equipment located externally, coefficient of heat transfer K = 0.7 W/m^2 K)
			H3	Mechanically refrigerated and heated
			H4	Spare
			H5	Insulated, coefficient of heat transfer K = 0.4 W/m^2 K)
			H6	Insulated, coefficient of heat transfer K = 0.7 W/m^2 K)
			H7	Spare
			H8	Spare
			H9	Spare

Table 10.9 Container code U

Container code	Group code	Type	Type code	Principal characteristics
U	UT	Open top container	U0	Opening(s) at one or both end(s)
			U1	Opening(s) at one or both end(s) plus removable roof in end frame
			U2	Opening(s) at one or both end(s) plus opening(s) on one or both sides
			U3	Opening(s) at one or both end(s) plus opening(s) on one or both sides plus removable top members
			U4	Opening(s) at one or both end(s) plus opening(s) on one side plus full openings on the other side
			U5	Full, solid side and end walls (no doors)
			U6	Spare
			U7	Spare
			U8	Spare
			U9	Spare

Table 10.10 Container code P

Container code	Group code	Type	Type code	Principal characteristics
P	PL	Platform (containers) • Platform based containers with incomplete structure • Fixed • Folding • Platform based containers with complete superstructure	P0	Platform
	PF		P1	Platform with two complete, fixed end walls
			P2	Platform with fixed posts, either free-standing or with removable top members
	PC		P3	Platform with folding complete end walls
			P4	Platform with folding posts, either free-standing or with removable top members
	PS		P5	Platform, open at the top and ends
			P6	Spare
			P7	Spare
			P8	Spare
			P9	Spare

Table 10.11 Container code T

Container code	Group code	Type	Type code	Principal characteristics
T	TN	Tank container for non-dangerous liquids	T0	Minimum pressure 45 kPa
			T1	Minimum pressure 150 kPa
			T2	Minimum pressure 265 kPa
	TD	Tank container for dangerous liquids	T3	Minimum pressure 150 kPa
			T4	Minimum pressure 265 kPa
			T5	Minimum pressure 400 kPa
			T6	Minimum pressure 600 kPa
	TG	Tank container for gases	T7	Minimum pressure 910 kPa
			T8	Minimum pressure 2,200 kPa
			T9	Spare

Table 10.12 Container code A

Container code	Group code	Type	Type code	Principal characteristics
A	AS	Air/surface containers	AO	–

Fig. 10.12 Norfolk Line container with container codes clearly identifiable

According to the two-character combinations **22** and **42** of the size code, the container on the left is 6,058 mm or 20′ long and the two containers on the right are 12,192 mm or 40′ long. All the containers are 2,591 mm or 8′6″ high and 2,438 mm or 8′ wide. The type code characters **G1** state that the container is a general-purpose container without ventilation but with vents in the upper part of the cargo space. According to the two characters **22** of the size code, the container is 6,058 mm or 20′ long, 2,591 mm or 8′ 6″ high and 2,438 mm or 8′ wide. The type code characters **P3** state that the container is a platform-based container with folding complete end walls.

According to the two characters **22** of the size code, the container is 6,058 mm or 20′ long, 2,591 mm or 8′ 6″ high and 2,438 mm or 8′ wide. The type code characters **R1** state that the container is a thermal container which can be mechanically refrigerated and heated.

According to the two characters **28** of the size code, the container is 6,058 mm or 20′ long, 1,295 mm or 4′ 3″ high and 2,438 mm or 8′ wide. The type code characters **P2** state that the container is a platform container with fixed posts, either free-standing (as here) or with removable top members.

According to the two characters **42** of the size code, the container is 12,192 mm or 40′ long, 2,591 mm or 8′6″ high and 2,438 mm or 8′ wide. The type code characters **U1** state that the container is an open-top container which may additionally be equipped with opening(s) at one or both end(s) plus removable top door rail in the end frame.

According to the two characters **45** of the size code, the container is 12,192 mm or 40′ long, 2,895 mm or 9′6″ high and 2,438 mm or 8′ wide. The type code characters **G1** state that the container is a general-purpose container without ventilation but with vents in the upper part of the cargo space.

According to the two characters **45** of the size code, the container is 12,192 mm or 40′ long, 2,895 mm or 9′6″ high and 2,438 mm or 8′ wide. The type code character **P** indicates that the container is a platform-based container with complete end walls. The standard still gives type code **P8** as being "spare".

Type code **P8** has been used here for a "collapsible flat" (to the right), i.e., the end walls are collapsible for empty carriage.

According to the two characters **45** of the size code, the container is 12,192 mm or 40′ long, 2,895 mm or 9′6″ high and 2,438 mm or 8′ wide. The type code characters **R1** state that the container is a thermal container which can be mechanically refrigerated and heated.

According to the character **4** of the size code, the container is 12,192 mm or 40′ long. The character **C** indicates that the container has a height of 2,591 mm or 8′ 6″ and a width ranging between > 2,438 mm and < 2,500 mm. According to type code **G1**, the container is a general-purpose container without ventilation, which nonetheless has vents in the upper part of the cargo space.

The character **4** stands for the length 12,192 mm or 40′. **E** states that the container has a height of 2,895 mm or 9′ 6″ and a width ranging between >2,438 mm and <2,500 mm. The type code characters **G1** state that the container is a general-purpose container without ventilation but with vents in the upper part of the cargo space.

Eiven the considerable number of containers available today, it is impossible to tell how long the current coding will suffice and when more characters will need to be introduced. **G1** state that the container is a general-purpose container without ventilation but with vents in the upper part of the cargo space.

10.2 Marking of Containers Carrying Dangerous Cargo

The container packing guidelines provides, in Sect. 4.4.1, with respect to the placarding of containers carrying dangerous cargoes, the following provisions:

> Placards (enlarged labels) (minimum size 250 mm × 250 mm) and, if applicable for maritime transport, "MARINE POLLUTANT" marks (minimum size of a side 250 mm) and other signs should be affixed to the exterior surfaces of a [container] or unit load or over pack to provide a warning that the contents of the unit are dangerous cargoes and present risks, unless the labels, marks or signs affixed to the packages are clearly visible from the exterior of the unit. This type of marking can be omitted on unit loads and over packs if the hazard labels, markings or warning symbols are clearly visible from the outside. If possible, the placards, hazard labels, markings or warning symbols on the outsides of the [container] should not be obscured when the [container] is opened.

Moreover, the container packing guidelines provides additional instructions for affixing placards, these being explained in more detail in the container packing guidelines at Paragraph 4.4.1.2:

> [Containers] containing dangerous cargoes or residues of dangerous cargoes should clearly display placards and, if applicable for maritime transport, "MARINE POLLUTANT" marks or other signs as follows: (1) a container, one on each side and, in the case of carriage by ocean-going vessel, also one on each of the ends of the unit; (2) a railway wagon [railroad car], at least one on each side; and (3) any other [container], at least one on both sides and on the back of the [container] and, in the case of a semi-trailer, also one on the front of the [container].

It has been proposed to mark the roof with a corresponding placard, in order to warn lifting gear or ground conveyor operators when picking up such containers. However, the regulations do not as yet contain any such obligation. Paragraph 4.4.1.2 of the container packing guidelines continues:

> Placards on the sides of a [container] should be affixed in such a position that they are not obscured when the unit doors are opened.

The "if possible ... not be obscured" contained in Paragraph 4.4.1.1 has since become an obligation, with the paragraph reading instead "should be affixed." Paragraph 4.4.1.3 provides guidance in relation to the handling of dangerous cargoes which present unusual or particular risks and hazards:

> Whenever dangerous cargoes present several risks, subsidiary risk placards should be displayed in addition to primary risk placards.

In this instance, the container at Fig. 10.13 is loaded with Class 4.1 flammable solids:

The container at Fig. 10.14 is loaded with Class 3 goods, i.e., flammable liquids, which are also marine pollutants (Fig. 10.15):

The regulations covering this area were amended in 2002. According to the previous regulations, the primary risk placards had to carry the correct class number, while the subsidiary risk placard was not supposed to carry any class numbers as indicated below (Fig. 10.16):

The way this requirement was fulfilled was open to choose and it was of no significance whether the class numbers were obscured by being painted or stuck over, or indeed cut off. According to the regulations in force since 2002, both primary

Fig. 10.13 Correct marking on a container side

Fig. 10.14 Correct marking on a container side

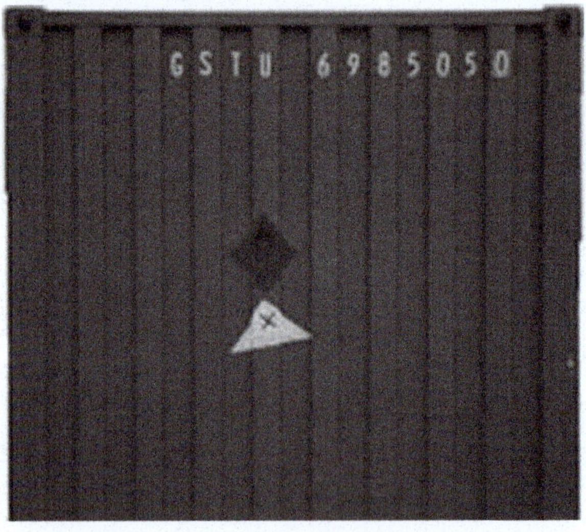

Fig. 10.15 Non-regulation affixing of a placard

Fig. 10.16 Previous variant of the primary and subsidiary risk placards

and subsidiary risk placards must carry the respective class numbers. To this end, Paragraph 4.4.1.3 of the container packing guidelines continues (Fig. 10.17):

> ... [containers] containing cargoes of more than one class, however, need not bear a subsidiary risk placard if the hazard represented is already indicated by the primary risk placard.

Fig. 10.17 New regulations: primary and subsidiary risk placards bear class number

For dangerous cargo containers carrying dangerous goods of only one class, the regulations in Paragraph 4.4.1.6 of the container packing guidelines apply:

> Consignments of packaged dangerous cargoes of a single commodity, other than cargoes of class 1, which constitute a full load for the [container] should have the UN Number for the commodity displayed in black digits not less than 65 mm high either against a white background in the lower half of the class placard or on an orange rectangular panel not less than 120 mm high and 300 mm wide, with a 10 mm black border, to be placed immediately adjacent to the placard (see annex 2). In those cases, the UN Number should be displayed immediately adjacent to the Proper Shipping Name.

In accordance with the IMDG code, the regulation applies to cargoes with a minimum mass of 4,000 kg, wherein (Figs. 10.18, 10.19 and 10.20):

> The container contains only one dangerous cargo, i.e., class 5.1 oxidising substances (agents), which increase the risk and intensity of fire, and are also corrosive. The (yellow) primary risk placard correctly carries the number 5.1, but the subsidiary risk placard does not carry the number 8 (also correct according to the old regulations), which has been obscured. According to the currently applicable regulations, this placard would also have to carry the class number 8.

In this example, the placards have been incorrectly positioned as they will be obscured when the door is opened. Paragraph 4.4.1.9 of the container packing code states the following:

> When solid carbon dioxide (CO^2 - dry ice) or other expendable refrigerant is used for cooling purposes, a warning sign should be affixed to the outside of the doors so that it is clearly visible to any person operating the doors. The sign should warn of the possibility of an asphyxiating atmosphere.

The name of the gas used as refrigerant should be inserted beneath the word **WARNING** (Fig. 10.21).

The IMDG Code makes recommendations for the use of wording for the "Container Packing Certificate." In this instance, the warning notice applies to solid carbon

Fig. 10.18 Examples of primary and subsidiary risk placards plus UN Number

Fig. 10.19 Correct: 4,000 kg consignment with hazardous materials, therefore UN number stated

Fig. 10.20 Incorrect: < 4,000 kg consignment with hazardous materials, therefore UN number not stated

Fig. 10.21 Warning sign for containers in which dry ice or other expendable refrigerants are used for refrigeration purposes

dioxide (CO^2 - dry ice) which is used for cooling purposes. To comply with the code, the container or vehicle carrying the load should be externally and conspicuously marked or labelled, such as at the door end, with the words:

DANGEROUS CO^2 (DRY ICE) INSIDE

VENTILATE THOROUGHLY BEFORE ENTERING

The container packing guidelines contains a reminder under Paragraph 4.4.1.10 that cargoes under fumigation may require special precautions:

> As [containers] offered for shipment under fumigation may require special precautions, they should only be accepted with the agreement of the carrier, and they should be identified to him prior to loading. [Containers] under fumigation are now included in class 9 of the IMDG Code.

The guidelines continue with the following instructions (Paragraph 4.4.1.11) pertaining to the marking or labelling of container undergoing or having recently undergone a course of fumigation:

> When a closed [container] or its contents has been fumigated and is to be shipped under fumigation, a warning sign should be affixed to the outside of the doors so that it is clearly visible to any person operating the door. An example of such a warning sign is given in annex 2. The sign should state the fumigant, the method of fumigation employed and the date and time when it took place. The sign should only be removed when the unit has been ventilated after fumigation, to ensure that no harmful concentration of gas remains.

The intention of this chapter was to provide some tips for marking dangerous cargo containers. In a real packing situation, obviously, the other regulations in the container packing guidelines should be consulted, especially the valid dangerous cargo regulations.

Annex

List of Current BIC Representatives (Correct as of September 2023):

Australia/New Zealand
Mr Michael Downes
SOUTHERN CONTAINER ADVISORY LTD.
14 Carlisle Street
Waimate 7924
South Canterbury
New Zealand
Tel: +64 3 689 8242
Mob: +64 21 255 9704
mike.downes@containeradvice.com

Benelux
Mr Bart Zuidhoek
BUREAU VERITAS N.V.
Mechelsesteenweg 128-136
B-2018 ANTWERPEN
Tel: +32 3 247 94 66
Mob: +32 3 247 94 99
bart.zuidhoek@bureauveritas.com

Brazil
Mr Alex André Rotmeister de Souza
CAMARA BRASILEIRA DE CONTEINERES
Rua Uruguaiana, nº 10 – Sala 1.508 – 15º andar

20050-090 RIO DE JANEIRO
Tel: +55 (21) 2263-1645
Fax: 55 (21) 2233-2333
cbcdiretoria@cbcconteiner.org.br

China
Mr Daohua Hu
BUREAU VERITAS
3/F, No.1288 Wai Ma Road, Huangpu District, Shanghai
200011, P.R.C.
Tel: +86 18 01750 2239
Fax: +86 21 6886 1316
daohua.hu@bureauveritas.com
blair.hu@bureauveritas.com

China (Taiwan office)
Mr Daohua Hu
BUREAU VERITAS
Tel: +86 18 01750 2239
Fax: +86 21 6886 1316
blair.hu@bureauveritas.com

Denmark/Finland/Norway/Sweden
Mr Svend Heerulff
BoxID Nordic
Atlasvej 3B
DK-8270 Aarhus-Hoejbjerg
Denmark
Tel: +45 21 66 60 33
code@bic-code.dk

Germany/Austria
Mr Clemens Bochynek
SGKV e.V.
Westhafenstrasse 1
DE – 13353 BERLIN
Tel: +49 30 206 13 760
Fax: +49 30 206 13 76 17
cbochynek@sgkv.de

Hong Kong (China)
Ms Janice Chow
Container Affairs Company

Unit A 38/F Tower 9 Bellagio
33 Castle Peak Road
Sham Tseng, N.T.
HONG KONG
Tel: +852 9208 6272
janice@omic.com.hk

Italy
Ms Camilla Lo Bianco
CENTRO INTERNAZIONALE STUDY CONTAINERS
Via Garibaldi, 4
16124 GENOVA (ITALY)
Tel: +39 010 86 08 852
Fax: +39 010 25 18 852
info@ciscoconsultant.it

Japan
Mr Takuya Horaguchi
THE JAPANESE SHIPOWNERS' ASSOCIATION
Kaiun Building, 2-6-4, Hirakawa-cho, Chiyoda-ku,
102-8603 Tokyo, JAPAN
Tel: +81-3-3264-7314
t-horaguchi@jsanet.or.jp

Korea
Mr BJ Chun
BOXJOIN CORPORATION
Room 1001, Daewoo The O Ville,
115, Hangang-daero, Yongsan-gu,
SEOUL 04376
Tel: +82 2 6264 1377
bjchun@boxjoin.com

Malaysia
Mr Vincent Loh
RITCHIE & BISSET (ASIA) Pte Ltd
Blk 437 #08-491, Choa Chu Kang Ave.4
Singapore 680476
Tel: +65 6271 8700
vincentloh@ritchie-and-bisset.com

Mexico and Northern Latin America
Mr Arturo Serratos

ESTRATEGIA INTERMODAL
Tecax # 71 casa 1
Col. Lomas de Padierna CP 14240
Alcaldía Tlalpan, CDMX
MEXICO
Tel:+52 (55) 3531 3911
direccion@estrategiaintermodal.com.mx

Netherlands
Mr Kees Noorman
ORAM
P.O. BOX 19405
1000 GK AMSTERDAM
Tel: +31 20 6222 111
noorman@oram.nl
Lie@oram.nl

Russia
Mr Alexander Fetisov
RUSSIAN MARITIME REGISTER OF SHIPPING
8, Dvortsovaya Nab
191186 ST PETERSBURG
Tel: +7 812 315 4698
fetisov.av@rs-class.org

Singapore
Mr Vincent Loh
RITCHIE & BISSET (ASIA) Pte Ltd
Blk 437 #08-491, Choa Chu Kang Ave.4
Singapore 680476
Tel: +65 6271 8700
Fax: +65 6271 3263
Mob: +65 9615 6256
vincentloh@ritchie-and-bisset.com

Spain
Mr Anton Elejabeitia
AENOR INTERNACIONAL, S.A.U.
C. Génova 6
28004 MADRID
Tel: +34 91 529 4930
Fax: +34 913 190 581
aeleja@aenor.com

Thailand
Ms Pilaiporn Thongsodsaeng
OMIC TECHNICAL SERVICES CO LTD
12-14 Soi Yen Akas 3, Yen Akas Road, Chongnonsri, Yannawa
10120 Bangkok
Tel: +66 2 2864120 Ext 221
tesco@omicnet.com

Turkey
Mr İlyas EMER
TURKISH STANDARDS INSTITUTION
Ostim 100. Yil Bulvari n°99
06374 Yenimahalle – ANKARA
Tel: +90 312 592 50 00
Fax: +90 312 5925013
ilyase@tse.org.tr

United Kingdom
Mr Andrew David Nutter
LRQA VERIFICATION LTD.
1 Trinity Park, Bickenhill Lane,
B37 7ES Birmingham
Tel: +44 (0)7896 816 867
Andrew.Nutter@lr.org

UAE/Bahrain/Iran/Iraq/Kuwait/Oman Qatar/Saudi Arabia/Yemen
Mrs Nayana Nandkumar
DUBAI SHIPPING AGENTS ASSOCIATION
Office n° 7 – Block B
Al Furdah Bldg. (Customs New Bldg.)
Port Rashid – Al Mina Road
P.O. BOX 27345, DUBAI
Tel: +971 4 345 3144
Fax: +971 4 345 6411
dubaisaa@emirates.net.ae

United States/Canada
Mr Kiki Shahani
P.W. SHAHANI ASSOCIATES, LLC
214 Coltsgate Drive
CARY, NC 27518
Tel: +1 908 892 5960
pwsa.bic@gmail.com

Vietnam
Ms Nguyen Minh Nga
OVERSEAS MERCHANDISE INSPECTION CO., LTD (OMIC)
45 Dinh Tien Hoang Street,
Ben Nghe Ward, District 1,
HO CHI MINH CITY
gm.vn@omicnet.com
minhnga@omicnet.com

Disclaimer: The author makes no representation or guarantee regarding the accuracy of the contact details provided, which are sourced directly from BIC. This information is subject to change.

Part IV
Container Load Packing

Chapter 11
Impact of Packaging and Packing Aids on Load Security

In some contexts, load securing can be defined quite simply as the physical securing of a consignment using friction and tight-fit methods. In our context, it should be taken to mean the "safety of loads and protection against potential damage by loads" and should cover the protection of the goods against all forms of damage and or protection of the environment against potentially damaging effects of the cargo. An all-inclusive approach such as this may encourage greater interdisciplinary thinking and consideration of the problems of load securing in context rather than as an isolated issue. Load securing, in its broadest sense, is dependent on a number of factors such as the characteristics of the goods that are being transported, the construction and or manufacturing features, the types packing and packaging used, the cargo transport units chosen for transportation, the aids and devices used during packing, the packing methods, the securing methods, the materials used to secure the load, the finish of the materials used to secure the load, and the transport methods and routes used etc. As always, the people responsible for the overall process play a critical role. Only when these people have been trained properly and have the right experience is it possible to minimise damage. Only the use of the correct and appropriate forms of load securing is able to prevent damage being caused to the goods that are being transported as well as to equipment, public property and the environment. Note regarding legal stipulations: the packing of containers and cargo securing in or on containers must be carried out in accordance with the container packing guidelines as published by the IMO.[1] The 'Scope' section of the container packing guidelines contains information about situations where the guidelines are not applicable. Accordingly:

> [The container packing guidelines] do not cover the filling or emptying of tank containers, portable tanks, or road tank vehicles, or the transport of any bulky cargo in bulk packaging.

---

[1] The exact title of the document is the IMO/ILO/UN ECE *Guidelines for Packing of Cargo Transport Units (CTUs).*

A. A. Olsen and F. Karkori, *Containerized Cargo Handling and Stowage*, Springer Series on Naval Architecture, Marine Engineering, Shipbuilding and Shipping 22,
https://doi.org/10.1007/978-3-031-63329-4_11

Moreover, the container packing guidelines do not apply to tanks and bulk cargo. Because of these omissions, it is worth stating the definitions of these containment types as defined by the container packing guidelines:

> **Bulk cargoes** refer to any cargoes which are intended to be transported without any intermediate form of containment in bulk packaging or portable tanks.
>
> **Cargo** refers to any goods, wares, merchandise and articles of any kind which are intended to be transported.
>
> **Dangerous cargoes** refer to any packaged, dangerous, hazardous or harmful substances, materials or articles, including environmentally hazardous substances (marine pollutants) and wastes, covered by the International Maritime Dangerous Goods (IMDG) Code. The term "dangerous cargoes" includes any empty and unclean packaging.
>
> **Over pack** refers to an enclosure used by a single shipper to contain one or more packages and to form one unit for convenience of handling and stowage during transport. Examples of over packs are a number of packages either:
>
> (1) *Placed or stacked on to a load board such as a pallet and secured by strapping, shrink-wrapping, stretch-wrapping or other suitable means; or*
> (2) *Placed in a protective outer packaging such as a case or crate.*
>
> **Packaging(s)** refers to the receptacles and any other components or materials necessary for the receptacle to perform its containment function.
>
> **Packages** means the complete product of the packing operation, consisting of the packaging and its contents as prepared for transport.
>
> **Unit load** refers to a number of packages that are:
>
> (1) Placed or stacked on and secured by strapping, shrink-wrapping or other suitable means to a load board such as a pallet.
> (2) Placed in a protective outer enclosure such as a pallet box.
> (3) Permanently secured together in a sling.

Unlike conventional methods of loading, closed containers offer better protection against robbery, pilferage, and climatic conditions. For moisture-sensitive cargoes, namely the vast majority of general cargoes, measures must be taken to ensure the goods to be loaded are not stored in the open atmosphere, whether for long- or short-term periods, and that they are only packed in roofed areas. Subsequently the basic requirements for preventing damage and thus protecting the load include:

- Container-suitable construction of loads, goods and or packaging, including container-suitable palletisation and unitisation.
- The appropriate packing of the load into the container, and if this is not sufficient; and
- Supplementing the packing measures by the use of suitable load securing mechanisms.

If these three rules are observed, the container is suitable to carry out the protective function for which it was designed. Failure to comply with these basic rules, however, may result in the container endangering its own cargo load. This section of the book has been written to demonstrate the basic principles required to transport cargo without causing damage to it, anyone or indeed anything else. In short, we can refer again to the 'Scope' section of the container packing guidelines, which states:

> These Guidelines are not intended to conflict with, or replace or supersede, any existing regulations or recommendations which may concern the carriage of cargo in [containers].

This is intended to stress the fact that other regulations still retain their validity. For example, the German railroad regulations states in clause Sect. 1:

> The cargo must be in a state that does not permit it to present a danger to the safety of the railroad.

This paragraph affects, in the first instance, the manufacturers, shippers and packers of goods. Other areas of law have similar regulations, including insurance law. Both the nature and the state of the packaged load and its packaging must meet all requirements for correct transportation. As we have already seen, and we will continue to see, this is very often not the case. Load securing considerations present themselves as an issue as early as the design phase of the product. In this sense the manufacturer intends to bring their products to market. To do this, the goods must be transported from the manufacturer's factory to the retailer, or where the goods are bought directly by the customer from the manufacturer to the customer's location. This might entail a significant distance and require shipping using one or more modes of transportation. Since the manufacturer has a personal stake in getting their products to their customers in an undamaged state, it makes sense for the producer to enable or facilitate the correct securing of their load by incorporating appropriate points for securing. These are often referred to as *lashing points*. In some jurisdictions, this is a legal requirement. The *German Law on the Safety of Equipment*, for instance, requires the manufacturer to do so. Once the load is received by the carrier, the load is no longer in the power of disposal of the shipper. This means all future handlers of the cargo are responsible for ensuring the consignment is in a condition suitable for transportation. If faults are detected, measures must be taken to ensure that the goods are made suitable for transport (again). In the case of hazardous goods, damaged packages must not be loaded under any circumstances. Moreover, instructions to the contrary should never be followed.

11.1 The Influence of Packaging Receptacles and Packaging Aids on Load Securing

Choosing the right type of receptacle and packaging aids has a direct influence on the security of the load, the prevention of damage and the costs of storing, handling and transportation. Therefore the main objectives of transport packaging are to:

- Protect the cargo from loss, damage or depreciation.

- Protect people, the environment and the transportation vehicle from potential damage caused by the cargo; and
- Enable elevated levels of mechanisation and rationalisation of storage, handling and transportation processes.

These tasks can only be achieved by careful selection of packaging methods and packaging receptacles and aids, as well as using the correct packing method. Modifications are easy to implement for cartons, crates and cases. However, fibre, plastic and steel sheet packaging can only be modified gradually since many filling plants and packaging lines are designed specifically for use with special receptacles. Thus, solutions must be found that take account of the needs of all concerned. This can only be achieved in conjunction with manufacturers, industrial associations and standardisation committees. This process also requires considerable investment. The design of fibre drums, barrels, canisters, pails, corrugated barrels, quick lever-closure drums, etc. must be revisited and modified in the long-term, both with regard to the fundamental design and the dimensions of the receptacles.

The risk of sustaining damage during transport is disproportionately large for some of the packaging receptacles commonly in use. Furthermore, transport using some of the existing packaging receptacles requires considerable outlay in terms of materials and personnel if the goods packed in these containers are to reach their destination undamaged. The closures on fibre drums, the handles on buckets, pails, canisters, and other similar containers can cause damage to cargo stored on top of or next to them as a result of high point loading. Professional loading requires the use of additional dunnage and increases the amount of stowage space lost during packing as well as increasing personnel costs. As a result of loads caused during transportation, drums of this type ride up over the corrugations of neighbouring drums. This encourages the formation of transport gaps and the damage associated with these. If the levers of quick lever-closure drums extend beyond the edge of the drum, these may snag. The consequences of this include unintentional opening of the drums and or damage to the drums themselves.

Transport damage caused by gaps formed as a result of the design of the receptacles can only be prevented by employing further materials which leads to additional costs for materials and personnel, or by selecting a different type of drum with wider corrugations and recessed closures. It is obvious that, in the long-term, the selection of a better drum is the more cost-effective solution. Careless packing is one of the main causes of accidents and damage. Fundamental errors are often made as early as the design stage and later in the production of the goods. This is because storage, handling and transport are rarely taken into consideration at these preliminary stages. The same applies to the production and use of various packaging receptacles and packaging aids. With that being said, in the majority of cases, however, it is the "packer" who makes the mistakes responsible for damage suffered during transportation. A package is only appropriate for carriage when it is able to withstand all of the manipulations likely to be expected during storage, cargo handling and transportation, and also in terms of keeping the environment protected from the goods. Not all packaging faults will cause damage, but the risk to the contents is increased considerably. Because the machine component itself has not been designed to be transported by forklift truck, it has been placed on a pallet (and not secured) for transportation within the company. Simply manoeuvring the component into a container is already fraught with a number of difficulties. In order to be able to secure it correctly, considerable outlay in terms of materials and staff hours will be required. If the packaging is designed to function as protection against climatic conditions or corrosion, then the damage to the plastic film that is likely to be caused by necessary securing measures is likely to have negative consequences for the cargo. What is required is a skid which can be transported using forklift, similar to the floor of a case, and to which the machine component can be bolted. Securing would then be easy. The skid plus machine component could be secured in place in the container using vertical bracing. The packing staff often give up when attempting to properly secure objects that are not suitably packaged. The load is simply placed in the container and left to its own devices. This case does not have any lower skids. In order to be able to comply with the sling here hoisting instructions given by the symbol on the case, or for transportation using a forklift, the forklift operator must force the forks under the case to lift it.

In practice, people fail to assign as much importance to the issue of the suitability of packaging for transportation as they should. The issue affects all types of transportation and is thus not specific to maritime transport alone. It is, however, of particular importance when loading containers for maritime transportation. The following basic principles may be used to assess the suitability of a load for transportation:

- Loads must be in a state which does not, in any way, affect safety during transport; and
- The load is secured appropriately for carriage if, when loaded and handled in the proper manner, it is unable to cause damage to persons, to handling or transportation equipment or to other cargo.

The term "safe for carriage" must be defined in such a way that it is clear to all persons involved in the process. The average lay person can assume that manufacturers of goods and the people and companies entrusted with their transportation are specialists in their field. If the terms shipping and carriage are used, they also cover the storage, handling and transportation of the goods as well as all associated preparation and final work. It must also be assumed that specialists are aware of the different demands made of small, large, display and sales packaging and are able to differentiate between these demands and the demands made of combination and transportation packaging and load units. Transportation packaging must be constructed in such a way that it is suitable for carriage in order to ensure that it meets all the demands of handling, storage and transportation. This applies to all aspects: protection of the goods, protection of the environment from the goods, or issues relating to handling, packing, stowage, securing, etc.

Unfortunately, in some areas, concepts such as ordinary commercial packaging, overseas packaging and seaworthy packaging are still often used. Concepts such as packaging suitable for transport or fit-for-purpose packaging are preferable, since such concepts are used in transport insurance law. The manufacturers and shippers of goods must, at the very least, inform themselves of the following issues and construct/select suitable packaging formats accordingly. All measures for protecting the goods from damage must focus on the loads experienced during storage, handling and transportation. Goods should always be packed safely if they require protection from complete, or partial loss or from damage, or if the goods themselves represent a risk to the environment. If these basic principles are adhered to and the relevant ISO, EN or DIN standards and other technical regulations are also observed, the best conditions for damage-free transportation can be achieved. In the area of hazardous goods, legislation has taken account of the above factors in the form of special packaging regulations. But what is the use of introducing packaging regulations for hazardous goods when they can (and must) be packed into a container with other goods which are less suitably packaged. Not all packaging faults will cause damage. However, the risk is increased considerably. Poor packaging has an additional psychological effect: bad packaging promotes carelessness and incorrect handling among handling staff. It will be difficult to stow other items on top of this crate since part of the contents extend above the top of the wooden crate. Packing a container with packages of this type causes additional expense since so much interlayer dunnage must be employed in order to stack other objects on top of the crate while ensuring that the contents of the wooden crate do not suffer damage.

In this chapter we have discussed some of the key points associated with the use of packaging and packing aids on loads intended for transportation. Without adequate packing, it is quite likely the goods inside a container will shift during transit. This can lead to many devastating consequences caused by both the shifting momentum of the cargo and the shifting centre of gravity of the container. In the next chapter, we will look at the use of pallets for securing cargoes in separate loads, the best practices associated with palletisation.

Chapter 12
Unitisation and Palletisation

Carelessness and poorly trained personnel are two of the main reasons for damage caused as a result of faulty palletisation and unitisation. The combination of packaging receptacles, pallet and stretch wrap does not automatically result in a so-called "palletised load." Only if the handling of pallets can be rationalised using ground conveyors can we really speak of palletised goods. These are handled by a number of companies at a cheaper rate. The types of errors that occur frequently are shown in Figs. 12.1 and 12.2.

The quick lever-closure type drums shown here are not suitable for transportation, since the lids are able to push each other open. This applies to both conventional methods of stowage and to palletisation. If this does not happen when the goods are wrapped, as in this case, then it will occur at the very latest when the load is jolted during transport. Pallets do not usually have diagonal bars which means that are very susceptible to asymmetric forces. In particular when used in containers, pallets can only really be used sensibly when they can be packed flush and have level sides and surfaces. Gaps make it more difficult to pack and secure the load. Shrink and stretch wrap must firmly bond the goods to the pallets. This bond should be able to withstand the normal loads caused by handling. That is not the case here. When loads arrive at a container packing station in such a pitiful state, the drums should be packed individually. It is obvious that drums or pails designed for use in containers can save on costs. Figures 12.3 and 12.4 demonstrates the potential danger that may arise when poorly packed pallets are not rigorously unpacked, and instead an attempt is made to load them despite their obvious deficiencies. Despite the fact that this is not permitted, a loader can be seen climbing onto the pallet lifted by the forklift truck in an attempt to stabilise the load. This attempt fails and the drums fall to the ground. If we refer to Fig. 12.5, the package appears to be a perfectly well shrink-wrapped pallet containing boxed goods. However, if we look closer, we can see wider dimensions of the load sloping towards the top, i.e., the fact that they extend beyond the bounds of the pallet, may mean that the goods are forced into a gap that is too small for them because they no longer have dimensions suitable for the means of transport.

A. A. Olsen and F. Karkori, *Containerized Cargo Handling and Stowage*, Springer Series on Naval Architecture, Marine Engineering, Shipbuilding and Shipping 22,
https://doi.org/10.1007/978-3-031-63329-4_12

Fig. 12.1 Example of palletisation which will actually increase freight costs

Fig. 12.2 Example of goods that are not safe for carriage

The cartons stacked on top restrict the options for packing other goods on top of this package. In the real world, many packers would slice open the shrink wrap and load these two cartons separately. This subsequently increases the probability of incorrect delivery. A likely reason for this type of unacceptable palletisation is that producers tend to apply their pricing/discounting system to round numbers, for example 50 or 100, instead of matching it to the actual capacity of loading aids, transport aids or means of transport. The most frequently seen error is pallets that are not packed flush. Whether the goods are too small for their pallet, extend beyond the bounds of the pallet or are placed diagonally on the pallet, all these instances encourage careless

Fig. 12.3 Example of an accident waiting to happen due to poorly loaded pallets

loading into vehicles and containers, makes load securing more difficult and can lead to the load being damaged (Fig. 12.6).

Poor palletisation is not just restricted to boxed goods. When loading containers with hazardous goods we often see inadequately shrink-wrapped or poorly strapped pallets. Referring to Fig. 12.7, we can see these load units cannot be safely stacked:

Under the slightest acceleration, the inadequately wrapped film will not be able to hold the plastic canisters and tin drums in place on the pallet. Alternatively, looking at Fig. 12.8, these load units are well packed, the surfaces are level, and they can be stacked.

Referring to Figs. 12.9 and 12.10, the sides are not perfectly flush with the pallet base. There are still gaps that should be filled. Goods that are palletised this badly cannot be stowed safely and without gaps, as the regulations require, without additional outlay in terms of work and materials.

In road and railway transportation, actions that contravene the general guidelines on palletisation can be compensated for by additional outlay on stowage and securing. In maritime transportation and when dealing with hazardous goods, packages of this nature are clearly in breach of the law. The general provisions of the IMDG code, states that the unitisation and labelling of items and unit loads to be transported must be carried out in accordance with the regulations. In Sect. 10.18 of the IMDG Code, unit loads are defined as a number of packages that are:

Fig. 12.4 Example of stowing two cartons too many!

Fig. 12.5 Example of incorrectly packed pallets containing boxed goods

Fig. 12.6 Example of poor palletisation

Fig. 12.7 Example of a load unit that is not suitable for transportation

- *Placed or stacked on and secured by strapping, shrink-wrapping or other suitable means to a load board such as a pallet; and*
- *Placed in a protective outer enclosure such as a pallet box; and*
- *Permanently secured together in a sling.*

Fig. 12.8 Example of a load unit that is not suitable for transportation

Fig. 12.9 Example of deficiencies when forming load units

Fig. 12.10 Example of unacceptable packing of pallets with drums

In accordance with Paragraph 10.18.2, packages that contain hazardous goods, which are permitted to be transported in accordance with the code, may only be transported in unit loads if the following conditions are fulfilled:

- *It may occur that the packages in a unit load will need to be separated. In this event, it must be ensured that the individual packages can be handled safely.*
- *The unit loads should be compact, have as regular a form as possible, and vertical sides. The top of the unit load should be level. It must be possible to stack the unit loads. They must be constructed and secured in such a way that it is unlikely that the individual packages can become damaged.*
- *The unit loads must be sufficiently strong to withstand repeated loading and stowing operations and they must be able to bear unit loads with a similar specific mass which are stacked on them to a height commonly occurring during transportation.*
- *The material used to bind a unit load together must be compatible with the substances contained in the unit load and must remain effective under the influence of moisture, extreme ambient temperatures* and sunlight.

Referring to Figs. 12.11 and 12.12 without the additional use of appropriate corner rails or cover boards, the straps alone will not be able to hold the plastic receptacles in place for long. It is only possible to stack further unit loads on top of these with additional effort. When stowing the second or third layer, the goods stowed below must be protected from potentially damaging pressure by the use of additional measures. Deficiencies can also be observed in the unit loads shown at Figs. 12.13 and 12.14. Changing proportions lead to increased freight costs since the outer dimensions are measured. If the pallet placed on top of these drums can be brought into the correct

Fig. 12.10 (continued)

Fig. 12.11 Example of a unit load with serious deficiencies

position, it will no longer be able to bind the load together securely. The tape wrapped around these cartons and the stretch wrap round the cases must be used correctly. Overall, the only recommendation that can be made here is to separate the packages and load them individually. If the lower pallet is intended to be part of the unit load, then it should be firmly bonded to the rest of the load. That is obviously not the case here. Without the pallet, this "pack of three" cannot be handled by forklift truck and thus cannot be loaded rationally. The stretch wrap itself is unable to provide sufficient stability. The benefits afforded by the efficient handling of unit loads are lost if gaps are left in the cargo. For the purposes of load securing, it would be more effective if each of the cases had been strapped separately. Pallets should always be packed in such a way that the goods being transported, whether it be in cartons, sacks, pails, drums or canisters, are always flush with the pallet. Furthermore, only pallets should be used which are suitably dimensioned for the intended transportation method and the handling equipment that will be used.

The trays containing empty cans are designed for use with the pallets used in Fig. 12.15a. The pallets themselves are compatible with the dimensions currently used for road and rail vehicles, swap bodies and inland containers in Europe. Thus, two pallets can be loaded next to each other widthways or three pallets lengthways. For the purposes of overseas transportation, there are currently very few containers with an internal width greater than 2.4 m. Thus, these pallets are not suitable for

Fig. 12.12 Example of a unit load with serious deficiencies

Fig. 12.13 Example of a unit load with serious deficiencies

Fig. 12.14 Example of a unit load with serious deficiencies

transportation in overseas containers. The European transportation market now has a small number of CPCs or "Cellular Palletwide Containers."

The packing receptacles in Fig. 12.15b are not designed for the pallet, and the pallets themselves are not designed for use with this container. This makes loading and unloading more difficult and increases the amount of securing that is required. The risk of damage also increases. It would have been more sensible to use pallets or pallet-like bases sized 116 cm × 116 cm for transport in the container. This would have allowed two units to be loaded into the container next to each other in normal steel corrugated containers. The goods themselves should be packed in modular unit load packaging such as cartons, canisters, drums, etc. which would allow a clean block to be formed on the pallet. Here the requirement for harmonised dimensions has been implemented and not just in terms of width, but also in terms of height. The telescope cartons at Fig. 12.15c fit tightly into the container and need no additional securing. The stability of the cartons is, however, a point of criticism. Whether these are likely to survive the rigours of transportation is a matter of doubt. Interlayer dunnage with a large surface area would improve the situation. It may even be possible to support this dunnage for the lower and middle layer. If the contents are heavy, it may be necessary to find out whether the four lengthwise slats on the pallet-like bases would cause pressure damage on the bottom of the cartons (Fig. 12.16).

This example also shows that the dimensions of packaging receptacles can be designed to fit the internal dimensions of the containers themselves. Transportation modules of this type can help save costs (Fig. 12.17).

The strapping as shown above is faulty. This method is unable to provide a stable unit load. The unpleasant tendency of corrugated drums to ride up over the corrugations can be prevented by appropriate unitisation on pallets, but this will still not cure the tendency of pallets to ride up against each other (Fig. 12.18).

Four drums with a maximum diameter of 585 mm can be stowed next to each other in standard containers. From the loading point of view, the most sensible option is to use drums with wide corrugations and a diameter of 580 mm, loaded onto pallets

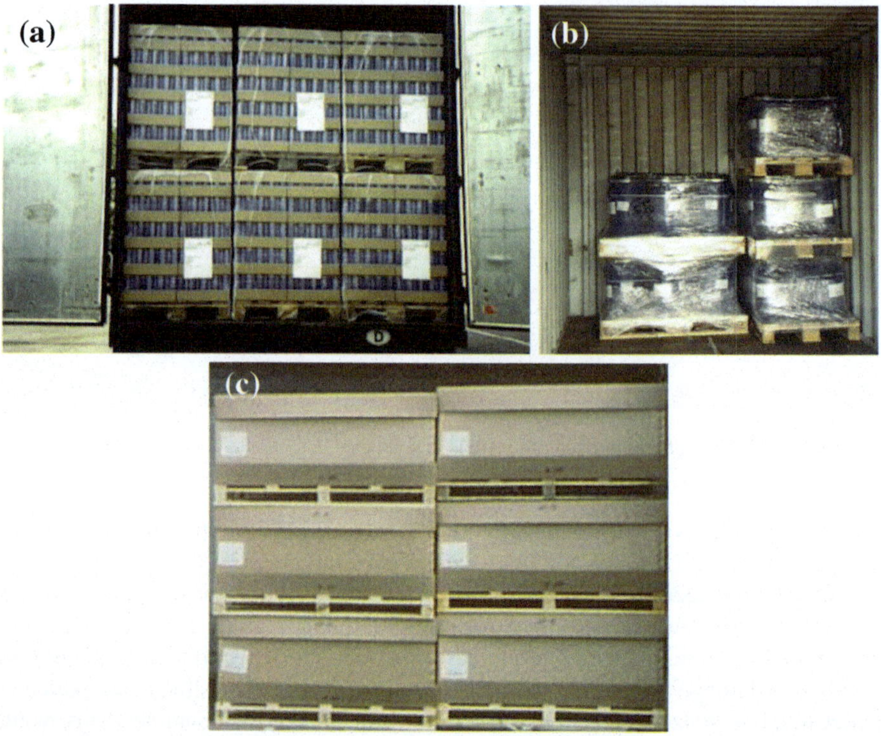

Fig. 12.15 **a** Pallets that are appropriately dimensioned for transportation by lorry. **b** Pallets that are not designed for the container. **c** Telescope cartons with pallet-like bases designed for use in the container

with dimensions of 1,160 mm × 1,160 mm. Formation of a unit load in the format shown above is acceptable, but its quality should not be overestimated. Since the strapping is vertical, it is unable to provide resistance to shifting. As a result, the top wooden board can easily shift. Diagonal strapping would be ideal but is extremely difficult to implement. Figure 12.19 below makes this problem clear:

The "lid" used for this unit load is an inverted pallet. In this case, only two straps have been used for unitisation but at least the drums were wrapped in shrink film first. Nevertheless: simply as a result of the transportation by road from the shipper to the container packing station, the "lid" has already shifted. In series production it would make sense to affix a "lozenge" to both the pallet and the lid which would ensure a 100% tight fit for the drums after strapping (Figs. 12.20 and 12.21):

In order to reduce the risk of torn strapping, it would make sense to introduce pallets and lids which have grooves cut specifically to take straps. This example also requires additional effort in order to guarantee that the plastic canisters reach the recipient in an undamaged state. The photographs below provide particularly vivid examples of poor packaging and unitisation: wooden feet that extend beyond the pallet or are higher than they are wide always cause problems during loading, packing

Fig. 12.16 Load dimensioned to suit the internal measurements of the container

Fig. 12.17 Example of poor unitisation

Fig. 12.18 Example of acceptable unitisation

Fig. 12.19 Diagonal shifting of the "lid" of a unit load

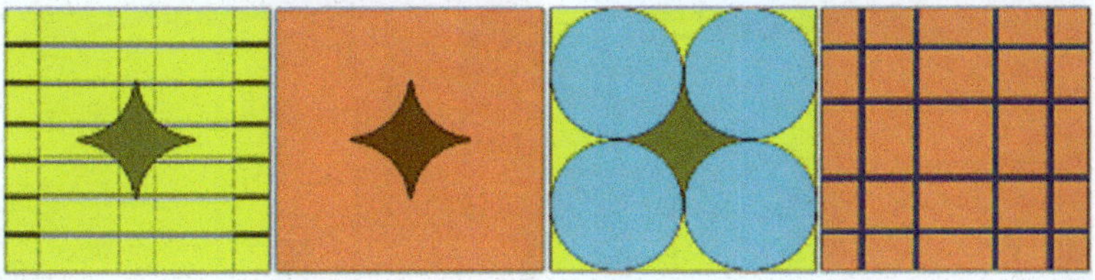

Fig. 12.20 Pallet and lid with "lozenge" can guarantee a tight fit

Fig. 12.21 Side view of the strapped pallet

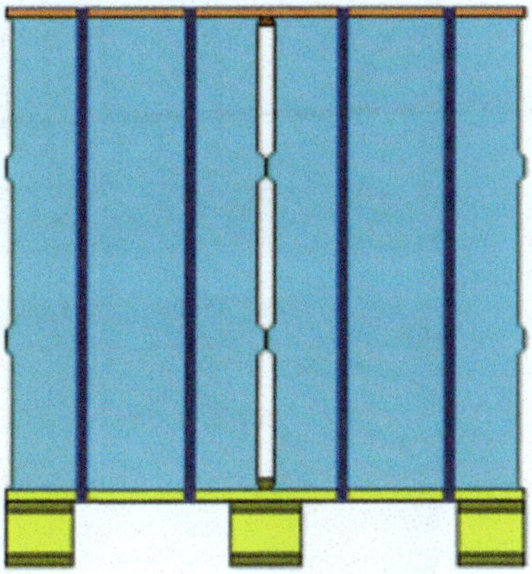

and securing. The loaded surface area is too small and results increased loads on the cargo stowed underneath. Even experienced stowage and loading personnel can only partially compensate for this by using additional dunnage. The use of steel strapping to hold packages in place should be avoided. The use of steel strapping without any edge protectors is particularly critical. Even the slightest of lateral forces are enough to collapse the feet of this plywood case. This combination package shows an "ALTO" arrow which simply indicates the top of the item, and not "transport this way up." This selection of frequently observed packaging faults is just the beginning. There is also an entire range of other weaknesses that indicate that the transportation of goods in general, and the packing of containers in particular, are operations that do not always run as smoothly as they could. The requirement laid down in all national and international legislation, regulations and agreements, namely that goods should be "packed safely" is unfortunately all too frequently not put into practice (Fig. 12.22).

These unit loads have fallen apart when being transported from the warehouse to the loading ramp on the company premises—despite being very carefully deposited by the forklift operator. Companies very rarely respond appropriately and immediately when confronted with inadequately packed goods, despite the fact that they have the opportunity to do so. The shipper is responsible for all consequences caused by a lack of packaging or by inferior quality packaging. In particular, the shipper is responsible for compensating companies for damage caused as a result of such faulty

Fig. 12.22 Example of very poorly stowed and strapped loads

packaging. A company which accepts goods for carriage that already show obvious evidence of damage is entitled to demand that an entry is made in the consignment note by the shipper documenting the state of the goods explicitly. Pallets that are not loaded flush require additional securing. Poorly applied steel strapping requires particular care when handling these loads using a forklift truck since they are easily torn away when applied in this manner. It is exceedingly difficult to assess whether

Fig. 12.23 Example of crates fit for writing off due to insufficient diagonal bracing

or not the damage to the carton has also resulted in damage to the goods inside it or will do so subsequently. If the lack of packaging/faulty packaging is not entered in the shipping documents by the shipper or the party delivering the consignment, then the company responsible for the onward transportation is required to document these faults. Every wooden crate that does not have sufficient diagonal bracing should be classed as faulty packaging. In this example (Fig. 12.23) the lack of diagonal supports has already caused deformation of this wooden crate. Depending how sensitive the packed goods are, this deformation could cause damage. In addition, this example shows evidence of loading and securing errors.

As observations in many container packing stations have shown, pallets made of chipboard are not suitable for overseas transportation purposes. They are too weak and are very susceptible to damage. As a result of the small surface areas of their feet, they also apply high loads to the objects they are standing on and can thus often only be used in conjunction with materials that are able to distribute this load. Chipboard pallets are easily damaged even when subjected to incredibly low loads. Both the sharp edges that remain and the reduction in surface area that results will further increase the loads on other cargo. This pallet (Fig. 12.24) is too big for the carton. This causes gaps when packing the container. Since chipboard pallets are extremely sensitive to bending forces, the steel strapping may break off parts of the pallet. This means that the carton is no longer bonded securely to the pallet. In order to get around this problem, the carton and its pallet were loaded on their side. But this is not a particularly workable solution because this can cause problems for handling staff and may even damage the contents.

Such poorly packed pallets can only be loaded safely with a high outlay in terms of personnel and materials. The holes in the lower telescope carton indicate that an inexperienced forklift operator was entrusted with handling this pallet. This unit load (Fig. 12.25) clearly shows the lack of understanding that has gone into unitising the load:

The consequence of this poor unitisation will be that the packages will be split and loaded into the container individually. If a company is not prepared to load packages of this type individually, considerable effort must be put into correct load securing.

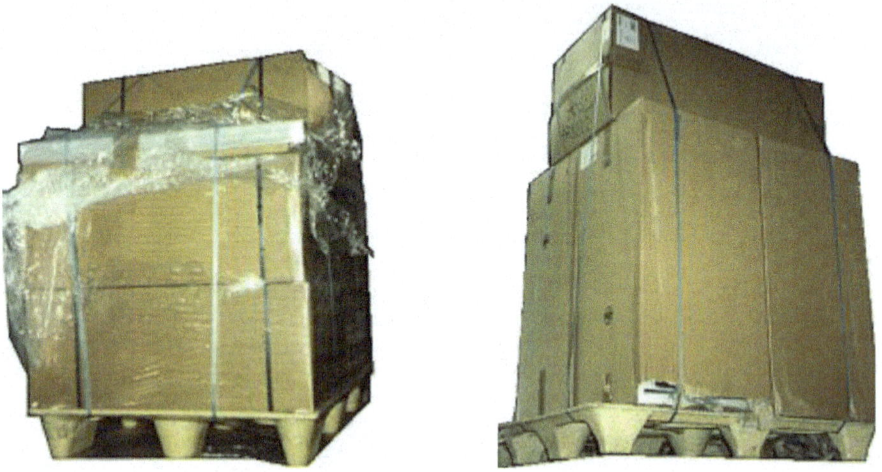

Fig. 12.24 Example of a unit load that is not fit for transportation

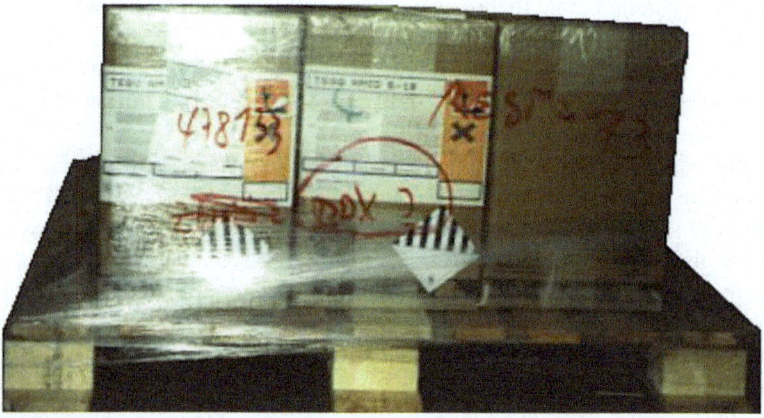

Fig. 12.25 In accordance with the IMDG Code (Load unit that contravenes the regulations)

With many packages, gaps remain when the packages are loaded next to each other. In other cases, the stretch wrap is too weak or does not form a firm bond with the pallets or other bases. Other unit loads cannot be stacked or have faulty strapping, etc. Loads of this type make it impossible to pack the container tightly without additional work and materials. The process of packing a container is made more complicated if unit loads are delivered which cannot be stacked. If unit loads are so heavy that they cannot be packed on the top layer or cannot be fitted into this space, then the packages must be split and loaded manually. The faults are no stable bond between the stretch wrap and the pallet base, the strapping cuts into the boxes, the surface is not level and thus does not allow anything to be stacked on top of it. Since the sides

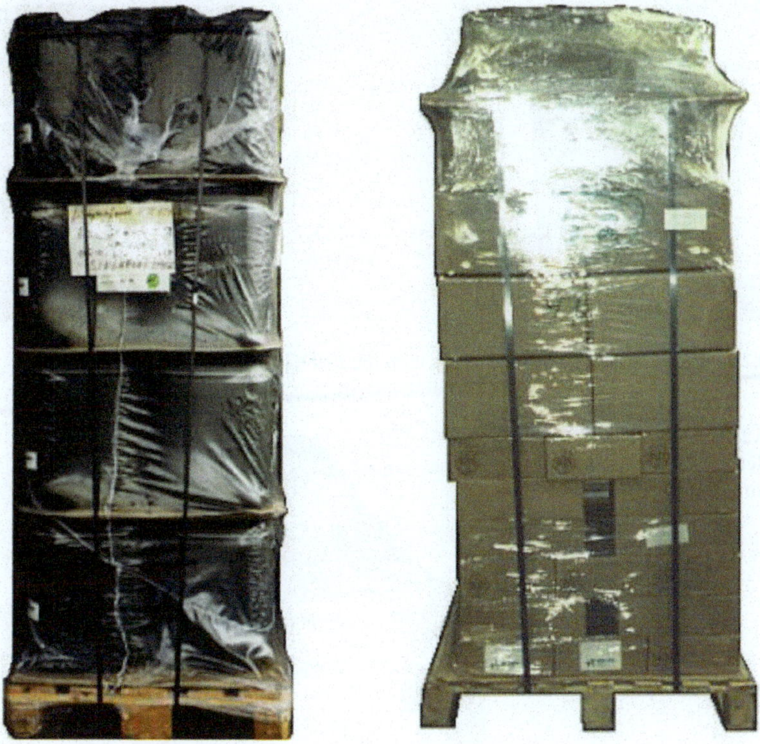

Fig. 12.26 Example of poor unitisation hindering the correct packing of containers

of the unit load are not flush, additional work is required in order to fill the gaps (Fig. 12.26).

According to the IMDG Code, the unit loads should be compact, have as regular a form as possible, and vertical sides. The top of the unit load should be level. It must be possible to stack the unit loads. They must be constructed and secured in such a way that it is unlikely that the individual packages can become damaged. In Fig. 12.27, it is clear the requirements of the IMDG Code have not been met. Nor has the following unit load met them: The main faults here are there is no firm bond to the pallet, the pallet is not loaded flush, and therefore the unit cannot be stacked safely. The pallet shown at Figs. 12.27 and 12.28) carrying hazardous goods is not packed flush. The bond to the pallet is not secure enough and there is a large gap in the top layer. Unit loads that are irregularly shaped such as those shown at Fig. 12.29 makes packing work more difficult. It may be necessary to split this unit and to load the packages individually.

In Fig. 12.30 we can see that one of the drums was already dented when the pallet was delivered by lorry as the load extends beyond the edge of the pallet. It is likely part of the strapping had worked loose causing the lid to slightly shift. Subsequently, it will only be possible to transport this load without causing further

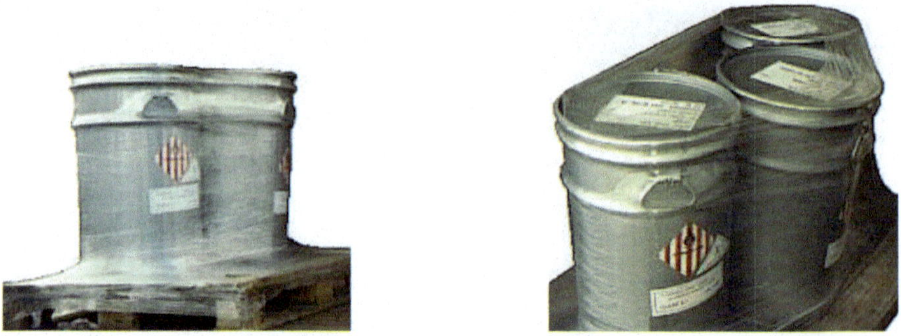

Fig. 12.27 This unit load does not comply with the IMDG Code

Fig. 12.28 No firm bond to the pallet; pallet not loaded flush, unit cannot be stacked

damage by investing a considerable amount of effort into load securing. Since this load comprises class 8 hazardous goods for maritime transport, this constitutes a violation of the law. The violation of the law is even more obvious in the example given in Fig. 12.31.

In this instance, this cannot be termed a unit load as the drums must be loaded individually. There are no bounds to some people's imagination when it comes to devising

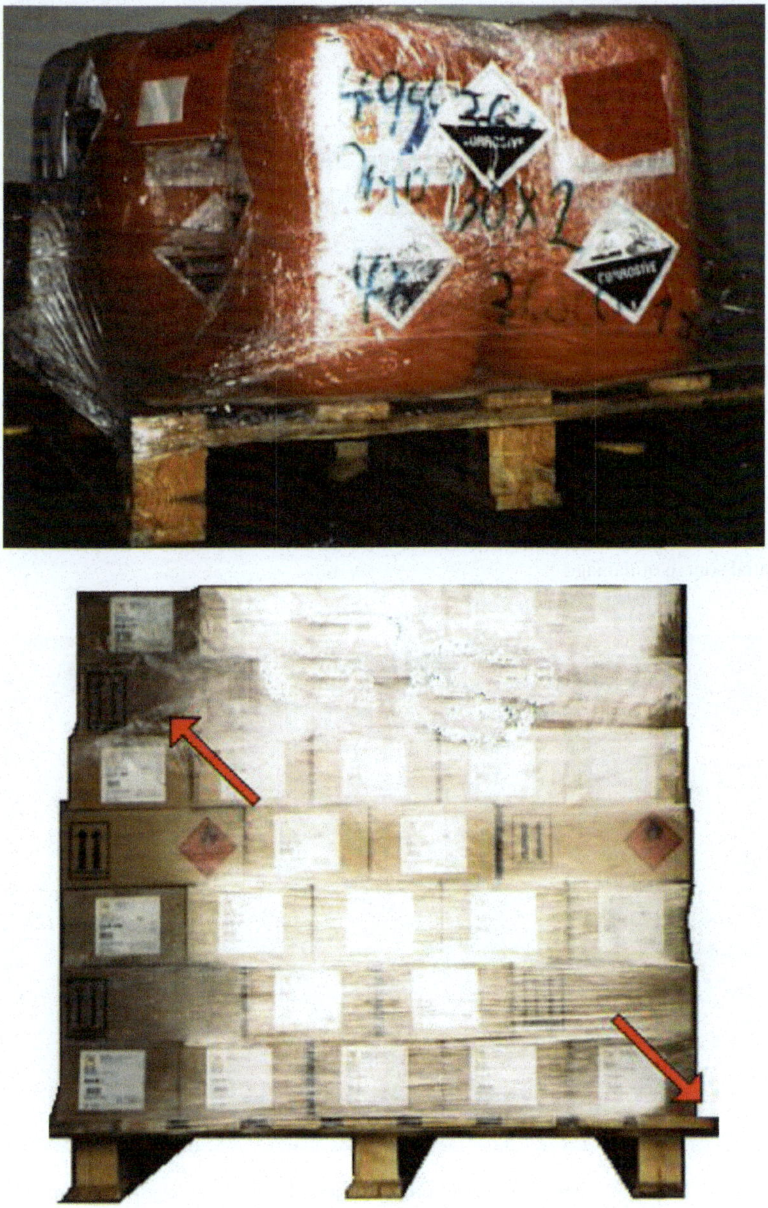

Fig. 12.29 Example of a load unit that does not comply with the regulations

Fig. 12.30 Example of where the projection of the load beyond the edges of the pallet may present a potential source of damage

Fig. 12.31 Example of where the projection of the load beyond the edges of the may present a potential source of damage

methods of forming unit loads. The few examples provided below (Fig. 13.34) should serve to illustrate this problem (Fig. 12.32).

Damage is almost inevitable if loads are packed into a container in this state, unless of course, a considerable amount of work and materials are invested during handling

Fig. 12.32 Example of where the projection of the load beyond the edges of the pallet may present a potential source of damage

to repack, improve the packaging, and secure the load in place safely. In place of the outlay required to unpack the pallets, dispose of the wrapping film, and to then pack the items properly and secure them in the container etc. it would be worthwhile to procure a (used) box pallet or case and load the pails into that instead. It should also be borne in mind that such a breach of regulations would incur a fine if the authorities

Fig. 12.33 Example of faulty formation of unit loads in violation of the loading guidelines

Fig. 12.34 Example of incomplete strapping

were to carry out an inspection. In Figs. 12.33 and 12.34 we can see that the plastic drums are not firmly secured to their pallet by the wrap. The gaps between the unit loads are an obvious violation of the regulations for the transportation of hazardous goods.

If we look at Fig. 12.35, we can clearly see the strapping is missing on two sides and on the lower and middle sections of the load. The existing strapping is badly placed. The edge protectors used are only partially effective in this case and they are missing on the other two sides of the pallet. If the film that has been used is a shrink cover, then the packers forgot to put the pallet in the shrink oven as well.

Fig. 12.35 Example of a unit load that does not comply with the IMDG code

The use of weak packaging hinders the efficient packing of containers. Additional measures for damage prevention are required. Not only that, but no edge protectors were also used when strapping the load. This rectangular dunnage has been used incorrectly: here a groove can be seen that has been cut specifically to take steel straps or plastic strapping in order to ensure that the strapping does not break or tear when it comes into contact with the floor. Used incorrectly as we can see here, the dunnage is likely to collapse. It is also increasing the load on the surface it is standing on. The example given below at Figs. 12.35 and 12.36 shows the effort that must be invested in order to counteract the problems of faulty packaging and how unsatisfactory the ultimate solution is:

Referring to Fig. 12.36, a number of mistakes have been made in packing this container: the pallet holding the drums has not been loaded hard up against the right-hand container wall. A wooden pallet has been placed vertically between the chipboard pallet and the drum pallet, but this does not fill the gap completely. The drainage cocks for the drums have been placed on top of the drums and stretch wrapped in place. The only feasible way of stacking other items on top of these drums would be to create a level surface using dunnage placed lengthways and then width-wise. The gap behind the drum pallet must be filled in manually. Since the packer must ensure that he does not stand on the drainage cock, he is unable to handle the heavy carton safely and it is distorted as a result of being accidentally placed on the edge of the drum. The stowage gap above the drum pallet is only partially filled here. Behind package there is a gap because the drainage cock of the second drum is located here. A gap has also been left for drainage cock in front of package and next to package. If the drainage cocks had been shrink-wrapped at the bottom of the drums, a wooden board could have been used as interlayer dunnage to allow the

Fig. 12.36 Example of a unit load that does not comply with the IMDG code

remainder of the stowage space in the container to be filled out more efficiently. It is correct and sensible to ensure that the hazardous goods drum pallet is packed in the vicinity of the container doors.

12.1 Securing Goods in Packaging Receptacles

The basic concept, namely attaching a suitable base in order to stabilise a case that runs the risk of tipping, is in itself not a bad idea, but only if the package is to be transported on an open flatbed lorry. It is unsuitable for use within a closed mode of transport. For example, the case in Fig. 12.37 can only be secured with considerable effort, using wooden bracing or a considerable amount of material to fill in the gaps. Lifting this object using a forklift truck from the side is only possible if the forks of the forklift truck are positioned close together. Even a slight uneven road surface could

Fig. 12.37 Example of poorly constructed case

cause the case to topple off the forks. If the freight costs are calculated according to volume, money is being wasted since the dimensions used to calculate the volume are the extremities of the case. It would have been more sensible to pack the contents into a stable case with a floor surface the same size as the special construction shown here (Figs. 12.37 and 12.38). If the load is firmly bolted inside a case of this type, it can easily be packed into a closed container and secured. This method would not incur greater freight costs than the existing special construction. Sometimes, the correct securing of the packed goods inside the packaging receptacle is overlooked. Even a sturdy box or case cannot ensure that the goods are not damaged in transit if the contents of the case are not securely packed. Here, the forces of inertia during road transport were sufficient to cause the load to penetrate the case. For customs-related considerations, it is often advantageous to disassemble goods and transport them in cases or other receptacles. The practice is very often adopted when shipping cars. Terms like CKD and SKD, meaning completely knocked-down cars and semi knocked-down cars, are used to describe vehicles that are packed in containers, crates and cases.

Fig. 12.38 Example of a poorly constructed case designed to accommodate two disassembled tractors for maritime transport

In this chapter, we have looked at best practice when unitising and palletising goods. We have also seen what worst practice looks like, and considered the potential consequences when goods are not properly secured on their pallets ready for transportation. As we have said various times before, it is of crucial importance that packers properly secure their goods on pallets as failure to do so can cause all manner of unintended consequences. In the next chapter, we look at how goods should be marked according to their dimensions and weight, whether they are hazardous (and therefore subject to the IMDG code), and whether they require specialist handling due to their fragile nature.

Chapter 13
Marking of Goods

Experts are generally in agreement that the marking of the goods is an essential part of the packaging. This opinion is reinforced by a series of court judgments. For example, in a judgment by the Duisburg Landgericht (regional court) in Germany, a company's claim for damages was rejected because the markings (handling symbols) necessary for safe handling of a package were missing. Worldwide, various domestic traffic regulations are almost unanimous in requiring that the shipper of a package must mark their packaging in such a way that the marking is permanent, clearly visible and cannot be misunderstood. The markings must correspond to the specifications in the consignment note. All previous addresses and labels must be removed or clearly negated. In some areas, it is also necessary to specify both the shipper and the recipient. Part 1 of the current DIN standard 55 402 entitled Shipping marks for packages describes the pictorial markings for the handling of goods, and Part 2 provides a directive for export packaging. Adherence to such standards should ensure uniform marking of the packages to be shipped. The standard does not apply to advertising and required information that is regulated by the guidelines of various transport service providers, such as postal services, railway services or by individual countries. Further information can be found in ISO/DIN 780. Markings should always be restricted to the necessary information and no more. They should be applied without a border and must be easily legible and permanent. They must be able to withstand the rigours of transportation. Monochrome (black and white) presentation should be used wherever possible. It must, however, be ensured that the markings contrast with the colour of the package. Where adhesive labels are used, the standard recommends black printed on a white background. The colours not permitted by the DIN standard are those that are reserved for hazardous goods, which includes for example, orange and bright red. The hazardous goods regulations previously demanded that a border be placed around the symbol. This is no longer the case. If goods are relabelled, both the pictorial marks and the lettering of the previous markings should be made illegible. This is intended to eliminate all risk of errors and misdirection.

A. A. Olsen and F. Karkori, *Containerized Cargo Handling and Stowage*, Springer Series on Naval Architecture, Marine Engineering, Shipbuilding and Shipping 22,
https://doi.org/10.1007/978-3-031-63329-4_13

The specifications in the shipping documents must also match the specifications on the packaging themselves. The problem for the practitioner here is that even if an incorrect or incomplete marking is applied, they are not permitted to correct the markings or add markings without the authorisation of the shipper. Otherwise, this could lead to issues with customs, foreign trade banks and similar interested parties. Those who are responsible for the packing and labelling of goods must be familiar with the various standards, or have them available for reference, and must always act in accordance with them. For those who "only" deal with the transportation of the packages, it is important that they are able to read and interpret the markings on the packages; they are not required or expected to attach or modify these markings.

13.1 Labelling in Accordance with the Standards

In accordance with the various standards, package labelling should consist of the following items of information with the sequence adhered to unless special directives apply for particular circumstances (Fig. 13.1):

The required marks include: (1) the shipping mark. The shipping mark of a package includes the identification mark, the identification number, the address and the total number of packages in the consignment, as well as the sequential number of the package within the consignment; (2) identification mark. The identification mark may be the first few letters of the name of the shipper or recipient or the company name of the recipient; (3) identification number. The identification number may be, for example, the order number, consignment note number or bill of lading number among other things; (4) address. If the package is to be received at a port, the name of the port only should be specified. If the package is to be transported further once it has reached port, then the address should be specified as the destination via the destination port; (5) the total number of packages in the consignment, sequential number. The total number of items in the consignment must be specified plus the sequential number of the package within the consignment; (6) information markings. Information markings include information about the country of origin, and the weight and dimensions of the package; (7) statement of origin. The statement of the origin must be in accordance with the requirements of the relevant countries. The statement of the country of origin is often mandatory. In some cases, however, it is not desired; (8) weight. As of a gross weight of 1,000 kg, packages must display weight information. With regard to ease of transport, handling and storage, however, the relevant standards also recommend indicating the weight in lighter packages; (9) dimensions. The standards state that the dimensions of a package should be specified in centimetres; (10) any relevant handling symbols. In order to guarantee the correct handling of packages and goods, pictorial symbols are used. The use of textual information to enhance the pictorial symbols should be avoided. The pictorial symbols are to be applied without a border. In the past, the hazardous goods regulations required the use of pictorial symbols with a border to identify hazardous goods. This is no longer the case. The authorities will generally tolerate a border if there is one.

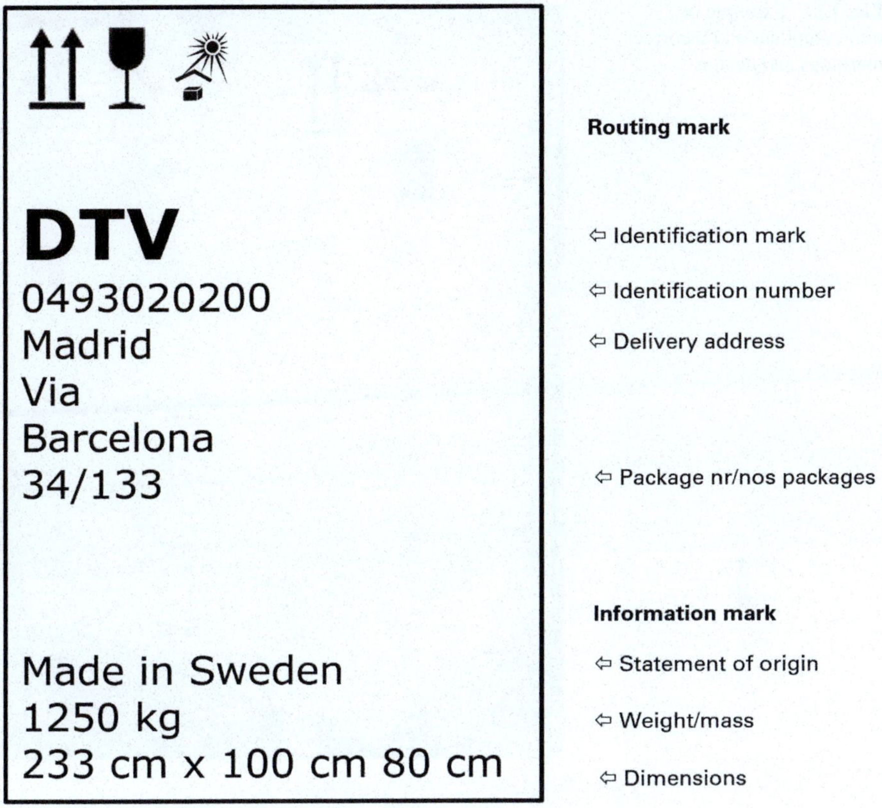

Fig. 13.1 Example of a marking that conforms to the current standards

The shipping mark and the statement of origin should be arranged one below the other. The weight and the dimensions should be applied to the bottom right of the package, and handling information preferably to the top left. For cuboid packages, the handling information and the markings should not be separated. They should be applied to at least two adjoining sides. The only exceptions to this rule are pictorial markings that are tied to a specific location, for example, those showing the centre of gravity. For cylindrical objects, it is sensible to apply the labelling to two opposite sides. It is sufficient to mark bagged cargo and bales on one side only. Referring to the example provided in Fig. 13.2, the pair of arrows should have been on the front of the crate. All markings should also have been duplicated on an adjacent side. As a further observation regarding the case, it is constructed in such a way that the lid boards can lift easily if slinging equipment with multiple cables and a single hoisting point are used. Here, either metal corners should be used, or the lid boards should be recessed or rabbeted. Furthermore, the air vent protrudes by approximately 1.5 cm. This may be torn off or may damage other goods during packing.

Fig. 13.2 Example of non-compliant and incorrect markings and design

13.2 Approved Marks and Labels

Information that is not required for shipping should not detract from the effectiveness of the markings. Nor should they be applied to the same sides as the markings. Unless otherwise specified in other regulations, all markings are to be affected using Arabic numerals and in capital letters. The size of the letters will depend on the amount of space available and is specified in the standards. Clarity and legibility must be ensured. The letters and numbers of the shipping mark should contrast with all other markings. All letters and numbers should, however, be smaller than the specified handling instructions. If we look at Fig. 13.3, we can see that neither the size nor the location of the labelling meets the requirements of the standard.

Unfortunately, the existing regulations are rarely applied in practice. In the event of doubt about the meaning of a marking, information should be obtained as soon as possible from a responsible person or recognised organisation. This is particularly important for hazardous goods markings as well as the "sling here" and "centre of gravity" markings.

Fig. 13.3 Example of markings that do not comply with the current standards

13.3 Meanings of the Pictorial Symbols According to DIN, ISO and IEC

The European meanings of the pictorial symbols in accordance with European standards and the English meanings as defined by the International Organisation for Standardisation and IEC are described below.

13.3.1 Keep Dry

Since goods of this type should also be protected from high levels of humidity, they must be transported in closed vehicles or containers. If it is only possible to transport particularly heavy or bulky goods on platforms, flats or in open top containers these must be covered carefully by way of tarpaulins. Load securing issues will not be discussed here. The important consideration is that the cover must be water-tight and that it is not possible for condensation to form below the tarpaulin. This can only be achieved with considerable effort. More information and hints about using tarpaulins can be found elsewhere. The cover must be as securely lashed as possible to make sure that it cannot be detached or destroyed by stormy conditions (Fig. 13.4).

Fig. 13.4 Photograph of a tarpaulin-covered case with the umbrella symbol

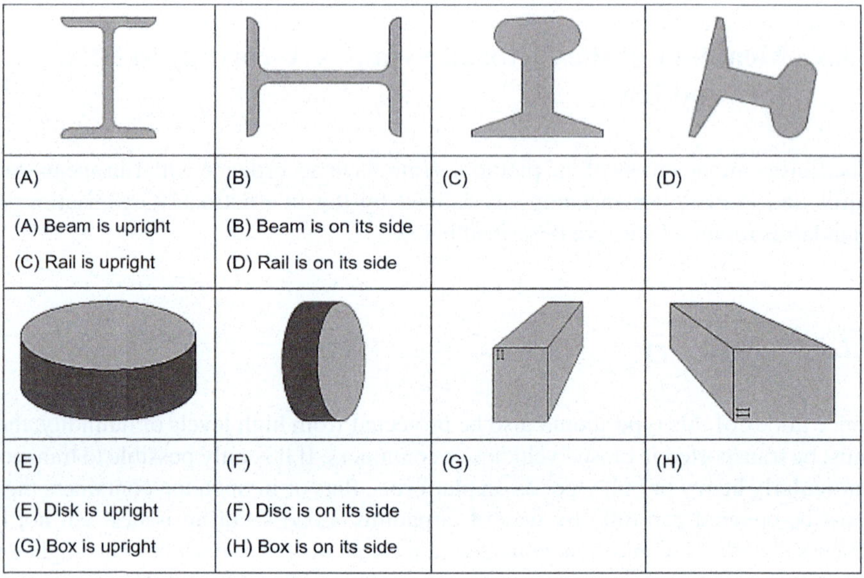

Symbol	Definition according to DIN	Definition according to ISO
	Keep dry	Keep dry

13.3.2 *'Protect from Heat' and 'Keep Out of Direct Sunlight'*

In order to comply with the requirements of this symbol, goods traveling to certain climatic regions may be loaded into insulated containers or vehicles, or at the very least, an appropriate distance to the vehicle or container walls and roof should be maintained. It is important to remember that any gaps must be filled in. When goods are imported from abroad, symbols are often used which are not ISO-compliant. In the vast majority of instances, these symbols can be understood by the storage and handling staff and leave little room for misinterpretation. In the example provided below, the sun as a source of heat is easy to recognise. Identifying the second symbol as a source of heat is somewhat more difficult.

Symbol	Definition according to DIN	Definition according to ISO
	Protect from heat Keep out of direct sunlight	Keep away from heat

13.3.3 *Do Not Use Hooks*

Any other kind of point load should also be avoided with goods marked with this symbol. However, this symbol does not ban the so-called plate hooks that are sometimes used to manually move appropriately marked, robust bags made of natural materials such as sisal, hemp, jute, and so forth.

Symbol	Definition according to DIN	Definition according to ISO
	Do not use hooks	Do not use hooks
	DIN does not recognise this symbol	ISO does not recognise this symbol

13.3.4 Centre of Gravity

This symbol is intended to provide a clear indication of the position of the centre of gravity. It only makes sense to use this symbol if the centre of gravity is not in the middle of the object. If there are no symbols giving the location of the centre of gravity, the warehouse and handling staff can safely assume that the centre of gravity is in the middle of the load. If a load with a centre of gravity which is not in the middle of the object is not marked up accordingly, the shipper/packer of the load can be held responsible in the event of accidents or damage. The centre of gravity is deemed to be clearly identified if the symbol is applied to two adjacent sides (at right-angles). The DIN standard, however, recommends that it be applied to three surfaces, namely two adjacent sides and the top. This is particularly relevant for large and heavy packages as the operator of the lifting equipment is then also able to see that the centre of gravity is not in the middle of the package. Exports destined for countries of the former Soviet Union should use the Russian symbol for the centre of gravity. There, the cross is placed below the circle. Containers with a centre of gravity that is outside the specified tolerance of $\pm 2'$ ($\approx$60 cm) for 20′ containers and $\pm 3'$ ($\approx$90 cm) for 40′ containers must always be marked with the centre of gravity symbol. Permits must also be obtained from the transportation and handling companies with regard to the special location of centre of gravity (Fig. 13.5).

Symbol	Definition according to DIN	Definition according to ISO
	Centre of gravity	Centre of gravity
	Russian symbol showing the centre of gravity	

Fig. 13.5 Graphic showing the marking of the centre of gravity on three sides of a load, in accordance with the current DIN standard

Fig. 13.6 Example of a package showing the outdated "top-heavy" symbol

13.3.5 Top Heavy

For some decades now, the DIN standard has not included the "top-heavy" symbol. This was symbolised by a weight balanced on an inverted triangle. It formerly had to be used for objects with a centre of gravity above the middle of the object. As can be seen from the example given below the information that this symbol should no longer be used has not yet reached all shippers (Fig. 13.6).

Symbol	Definition according to DIN	Definition according to ISO
	Top heavy	Top heavy

13.3.6 Clamp Here

The statement that the package may be clamped at the indicated points is logically equivalent to prohibiting clamping anywhere else. This is a very useful piece of information in practice. The example gives the specified maximum pressure as 0.77 kPa (kilopascal) or 7.7 hPa (hectopascal) or 0.00077 MPa (megapascal). Expressed in the unofficial, yet still widely used units of pressure, this is 0.0077 Bar or 0.077 kg/ cm^2 or kp/cm^2. For handling staff, the specification 0.077 t/m^2 may also be helpful.

Symbol	Definition according to DIN	Definition according to ISO
	Clamp here	Clamp here
Symbol	Modified symbol as used in practice	
	Only clamp with the specified pressure	

13.3.7 *Sling Here*

If the 'sling here' symbols for a load with a central centre of gravity are equidistant from the middle or are equidistant from the centre of gravity for an off-centre load, then the package will be level when suspended if slinging equipment of the same length is used. If this is not the case, the slinging equipment must be shortened appropriately where the distance to the centre of gravity is shorter. If one were to attempt to use slinging equipment of the same length the case would not be level. Thus, the case would tilt considerably to the left and may fall. With certain goods that must be slung in a particular way, it makes sense to provide the warehouse and handling staff with additional information to help them handle the object properly. It may also be sensible to include information about restrictions on the maximum amount the load can be tilted or indications that the load is not to be tied (Figs. 13.7, 13.8, 13.9 and 13.10).

Symbol	Definition according to DIN	Definition according to ISO
	Sling here	Sling here

13.3.8 *Fragile Goods*

This symbol is used for fragile goods. Goods marked with this symbol should be handled carefully and should never be rolled over or tightly tied. In the case of certain

Fig. 13.7 Graphic of case with a centre of gravity in the middle of the load

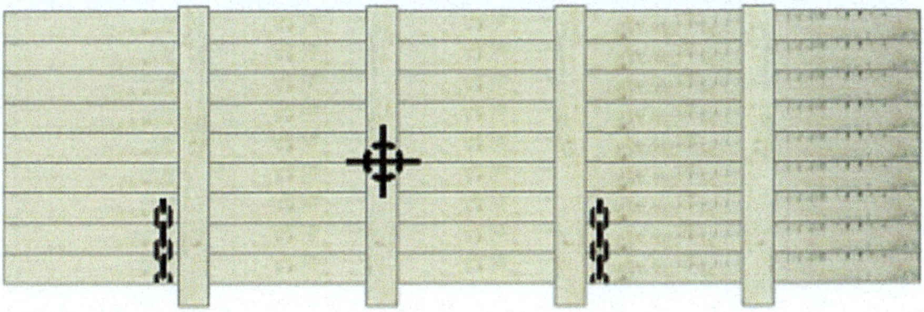

Fig. 13.8 Graphic of a case where the centre of gravity is not central

Fig. 13.9 Graphic of a case whose centre of gravity is not central and with asymmetric slinging instructions

hazardous goods, packages of this nature should not have anything stacked on top of or over them. More information about this can be found in the General Provisions of the relevant regulations or in the special notes for the various classes or special goods.

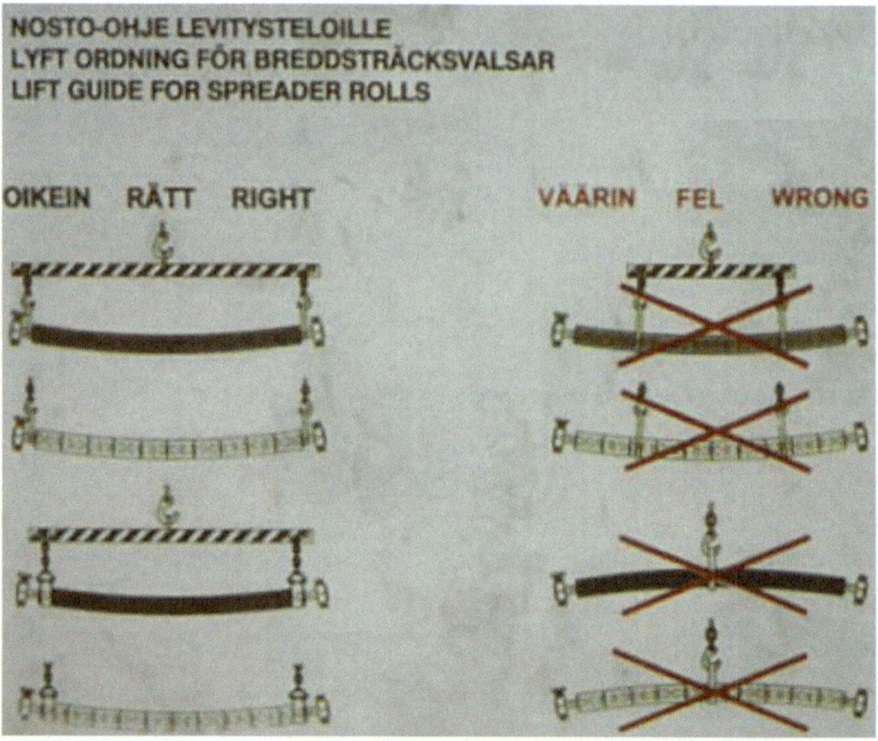

Fig. 13.10 Example of signage showing additional slinging information using special markings

Symbol	Definition according to DIN	Definition according to ISO
	Fragile—Handle with care	Fragile—Handle with care

13.3.9 Keep Upright

In this instance, the package must be stored, handled and transported in such a way that the arrows always point upwards. Rolling, swinging, severe tipping or walking over goods marked in this way must be avoided. The top box shows that the package is marked correctly. However, the arrow indicates that the marking has been applied at an inappropriate location and that the package has already been damaged by incorrect handling. The strapping used is not suitably placed for handling with a forklift truck. If a package does not have this pair of arrows, it may be handled or transported lying on its side or even upside down, which could damage the contents. The pair of arrows does not mean that an object must be stowed on the top layer, nor do they mean "open here". The latter of these two meanings is indicated by a single arrow or other similar symbol. A package that is marked with an arrow like the ones shown does not have to be transported upright. The correct symbol should clearly indicate the intended meaning. However, applying all sorts of nonsensical symbols is a practice that should be avoided. One can only really expect long-term observance of the handling symbols if they are used sparingly and sensibly. The shipper of this case probably wants it to be transported upright. But this is not what the symbols mean. It is also better to apply the handling symbols next to each other and on two adjacent sides (Figs. 13.11 and 13.12).

Fig. 13.11 Example of a pallet box and the appropriate handling symbols

Fig. 13.12 Example of the use of incorrect handling symbols

Symbol	Definition according to DIN	Definition according to ISO
	Keep upright	Keep upright

13.3.10 *Do Not Roll*

The package may not be rolled, swung, or tilted to any great degree during handling. The symbol should always be observed.

Symbol	Definition according to DIN	Definition according to ISO
	This is not yet incorporated in the standard	Do not roll

13.3.11 Tear Off Here

This marking is intended only for the receiver. Generally speaking, warehouse and handling staff should not open packages.

Symbol	Definition according to DIN	Definition according to ISO
	Tear off here	Tear off here

13.3.12 No Hand Truck Here

If this symbol is missing on one or more sides of a package it is assumed that the hand truck may be used on the unmarked sides of the package. Although the "hand truck here" symbol does not officially exist, a symbol of this nature is often used. The use of this symbol to provide staff with clear information is to be welcomed. Thus, for example if a carton containing a refrigerator may be moved by hand truck from two sides, but not from the other two sides, then it would make sense to use both of these symbols and apply them to the relevant sides.

Symbol	Definition according to DIN	Definition according to ISO
	No hand truck here	No hand truck here
	Unofficial Hand truck here	Unofficial Hand truck here

13.3.13 Do Not Destroy Barrier

A barrier layer which is (virtually) impermeable to water vapour and contains desiccants for corrosion protection is located beneath the outer packaging. This protection is no longer afforded if the barrier layer is damaged. Since the corresponding symbol has not yet been approved by ISO, puncturing of the outer layer should also always be avoided for any packages bearing the words "Packed with desiccants". Depending how sensitive the goods are, the length of the journey, the climatic conditions, the onward carriage and standing time of the goods, an appropriate form of conservation should be employed. Goods can be protected from corrosion using coatings, spraying, painting, dipping, desiccants, volatile corrosion inhibitors (the VCI method) as well as other effective measures. Attention should be paid to all critical issues before doing so. If in doubt, advice should be obtained from professional packers, or the job should be handed over to a reputable packing company. Sealed packages, and in particular those in which desiccants have been included, should not contain hygroscopic materials. Packages of this nature should be carefully protected from damage. Referring to Fig. 13.4, we can infer from the lack of diagonal bracing that the crate may suffer deformation during handling thus causing damage to the contents. The symbols indicating a sealed package are missing. This can easily be damaged during handling. A case would have been a more suitable receptacle. The single arrows do not oblige the handling staff to keep the crate upright at all times.

Symbol	Definition according to DIN	Definition according to ISO
	Do not destroy barrier	Do not destroy barrier

13.3.14 *Electrostatic Sensitive Device*

Touching parts that are marked like this under relative humidity conditions of less than 70% should be avoided if wearing electrically insulated shoes or when standing on an insulating surface. Low levels of relative humidity must be expected in hot and dry summer conditions and on very cold winter conditions.

Symbol	Definition according to DIN	Definition according to ISO
	Electrostatic sensitive device	Electrostatic sensitive device

13.3.15 *Permitted Temperature Range*

In accordance with the regulations, the symbol should either have an additional °C (°F) for a single temperature or °C (°F) maximum and °C (°F) minimum to provide the upper and lower temperature limits of a range. The relevant temperatures or temperature limits should also be noted in the shipping documents. Transportation of goods that are marked like this are generally carried using refrigerated containers or vehicles, or at the very least, insulated containers or well insulated vehicles.

Symbol	Definition according to DIN	Definition according to ISO
	Permitted temperature range	Permitted temperature range

13.3.16 *Temperature-Controlled Load*

The symbol for temperature-controlled cargo has been included in the hazardous goods regulations for a number of years. The relevant hazardous goods regulations provide information about the necessary markings and their meanings.

Symbol	Definition according to DIN	Definition according to ISO
	Temperature-controlled load	

13.3.17 *Do Not Use Forklift*

This symbol should only be applied to the sides of a package where a forklift truck cannot be used. Absence of the symbol on other sides of the package amounts to permission to use forklift trucks on these sides.

Symbol	Definition according to DIN	Definition according to ISO
	Do not use forklift truck here	Do not use forklift truck here or: Use no forklift

13.3.18 *Safe to Use Forklift*

The following symbol is not part of either the DIN or the ISO regulations but is sensible and should be used as required.

Symbol	Definition according to DIN	Definition according to ISO
	Use forklift truck here or: Forklift here	

13.3.19 Stacking Limitation

A sensible combination of this and the ‘safe to use forklift’ symbols would be, for example, on a long case that as a result of its construction can be lifted with a forklift truck from its long sides but not from either of the ends. However, other combinations or variations are entirely possible. The stacking limitation should be specified in kilograms (kg). Since a marking of this nature is only ever applied to packages that are able to take very little load, these are usually goods that are packed in the upper layer. If the limit is specified as 0.00 kg it should be clear to both warehouse and handling staff that the goods may not have packages stacked on top of them. The recommended use of the symbol should be clearer than any proprietary labels. The use of ‘Fragile’ on its own is no guarantee that other items will not be stowed on top of this package. The marking ‘Please do not load’ is more likely to cause confusion, as it is not clear what is meant. For the carriage of homogenous loads, the industry has developed a number of its own symbols which are very clearly understood and refer to the number of layers or tiers that may be formed using certain packages. If the case should not be stacked, it would be better to use the appropriate symbols instead of writing ‘do not stack’ by hand on the outer packaging. The hazardous goods regulations state that the markings on a package should be applied in such a way that they are guaranteed to survive three months exposure to sea water. The required hazardous goods markings and their meanings can be taken from the appropriate regulations. General notes can also be found in the appropriate sections of the container packing guidelines (Figs. 13.13 and 13.14).

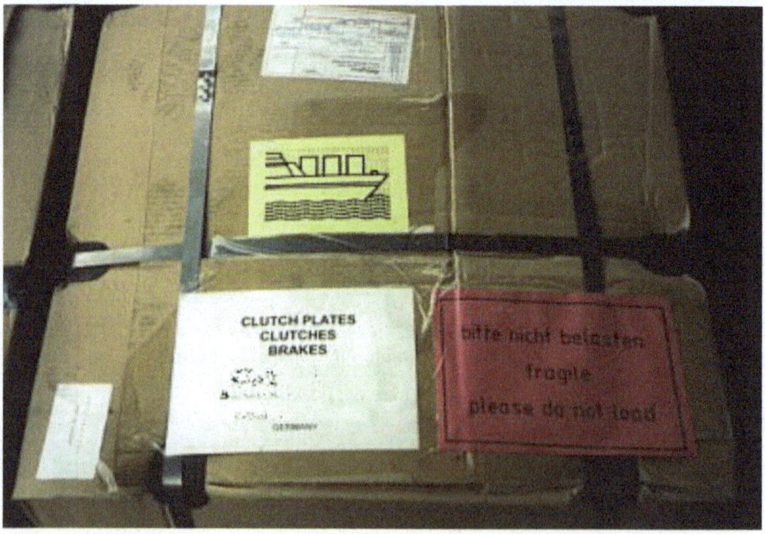

Fig. 13.13 The recommended use of the symbol should be clearer than any proprietary labels

Fig. 13.14 Cargo is at risk if the handling symbols are missing

Symbol	Definition according to DIN	Definition according to ISO
	Stacking limitation	Stacking limitation
	Only transport in the upper layer Or: Top layer Or: Do not stack	

Symbol		Definition according to DIN and ISO
		Only stack six layers high

13.3.20 Protect from Heat and Radioactive Sources

The required hazardous goods markings and their meanings can be taken from the appropriate regulations. General notes can also be found in the appropriate sections of the container packing guidelines. Since the term heat is not clearly defined, the packages should be stowed in as cool a location as possible. In addition to this, sensitivity to the temperatures that can be expected should also be checked. The presence of radioactive sources should in theory be known as a result of adherence to the regulations or as a result of the correct labelling of packages and receptacles. If refrigerated containers or refrigerated vehicles are not to be used, it is recommended that at least insulated containers or vehicles are used. Under no circumstances should loads of this type be stacked close to the walls or roofs of normal containers or vehicles.

Symbol	Definition according to DIN	Definition according to ISO
	Protect from heat and radioactive sources	Protect from heat and radioactive sources

13.3.21 Do Not Expose to Magnetic Fields

Certain packages such as computers and accessories, should not be exposed to magnetic fields. Industries in many countries has developed easily understood symbols to indicate this.

Symbol	Definition according to DIN	Definition according to ISO
	Do not expose to magnetic fields	

13.4 Tags and Detachable Labels

When transporting items such as steel bars and pipes, for instance, the labels are very often in the form of sheet metal tags that are hung on the cargo. When handling loads of this type, it must be ensured that these tags do not become detached, as this could lead to problems with customs authorities or incorrect delivery. If a tag of this type is detached or such a detached tag is found, it must be taken to the appropriate office or person.

In this chapter, we have looked at how goods should be marked according to their dimensions and weight, whether they are hazardous (and therefore subject to the IMDG code), and whether they require specialist handling due to their fragile nature. This is important information that must be clear and legible so that all container handlers are acutely aware of how to safely handle the cargo. In the next chapter, we will begin to examine how to pack goods so that they comply with the guidance discussed thus far in this book.

Chapter 14
Packing and Stowing Methods

Unlike the transportation of general cargo on conventional ships, containers represent small, closed loading areas or loading platforms of a specific dimension. The first definition applies primarily to box containers, whereas the second applies to platform containers. The height and width dimensions may be exceeded to a certain degree. The term "stowage methods" remains a frequently used term in the container handling and shipping industry. But the use of containers and other cargo transport units demands further distinctions in the use of terms. This is because not only do the containers themselves have to be loaded and unloaded but also, the containers have to be positioned on land or on various means of transport. It is therefore necessary for regulations, such as the container packing guidelines, to define terms to distinguish between the various activities. The section on "Definitions" in the container packing guidelines provides the following statement:

> Packing means the packing of packaged and/or unitised or over packed cargoes into [containers].

Before we move onto the techniques for packing containers, it is probably worth outlining some of the key terms used by those in the container handling industry. It is not unusual for terminal staff to refer to the "stuffing" of containers in this context. It is also a term frequently used at "packing centres". For safety, it is essential that all dead space in the container is filled. "Unpacking" refers to the removal of cargo from the container. Terminal staff often refer to this activity as "stripping". "Stowage" means the positioning of packages, IBCs, containers, swap-bodies, tank containers, vehicles or other containers onboard ships, in warehouses and sheds or in other areas such as terminals. It should be obvious that every container load needs to be packed carefully. Where necessary, packing should be supplemented by appropriate load securing measures. Unfortunately, not everybody thinks that the need for careful packing is obvious. Packages with different dimensions often present packing personnel with problems which are difficult to resolve. If the dimensions of

A. A. Olsen and F. Karkori, *Containerized Cargo Handling and Stowage*, Springer Series on Naval Architecture, Marine Engineering, Shipbuilding and Shipping 22,
https://doi.org/10.1007/978-3-031-63329-4_14

the packages do not meet the requirements of modularity, any gaps which remain must be filled in rigorously.

On arrival in port, random checks are occasionally performed to ensure that the load has been packed and secured correctly. However, these checks are typically limited to a small number of containers carrying dangerous goods. Those spot checks are arranged and performed by the harbour police or by the shipping company. It is unusual or even impossible for a load to be checked or corrected during a sea voyage. Loading and securing must therefore always be performed correctly. Stowage and securing errors made prior to acceptance of the container or which arise during the voyage can thus have more serious consequences than errors with a non-containerised load. Time is also a significant factor: on a long sea voyage, even minor initial deficiencies can develop into serious losses. The proportion of containers considered inadequate is alarmingly high—around 70%!

There are a number of national and international regulations relating to shipping and transport which provide important guidelines on how to handle and package goods, as well as how to pack and secure them. These legal stipulations are supplemented by a number of civil-law regulations and contracts, in particular special conditions for individual means of transport. Further information can be found in the recommendations of special organisations or various companies. Loads can only be transported safely around the globe by observing the fundamental issues. The only way to prevent damage from arising is by packing and unpacking in a professional manner and by using appropriate packing and securing methods. All preparation and final work must be fully integrated in operational organisation procedures. The remaining units in this section of the handbook address and explain a number of important issues.

14.1 Preparatory Work

When producers pack their own goods into cargo transport units, they should do so with trained staff and sufficient expertise. If external companies are entrusted with the storage, handling or transport of goods, they must fulfil their duties as carefully as if they were dealing with their own goods. They must ensure that nothing is spoilt or damaged and that articles and materials entrusted to them cannot have a damaging effect on other goods or on the environment. This obligation to exercise due care entails an important principle concerning the provision of suitable storage areas and handling and transport equipment:

> Open or closed areas intended to receive goods must have facilities and characteristics suitable for transporting or storing these goods. They must be in a state or be brought into a state which allows correct and safe transportation.

This results in a series of obligations and activities for all staff involved in the supervision, storage, handling and transport of goods to ensure smooth shipment of those goods. These include, most importantly, the accurate planning and sophisticated organisation of all processes. Otherwise, errors can propagate to every area of goods handling. Good organisation involves not only clear procedural instructions but will also ensure that these instructions are observed and monitored and that the staff receive regular training. This section provides a brief summary of important preparatory work. More detailed information on some of the specific issues outlined there will be found in separate sections. The actual work procedures and final work is dealt with in separate sections. Preparatory work for the orderly acceptance or handing over of a load includes particularly the following tasks: **the selection of suitable facilities for storage, handling and transport of goods and adequate stowage and storage space** is the particular task of all persons involved with the logistics for vehicles, containers or devices, and of shed or warehouse managers. They must ensure that suitable means of transport, ground conveyors and other equipment are available in the right place at the scheduled time. Outdoor short and long-term storage areas, covered hangars, sheds and warehouses should be marked out beforehand and dedicated to receiving specific goods. Staff under instructions have little influence in these considerations. Exceptions to this arise if staff are permitted to freely select from a series of empty containers. The loading staff in ports have no influence on the stowage locations for containers on board ships. These locations are specified by the stowage centres of the shipping companies or by the ship masters. **Checking or establishing fitness for loading** is an important prerequisite for being able to begin packing. Irrespective of the type of cargo transport unit used, fitness for loading must be established. The same also applies for short and long-term storage areas. They also must be in an adequate state to allow the goods to be stored safely and without damage. If this is not already the case, they must be brought into this state. Different properties of the different goods which constitute a load necessitate different requirements. Only a few general principles will be dealt with here. Although some concessions can be made under certain circumstances, they will also have to be supplemented by further measures under other circumstances. The term "fitness for loading" implies that work can be begun without delay. It may be necessary to line or pad out the floor of the area with boards, planks, squared lumber etc. It must be possible to handle the goods by lifting gear or ground conveyors without difficulty. Fitness for loading also means that permitted stacking loads, wheel loads or axle loads are (can be) observed. Otherwise, the result could be damage to vehicles, stowage or storage areas or the goods that make up the load. Sometimes, the permissible load stated is clearly visible. Sometimes, it can easily be ascertained. In some cases, it must be taken from the relevant regulations (Fig. 14.1).

The degree to which a container must be **clean** must always be considered in relation to the particular goods to be stored or transported. The requirements for stowage space and holds will naturally differ for coffee, spices and other foods and foodstuffs compared with, for instance, granite blocks, steel, machine parts or hazardous materials. The minimum requirement is that the stowage space is swept out. In this context, it is important not only that the floor is swept out, but also that any other dusty or contaminated parts, such as side and end walls, doors, roofs, covers, stanchions etc. are cleaned appropriately. Basic cleaning can prevent not just soiling or contamination but also chemical reactions. Sweeping out is also valuable on a physical level,

WARNING!

Before loading a container, the shipper must ensure that it is in a fit state for loading. The acceptance shows that the shipper considers the container fit to take the load that is to be shipped. It should also be noted that the recipient is responsible for ensuring the container is unloaded and returned cleaned and in a fit state using conventional methods or steam-cleaned. Steam cleaning may only be used when the inside walls of the container have been aired sufficiently and have attained positive temperatures. The substances used to clean the container should be neutral in odour. All load related labels should be removed. If in doubt, please contact your nearest HAPAG-LLOYD representative.

Fig. 14.1 Example safety notice inside a typical shipping container

since it generally improves the friction factor of the floor—dusty floors exhibit a lower coefficient of sliding friction. **Absence or neutrality of odour** is always required when odour-absorbing goods are stored or transported, otherwise their value could be adversely affected. Ensuring absence or neutrality of odour also means not using any odour-tainted or odour-emitting load securing equipment, dunnage or segregating materials in places where they could cause damage. Loads which can (could) negatively impact on each other must not be loaded together. Odour-neutrality is not required for the means of transport and accompanying load with most dangerous cargoes, but it can be important for certain chemicals. If odour-neutrality is required, appropriate information must be provided to the relevant offices. This is especially important if a molecular influence from other goods would have a damaging effect. In many cases, it is sufficient to remove residue from previous loads and to air thoroughly. Sometimes, it is useful to spray odour-binding chemicals, which should be environmentally friendly and in no way health-threatening or even toxic. Special measures are required for storing, handling and transporting loads which are particularly odour-sensitive. The same applies if strong or badly smelling or dangerous loads have been shipped and the affected areas have to be prepared specially or restored before a load can be accepted. Extensive cleaning work, such as washing etc. can then be necessary. There are specialist companies on the transport market who offer these services.

Decontamination with special chemicals must be ordered specially and must be monitored. In certain cases, specialist companies with specially trained staff are to be called in. Storage space and cargo transport units must be dry and offer protection from moisture, unless otherwise contractually agreed or this is not customary practice in a given trade. Generally, however, they must not show any defects which could lead to damage by moisture. Wet patches are to be removed by applying sawdust, dry sand or other absorbent materials and sweeping them up. Certain circumstances make it necessary to use special binders. In the simplest case, a thorough airing with "relatively" dry air is sufficient. To prevent moisture damage from occurring at a later date, the storage and stowage areas and means of transport must be inspected for leaks etc. This also means checking pipework, roofs, tarpaulins, doors, ventilation openings etc. and in particular checking the seals. On rainy days, containers should not be packed outside. It is another mark of good organisation that either covered areas are provided, or work is stopped under these conditions. Long and short-term storage areas and cargo transport units must be free of any articles which could cause **mechanical damage**. They must therefore be inspected for materials which could damage the load, such as old load securing material, protruding mountings, incorrectly inserted roof bows or slats, sharp or pointed objects, etc. If there is a risk of damage, these items must be removed. This also includes removing old wires, wood splinters, nails, small stones etc. All protruding mountings, lashing rings, and sharp and pointed objects which cannot be removed, are to be covered. Carelessness during this work can not only cause significant damage to the goods themselves, but also consequential losses when the goods are further processed. An example of this is an instance of damage which occurred during the transport of newsprint in a container in which bags with plastic pellets had previously been transported. Due to inadequate cleaning, individual pellets became embedded in the paper rolls. When the rolls were subsequently processed, the granules brought a complete printing line to a standstill. The damage caused was enormous.

Inspections and protective measures are necessary before using cargo handling equipment, load-bearing equipment, mechanical aids, tools etc. They must not cause tears, pinching, pressure points, scratches, soiling or a reduction of quality or safety standards which depreciate the value of the goods. Defective equipment must be repaired so that it is suitable for use or be rejected. This includes checking the loading capacity of ground conveyors and lifting gear, load-bearing equipment, etc. The materials used must be adequate to meet the requirements for correct segregation and correct use of dunnage. They must be available in a sufficient quantity and be of an adequate quality. Any missing stock must be made up in good time, to prevent any delays. Unusable materials must be rejected. Load securing means and transport protection equipment which have been installed or supplied are to be inspected to ensure they are ready to use, complete and in perfect condition. This includes stanchions, trailer rails, roof bow mountings or slats, partitions, permanently installed lashing systems, air bags, etc. **The provision and selection of conventional load securing equipment** must be organised in good time. The most suitable materials are to be selected from those which are available or provided. Unusable parts must be rejected. Claims for incorrect or incomplete deliveries must be made in good time, to prevent time delays during purchasing or replacement delivery. Materials of this type could be Steel wire, wire rope clamps, shackles, turnbuckles, steel strapping, lashing belts, lashing ropes, wood, etc. If individual items load securing equipment are passed on to third parties or used by third parties, this must be documented by appropriate receipts, etc. **Preparation of stowage, packing and storage areas or spaces** must also include levelling any uneven surfaces. All surfaces, including container floors, must be prepared so as to enable work to be carried out without any delay. If goods are to be packed in several layers, it may be necessary to line, flatten or pad out the floor of the area with boards, planks, squared lumber etc. It must be possible to handle the goods by lifting gear or ground conveyors without difficulty. The **load-carrying capacity and the permissible loading capacity of the underlying surface** must be checked. Permissible and required stacking loads, axle loads, and wheel loads must be compared. Exceeding the permissible loading capacity or loads can result in damage to the operating equipment or cargo transport units. For some storage areas and containers, the permissible load or load-bearing capacity is clearly visible. It can sometimes be determined simply by dividing the load-bearing capacity or load limit by the loading area. The "General guidelines" of the UVV (a national German regulation) stipulate that permissible values must not be exceeded.

Necessary **protective clothing and equipment** must be requested and provided in good time. Basic safety clothing includes Safety helmets, steel-capped safety boots with impervious soles and gloves. Gloves are indispensable for protection against "frayed edges" in wire ropes, splinters on rough wood, etc. When dealing with dangerous cargoes or working materials, it can be mandatory to use special protective clothing and equipment. Dust masks, filter masks or gas masks, acid-resistant footwear, antistatic footwear, rubber aprons, rubber boots, rubber gloves etc. must then be worn. It must be noted that even very traces of alcohol in the breath of the user can render certain filters ineffective. When using chainsaws, "saw-proof" clothing, ear protectors and face protection must be worn. Employers must provide the relevant safety equipment; employees must use it. All necessary resources must be kept ready for **planning and documenting packing operations**. Computer lists, forms, marking pens and pencils, placards, rules, measuring tapes, etc. must be immediately accessible or available at all times, to prevent delays during handling. Electronic support facilitates documentation and accelerates the handling process. **Supervisory authorities and specialists** must be requested and informed in good time. These may be radiation protection experts, persons with appropriate qualifications for handling explosives or similar experts, who must be present during handling. Frequently, fire services, veterinary or plant protection officers, customs or similar authorities must also be brought in. In some cases, firefighting equipment must be to hand and the appropriate staff must be readily accessible. Provisions may need to be made for monitoring the handling process or the containers. In some cases, it is required that extensive **information is provided to the staff**. When handling dangerous goods, all staff and other persons involved must be instructed before work begins about the special properties and risks associated with the load, and what protective measures need to be taken. The names of all **supervisors** must be known to all persons involved in the monitoring process. A written record must be kept of the names of the appointed supervisors. In according with the container packing guidelines:

> **Responsible person** means a person appointed by a shore-side employer who is empowered to take all decisions relating to his/her specific task, having the necessary current knowledge and experience for that purpose, and who, where required, is suitably certificated or otherwise recognised by the regulatory authority.

It is essential that all safety instructions on the containers are observed. To make provision for emergencies, staff should be informed about the location of fire alarms, emergency exits, the nearest stretcher and all relevant emergency numbers. In addition, every employee must be acquainted with the reporting procedures and general procedures in the event of faults and accidents occurring. Trade association demands for safe workplaces must be met. These include adequately lit traffic routes and freely accessible escape routes. In Germany, for instance, one of the requirements is that:

> The employer must ensure that only ramps which cover the entire width of the container are used when driving onto containers with ground conveyors.

It is obvious that safety-relevant aids of this kind must be checked regularly:

> Drivers of ground conveyors must only drive onto containers and vehicles once all necessary ramps and bridging have been set up and secured against unintentional movement.

When driving onto containers, ramps are considered to be secured if they have been hooked into the containers or, for instance, lashed to the container with chains. Workers must use equipment to protect against falling when working more than 2 m above the ground or above the height of a single container.

14.2 Selecting and Checking Container Conditions and Standards

From an economic perspective, the most suitable type of cargo transport unit is the one which best harmonises the volume-to-payload ratio of the container with the stowage factor of the cargo. This provides the best utilisation of the space and weight constraints. Other aspects that must be considered when selecting a container include the type of product, weight, centre of gravity, substructure and load-bearing surfaces, type of packaging, permissible loading capacity and or stackability, load securing options, and other similar aspects. The internal dimensions, door dimensions and other similar information can be obtained from the owners of the containers or the relevant leasing companies or can be taken from their prospectuses beforehand. In all cases, the container should be inspected. Not only for reasons of cargo care, but also for reasons of liability and accident prevention. "EIR"s are usually used at the handover points. This abbreviation stands for *Equipment Interchange Receipt*. These are receipts for the handover of containers, trailers, chassis and similar cargo transport units and items of equipment. On the basis of a visual inspection, the EIR lists, or marks in the diagrams printed on the form, all defects found on a container. If an inspection reveals, for instance, that a container exhibits damage which has not previously been documented, the costs for restoring or repairing it must be borne by the person who was previously responsible for the container. Thorough inspection and true and accurate documentation are therefore in the interests of all parties involved. The container packing guidelines require a thorough inspection of cargo transport units before they are packed. A number of issues are listed in the guidelines as a guide for visual inspection. For exterior inspection, the container packing guidelines prescribe the following:

> 2.1.1 The structural strength of a container depends to a great extent on the integrity of its main framework comprising the corner posts, corner fittings, main longitudinal and the top and bottom end transverse members which form the end frame. If there is evidence that the container is weakened, it should not be used.
>
> 2.1.2 The walls, the floor and roof of a [container] should be in good condition, and not significantly distorted.

If containers are placed on an uneven surface, they become distorted. It is then only possible to open and close the doors with force. Fork-lift trucks are frequently misused in this way. Questionable practices such as these are best avoided. Unevenness of the ground or underlying surface must be levelled out before setting down the containers, possibly by using simple aids such as hardwood strips. The container packing guidelines go on to state (Fig. 14.2):

> 2.1.4 A container on international voyages should be affixed with a current International Convention for Safe Containers (CSC) Safety Approval Plate. A swap-body may be required to have a yellow code plate, fixed at its side wall (for details see UIC leaflet 596), which proves that it has been codified in conformity with the safety rules of European railways. Such swap-bodies need not be affixed with a CSC plate, but many of them will have one in addition to the yellow code plate.
>
> 2.1.5 Irrelevant labels, placards, marks or signs should be removed or masked.
>
> 2.1.6 A vehicle should be provided with lashing points for securing it aboard ships (refer to ISO 9367-1: "Lashing and securing arrangements on road vehicles for sea transportation on RoRo ships - General requirements - Part 1: Commercial vehicles and combinations of vehicles, semi-trailers excluded" and to ISO 9367-2: Lashing and securing arrangements on road vehicles for sea transportation on RoRo ships - General requirements - Part 2: Semi-trailers".)

14.3 Using Canvas Covers

When canvas covers are used, they should be checked as being in satisfactory condition and capable of being secured. Loops or eyes in such canvas which take the fastening ropes, as well as the ropes themselves, must be in good condition. When loading swap-bodies, it should be borne in mind that in most cases, the bottom and floor of swap-bodies are the main areas of their structural strength. Checking vehicles includes checking the loading area, the side walls, the locking equipment, the support legs, load securing devices, the supports used to guide the cover, the slats used to guide the cover, the canvas covers and the section where canvas cover is closed. When performing an interior inspection, the container packing guidelines stipulate the following:

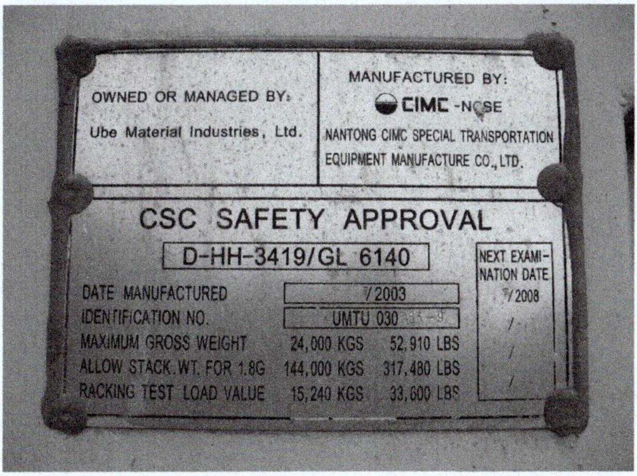

Fig. 14.2 Example of a typical CSC plate

> 2.2.1 A [container] should be weatherproof unless it is so constructed that this is obviously not feasible. Previous patches or repairs should be carefully checked for possible leakage. Potential points of leakage may be detected by observing if any light enters a closed unit. In carrying out this check, care should be taken to ensure that no person becomes locked inside a unit.

There was one incident in which employees locked a colleague inside a container during a check before a work break as a practical joke. During the break, the container was then removed. The employee, who suffered from claustrophobia, suffered a heart attack. Practical jokes of this kind must be severely sanctioned by the company.

> 2.2.2 A [container] should be free from major damage, with no broken flooring or protrusions such as nails, bolts, special fittings, etc. which could cause injury to persons or damage to the cargo.

> 2.2.3 Cargo tie-down cleats or rings, where provided, should be in good condition and well anchored. If heavy items of cargo are to be secured in a [container], the forwarder or shipping agent should be contacted for information about the cleat strength and appropriate action taken.

> 2.2.4 A [container] should be clean, dry and free of residue and persistent odours from previous cargo.

This paragraph in the container packing guidelines relates to the fitness for loading of cargo transport units. For further information on this subject, refer to the section on "Preparatory work":

> 2.2.5 A folding [container] with movable or removable main components should be correctly assembled. Care should be taken to ensure that removable parts not in use are packed and secured inside the unit.

This is particularly relevant to flat racks. Notices describing how to set up and fold these containers safely are generally attached to these containers. For further information, refer to the section "Other markings".

14.4 Before Packing

Section 3.1 of the container packing guidelines deals with actions required before packing. The following is an extract from the guidelines which relates to this specific paragraph:

> 3.1.1 Before packing a [container], careful consideration should be given as to how the unit will be presented during the packing operation. The same applies for unpacking. The [container] may be presented for packing or unpacking as follows: (1) loaded on a semi-trailer chassis together with a truck; (2) loaded on a semi-trailer chassis, but without a truck; (3) loaded on a rigid truck or chassis; (4) standing on the ground.
>
> – standing on its supporting legs (in case of class C swap-bodies); (5) loaded on a rail-car; (6) loaded on an inland barge; or (7) loaded on a seagoing vessel. Any of these configurations is possible. The actual packing or unpacking situation often depends on site and facility considerations. However, whenever the [container] is presented on a chassis or on supporting legs, special care should be taken in planning the packing or unpacking operation.
>
> 3.1.2 A [container] to be packed should rest on level and firm ground or on a trailer or a rail-car. If a [container] is on a trailer, care should be taken to ensure the trailer cannot tip while the container is being packed, especially if a lift truck is being used. If necessary, the trailer should be propped. Brakes should be securely applied, and the wheels chocked.

The container packing guidelines highlight particular hazards with illustrations and relevant comments.

14.5 Chassis Supports

It should be noted that the support legs of chassis are designed in such a way that they can carry the total permissible weight of the load or their own maximum payload. However, the support legs are unable to bear additional forces which may arise from, for example, driving into a container with a forklift truck. The container packing guidelines also provides a warning regarding inadequate support of semi-trailers (Fig. 14.3).

In this example (Fig. 14.13) there is a very real risk the pallet might collapse at the slightest horizontal movement of the semi-trailer. During loading and unloading, the position of the loading area in relation to the horizontal plane changes, as the vehicle rides up and down on the suspension (Fig. 14.4).

Thanks to the wheels and the handles, the support can be moved without a vast amount of effort. The wheels are effectively deactivated when the support is loaded

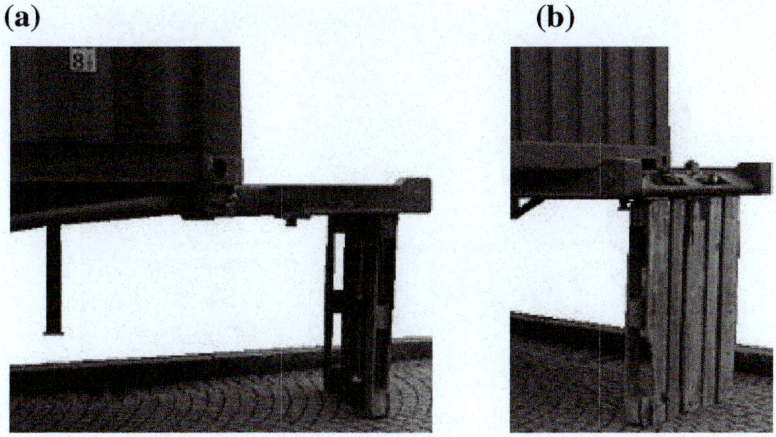

Fig. 14.3 Examples of dangerously inadequate support of a trailer chassis

Fig. 14.4 Example of a correctly supported chassis

against the spring and reactivated when the load is relieved. Approved supports are inspected and bear an appropriate approval plate with the necessary data on their permissible loading capacity. For reasons already given, the height of the chassis changes with the container during handling. The supports must therefore be regularly checked and adjusted. Any continuation of work under these conditions would represent a breach of regulations (Fig. 14.5).

> 3.1.3 When a swap-body standing on its supporting legs is packed, particular care should be taken to ensure that the swap-body does not tip when a lift truck is used for packing. It should be checked that the supporting legs of the swap-body rest firmly on the ground and cannot shift, slump or move when forces are exerted on the swap-body during packing.

Fig. 14.5 Photograph of a swap-body standing on its support legs

14.6 Swap-Bodies

A swap-body can be stored provisionally by stacking hardwood pallets with a sufficient loading capacity horizontally on top of each other or by using cross-beams made of squared lumber of a sufficiently large cross-section. These aids would also have to be checked regularly and their height adjusted. As this is only possible in practice if work is interrupted, such provisional solutions are not recommended. As a precautionary measure, swap-bodies should be provided with additional support if fork-lift trucks are to be driven into them. The container packing guidelines only advise that fork-lift trucks do not drive onto the swap-bodies too quickly. The picture below illustrates the danger of the legs bending. It is better to never drive onto a swap-body with fork-lift trucks—or even to move loads horizontally in other ways with heavy equipment, if the containers are only standing on their legs—but unfortunately this is common practice. Before driving onto a swap-body, it should always be checked that the supports are sound and are securely fastened. Here, not only is one "leg" deformed, but the diagonal support is even fastened with "ropes".

14.7 Driving into The Container

Devices for reducing the quantity of dangerous substances in exhaust gas include, for instance, exhaust purification systems with catalytic converters. For guidelines on the use of vehicles driven by Diesel engines, see the German *Technische Regeln für Gefahrstoffe TRGS 554* "Dieselmotoremissionen" (Technical Regulations for Hazardous Substances "Diesel engine emissions"). For information on securing the vehicle to be loaded or unloaded against unintentional movement, refer to Sect. 37 Paragraph 2 of the *German Accident Prevention Guidelines* (UVV) for vehicles or national equivalent.

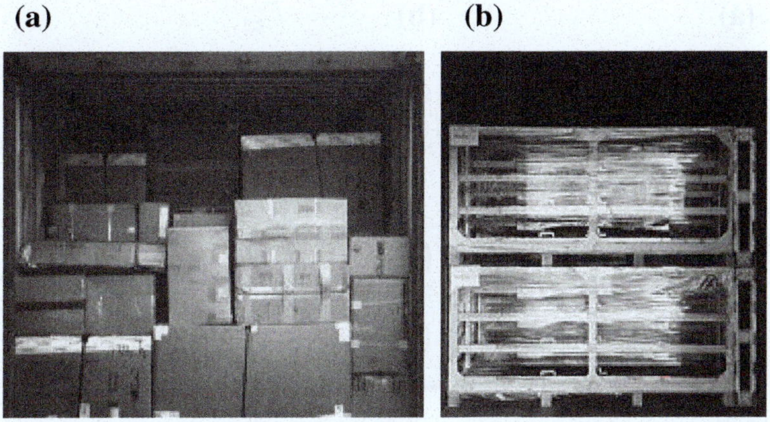

Fig. 14.6 Examples of not packed securely versus packed securely containers

14.8 Stowage Planning

Working and general stowage plans are prepared and used in normal maritime transport. The same practice should be adopted when packing other cargo transport units. The container packing guidelines even make this an obligation: Sect. 3.1 of the container packing guidelines deals with actions required before packing. The following are some of the points defined in the guidelines under Paragraph 3.1.4 (Fig. 14.6):

> Packing should be planned before it is started. This should make it possible to segregate incompatible cargoes and produce either a tight or secured stow, in which the compatibility of all items of cargo and the nature, i.e., type and strength, of any packages or packaging involved are taken into account. The possibility of cross-contamination by odour or dust, as well as physical or chemical compatibility, should be considered.

Furthermore, Paragraph 3.1.5 provides some fundamental guidelines as to basic stowage methods (Fig. 14.7):

> The planned cargo should not weigh more than the maximum payload of the [container]. In the case of containers, this ensures that the permitted maximum gross weight of the container (which includes the payload) marked on the CSC Safety Approval Plate will never be exceeded (see also annex 3). For [containers] not marked with their maximum permissible gross weight, tare weight or any other features, any of these values should be known before packing starts. According to CEN standards, a swap-body of class C (7.15 m to 7.82 m) will have a maximum gross mass of 16,000 kg and a swap-body of class A (12.0 m to 13.6 m) will have a gross mass of up to 32,000 kg.

Annex 3, which is referred to under this point, deals with the consequences of overloading containers. Precise planning of packing is indispensable. Planning becomes even more important when the amount of the load to be packed is sufficiently large, in that it ensures that the containers are utilised to their maximum in terms of space

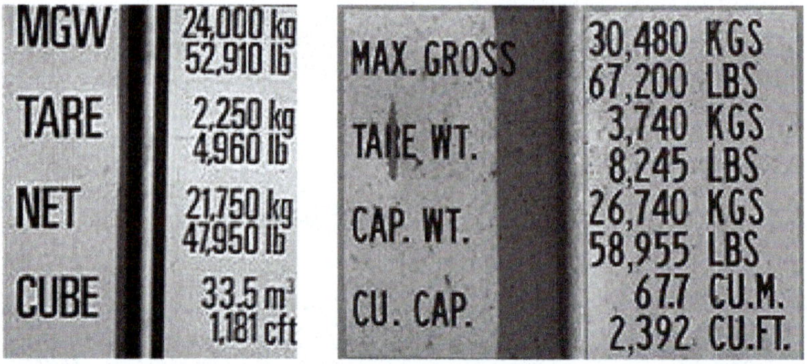

Fig. 14.7 Example of the typical information displayed on 20′ and 40′ containers

and weight. This is the only way to minimise transport costs. A detailed packing plan is indispensable for this. The volume-to-payload ratio of the containers can be determined from their volume and payload information, which can be obtained beforehand from prospectuses or documents provided by the container owners. The volume-to-payload ratios for the containers shown are calculated as follows:

$$20' \, container = 33.5 \, \text{m}^3 / 21.75 \, \text{t} = 1.54 \, \text{m}^3 / \text{t}$$
$$40' \, container = 67.7 \, \text{m}^3 / 26.74 \, \text{t} = 2.53 \, \text{m}^3 / \text{t}$$

This means that for every metric tonne of payload on the 20′ container, 1.54 m^3 of packing space is available, and on the 40′ container there is 2.53 m^3 available per metric tonne of cargo mass. The volume-to-payload ratio of a transport container therefore answers the question "how many cubic metres of stowage volume are available per metric tonne of payload?" The dimensions and masses of the items to be loaded can be used to determine the stowage factor of the load. Assuming that a consignment of 200 metric tonnes is to be loaded: the volume of the items is calculated to be 440 m^3. The net stowage factor of the cargo is calculated by dividing the cargo volume by the cargo mass. Net stowage factor of the cargo = 440 m^3/200 t = 2.2 m^3/t. This means that, if the load is packed with no wasted space, a volume of 2.2 m^3 would be required for every metric tonne of cargo. Since most loads are not 100% suitable for modular packing, certain losses will result from stowage or packing. Experienced practitioners can estimate these losses well or know from their own records what values are to be taken for calculation purposes. This information can also be obtained from relevant specialist literature. For our example, let us assume stowage loss of ten percent. The gross stowage factor will then not be 2.2 m^3/t but in fact 2.42 m^3/t. Some experts say, "cargo is stowed by a factor of 2.42". A comparison with the volume-to-payload ratios of the containers immediately reveals that economical transport can only be expected with the 40′ containers.

If a consignment of 300 m^3 volume and 230 t mass were waiting for loading and it were known that fifteen percent stowage loss is to be expected, the net stowage factor of the cargo would be 300 m^3/230 t = 1.30 m^3/t and the gross stowage factor 1.5 m^3/t. In this case, experts would say the "cargo is stowed by one and a half times" or the "cargo is stowed by a factor of 1.5". By comparing the numeric values, it can immediately be seen that only 20′ containers can be considered for economical transportation. Other aspects must of course still be considered, such as factors relevant to a particular transport route or to the destination countries, etc. These points are dealt with in the following and other sections of this book. Notes can also be found in the container packing guidelines under Paragraphs 3.1.6 and 3.1.7, respectively:

> Notwithstanding the foregoing, any height or weight limitation along the projected route that may be dictated by regulations or other circumstances (such as lifting gear, handling equipment, clearances and surface conditions) should be complied with. Such weight limits may be considerably less than the permitted gross weight already referred to.
>
> Stowage planning should take account of the fact that [containers] are generally designed and handled assuming the cargo to be evenly distributed over the entire floor area. Where substantial deviations from uniform packing occur, special advice for preferred packing should be sought.

This applies indirectly to the line load of containers. This is determined by dividing the payload of the containers by a calculation length. For 20′ containers, this calculation length is 2 m less than the container length, for 40′ containers 3 m less. It can also be expressed as follows: in order to obtain the calculation length, 1 m is subtracted from each end for 20′ containers and 1.5 m from each end for 40′ containers.

(a) Determining the calculation length for the line load for 20′ containers

−1.0 m	⇦	Residual length approximately 4 m	⇨	−1.0 m

(b) Determining the calculation length for the line load for 40′ containers

−1.5 m	⇦	Residual length approximately 9 m	⇨	−.5 m

The payload, maximum gross weight (MGW), net, capacity weight (CAP WT) or whatever is specified on the container as the permissible payload, is divided by the relevant calculation length. The result is the permissible line load.

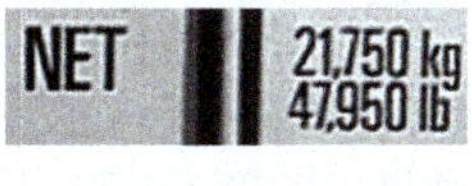

divided by	(a) 4 m	(b) 9 m	equals for the
	(a) 20′ container	(b) for the 40′ container	a permissible line load of:
	(a) 5,437.5 kg/m	(b) 2,971.1 kg/m	

It must be noted that the payload should actually be given in Newtons and the line load in Newtons per meter or daN/m or kN/m, if the line load is assumed to be a force. If the line load is regarded as a "line mass", the units kilogram per meter or metric tons per meter would be completely correct. For older 20′ containers with a payload of around 18,000 kg, the permissible line load is this around 4.5 t/m, while for older 40′ containers with an average payload of 27,000 kg it is −3.0 t/m. On more modern containers, the values are considerably higher: on some 20′ containers, as much as 6.75 t/m. If necessary, the current values must be determined for every loading operation:

> 3.1.8 When a heavy, indivisible load is to be shipped in a [container], due regard should be given to the localised weight-bearing capability of the unit. If necessary, the weight should be spread over a larger area than the actual bearing surface of the load, for example by use of properly secured baulks of timber. In such a case the method of securing the load should be planned before packing is started and any necessary preparations should be made.

For instance, a machine part weighing 20 t and 3 m long could not simply be loaded into the 20′ container described above without additional measures being taken. There would be a danger of it damaging the container floor. The machine part used as an example (20 t weight, 3 m long), exerts a load of 6.666 metric tonnes per running metre. This is calculated by dividing 20 t by 3 m. Since forces are involved, a correct calculation would have meant dividing the normal force of 19,620 N by the length of 3 m. This would have produced the correct values of 65,400 N/running metre, 6,540 daN/running metre or 65.4 kN/running metre. Comparison with the permissible line load of the container of 5,437.5 kg/m (5,334.2 daN/m or 53.342 kN/m) immediately reveals that loading cannot be carried out unless additional measures are taken. In this example, this means that the weight of the case has to be distributed over a larger number of bottom cross members. This can be achieved by laying squared lumber longitudinally across them. The required length of the squared lumber is determined by dividing the weight of the case (or more correctly the "normal force") by the permissible line load of the container. In the example, this would be 20,000 kg divided by 5,437.5 kg/m = 3.68 m (19,620 daN/5,334.1875 daN/m = 3.68 m). An older model of container would not have been permitted to be used for transport, due to its low permissible payload of 18,000 kg. A more modern model of container with a payload of 24,000 kg has a line load of 6 t/m. The box could be loaded into a container of this kind without requiring any further measures. If the box is to be loaded with very light but bulky items in the 40′ container in the example, the following calculation would need to be made: 20,000 kg/2,971.1 kg/m = 6.73 m. In the 40′ container described; the box would have to be placed on squared lumber which was at least 6.73 m long. As far as practical considerations are concerned, it should be noted that where possible squared lumber with a large vertical edge length should be used, e.g., 8 cm × 16 cm or 10 cm × 20 cm, etc. Half-sizes of this kind are generally less expensive. However, since there is a danger of them collapsing, they must be braced firmly laterally to prevent tilting. Preferably, this should result in a wooden lattice with a large height.

> 3.1.9 If the planned cargo of an "open topped" or "open sided" [container] will project beyond the overall dimensions of the unit, special arrangements should be made. It should be borne in mind that road traffic regulations may not allow such overhangs. Furthermore, [containers] are often loaded door-to-door and side by side, thus not permitting any overhang.

It can clearly be seen from this photographic representation (Figs. 14.8 and 14.9) that three slot rates must be paid for transport, since the two adjacent slots on the ship must be kept free. If the centre of gravity of the flat rack makes it technically possible to load it with an overhang on one side and it is also possible to load it in this state on the ship—after agreeing this with the shipping company—, only two slot rates will have to be paid. Over width road vehicles must be marked with red and white, diagonally striped plates. On board the ship, over height flat racks can only be stowed as the uppermost container either in the hold or on deck. Ships with flaps can segregate individual stowage areas from others and could also store the flat rack in another position. As one slot must be kept free above the flat rack, two slot rates must also always be paid (Figs. 14.10, 14.11 and 14.12):

> 3.1.10 The centre of gravity of the packed cargo should be at or near the longitudinal centre-line of the [container] and below half the height of the cargo space of the unit. (See also 3.2.5 and other relevant sections.)

> 3.2.5 The weight of the cargo should be evenly distributed over the floor of a container. Where cargo items of a varying weight are to be packed into a container or where a container will not be full (either because of insufficient cargo or because the maximum weight allowed will be reached before the container is full), the stow should be so arranged and secured that

Fig. 14.8 Photographic representation of a flat rack with overhang on both sides in the cell guides of a ship

Fig. 14.9 Photographic representation of a flat rack with overhang on both sides in the cell guides of a ship

Fig. 14.10 Photograph of an over height flat rack

Fig. 14.11 Example of the prescribed centre of gravity for loaded containers

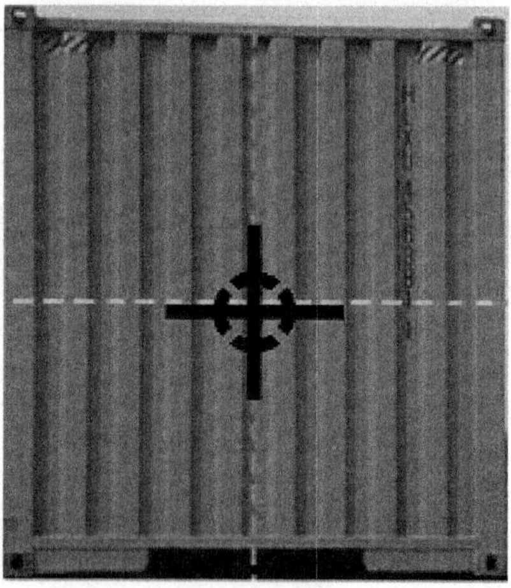

200 kg	200 kg	200 kg	200 kg	200 kg	200 kg	200 kg	200 kg	200 kg
200 kg	200 kg	200 kg	200 kg	200 kg	200 kg	200 kg	200 kg	200 kg
200 kg	200 kg	200 kg	200 kg	200 kg	200 kg	200 kg	200 kg	200 kg

Fig. 14.12 Ideal load distribution in a container with a payload of 18,000 kg * *not to scale*

> the approximate centre of gravity of the cargo is close to the mid-length of the container. If it is not, then special handling of the container may be necessary. In no case should more than 60% of the load be concentrated in less than half of the length of a container measured from one end. For vehicles, special attention should be paid to axle loads.

If necessary, the real centre of gravity of the container must be determined by calculating its moment.

> 3.1.11 When planning the packing of a [container], consideration should be given to potential problems which may be created for those who unpack it, e.g., cargo falling when doors are opened.

In the real world, dreadful incidents have occurred. To cite only one accident here: cotton bales weighing around 220 kg falling out onto a worker injured him so severely that he suffered serious fractures to his spine and hip in addition to internal injuries to vital organs. The young man only survived thanks to immediate medical attention and a number of operations. For the rest of his life, he is now confined to a wheelchair. And all this simply because the cotton bales in the container were badly packed and not secured.

> 3.1.12 Before a [container] is packed, it should be ensured that the personnel responsible for the packing are fully informed about all the risks and dangers involved. As a minimum requirement some sketches showing the basic rules of [container] packing should be available. The present Guidelines should also be readily available. If necessary, the shipper and the packing personnel should consult each other regarding any special feature of the cargo to be packed into the units. In particular, information on possible dangerous cargoes should be considered very carefully. Consideration should also be given to the provision of appropriate training for personnel involved in packing [containers].

Data sheets with a plan view and/or side view of different containers, in which the items to be packed are drawn in to scale, should be provided for planning packing (Figs. 14.13, 14.14 and 14.15).

The actual design of the packing plans is of little significance. What is crucial is that they contain all the most important information, e.g., the interior dimensions of the container, and that the stowage plan can also provide important evidence in the event of any subsequent (legal) disputes. When planning packing and illustrating it in the forms, the following points must be considered: How are an even load along the floor and an acceptable centre of gravity to be achieved? What goods will have

Fig. 14.13 Graphic of a blank form for a packing plan

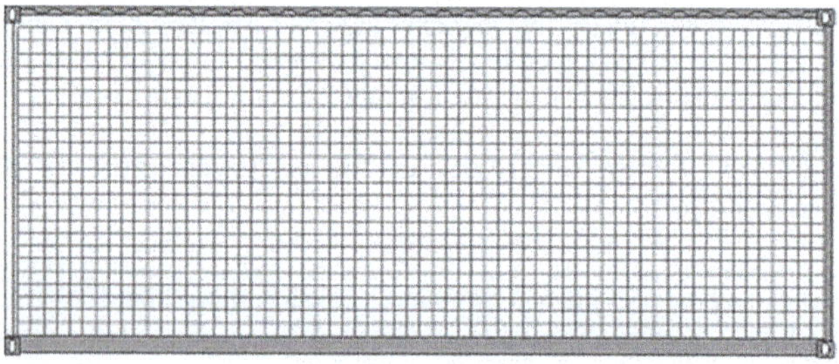

Fig. 14.14 Graphic of a blank form for a packing plan with dm^2 grid

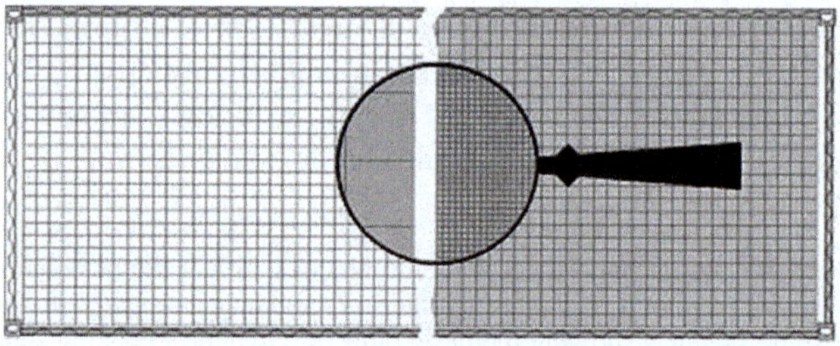

Fig. 14.15 Graphic of a blank form for a packing plan with dm^2 or cm^2 grid

to be brought in using a fork-lift truck? Will any other ground conveyors need to be used? Is manual handling necessary—or even desired? Do all items have a sufficient loading capacity? Can they be stacked? Will items have to be packed on top? Can the necessary basic stowage rules be met for all items? Are dangerous cargoes also to be loaded? Has unloading also been considered? And so on. It should be checked whether the stowage factor of the cargo has previously been determined using the weights and dimensions of the cargo and compared with the volume-to-payload ratio of the container. This is always good practice; moreover, this procedure gives all less experienced staff an insight into what is feasible. It goes without saying that the maximum payload of the container must not be exceeded.

> 3.1.13 When packing a [container], the shipper and persons responsible for packing should bear in mind that any failure to pack and secure the cargo correctly may result in additional costs that they will have to bear. If, for example in railway transport, a unit is found not to be properly packed and secured, the rail-car may be marshalled out of the train into a siding and the transport can only be continued once the cargo has been properly secured. The shipper may have to pay for this work, especially for the repacking and re-securing operation, as well as for the additional time during which the rail-car has been used. In addition, he may be held responsible for any delay of the transport operation.

> 3.1.14 Not all handling equipment is suitable for container packing. Lift trucks used for container packing and unpacking should have a short lifting mast and a low driver's overhead guard. If the lift truck operates inside the container, equipment with electric power supply should be used. Container floors are built to withstand a maximum wheel pressure corresponding to an axle load of a lift truck of 5,480 kg or 2,730 kg per wheel. Such an axle load is usually found on lift trucks with a lifting capacity of 2.5 t.

The short lifting mast and low driver's overhead guard alone do not say anything about the usefulness of the fork-lift truck. If these values are less than the height of the door, the fork-lift truck container can de driven into the container, but it is not automatically clear whether load can be stacked. This is where free lift is crucial. The higher the forks can be lifted without changing the total height, the more suited the fork-lift truck is for stacking and working in the container. More information is provided on axle loads and working with fork-lift trucks in other sections of this book.

> 3.1.15 If the [container] floor is at a different height level than the loading ramp, a bridging unit may need to be used. This may result in sharp bends between the loading ramp and the bridging unit as well as between the bridging unit and the [container] floor. In such cases the lift truck used should have sufficient ground clearance to ensure that the chassis does not touch the ramp when passing these bends.

This chapter has discussed some of the key points goods packers are expected to observe when packing goods into containers intended for transport. We also looked at some of the stowage methods that may be used to good effect. Needless to say, even the best planned and executed packing operation will lead to void space which, when considering the various forces that are exerted on goods during road, rail and sea transport, can cause goods to shift inside the confines of the container. To prevent this from happening, various forms of segregation, dunnage and stowage techniques can be employed. We will examine some of these in the following chapter.

Chapter 15
Segregation, Dunnage and Stowage

In this chapter we will discuss some of the key points associated with the segregation of cargoes inside containers, as well as the use of dunnage and various stowage techniques. It is well worth mentioning at the beginning that cargo stowed in containers must be secured on all sides to prevent shifting, in particular towards the door. This means spaces between packages and/or container walls must be filled. It is always advisable to secure the cargo by using the lashing points provided on the top and bottom side rails and on the corner posts of the container. Ideally, containers should be packed, and the cargo secured in accordance with guidelines for packing cargoes other than bulk cargo in or on containers for carriage by any means of transport at sea and ashore. The container packing guidelines, published by the IMO and ILO, crucial to establishing and maintaining good container packing practices.

15.1 Using Segregating Materials

The term "segregation" is used to describe the visible segregation of packages or batches of cargo. It can be useful to segregate loads in a transport container which are of the same type but destined for different recipients if the goods could otherwise be confused or subsequent sorting work might be required. Sorting involves considerable time outlay which can lead to additional costs. Segregating goods in good time is therefore critical for cost-effective delivery of goods to their final destination or to interim destinations. It is thus of great economic significance for conventional maritime transport and for storage. Segregation plays a less important role in the transport of cargo with containers and other cargo transport units. Segregation is unnecessary if there is no danger of goods being confused because they have different package dimensions, a different shape or colour or other different characteristics. Transport law stipulations require that the consignor, carrier or shipper of a cargo provide the goods for delivery in a suitable condition, meaning that the goods must

A. A. Olsen and F. Karkori, *Containerized Cargo Handling and Stowage*, Springer Series on Naval Architecture, Marine Engineering, Shipbuilding and Shipping 22,
https://doi.org/10.1007/978-3-031-63329-4_15

be both complete and in an undamaged condition. Such legislation may lead to an obligation to segregate the goods. In principle, two different methods are used for segregation:

- Spatial segregation; and
- Optical segregation using appropriate materials.

Spatial segregation of batches of cargo or individual packages is relatively easy to carry out on shore. On open spaces, and in sheds or warehouses, the storage spaces are indicated by appropriate location specifications, chalk marking or similar markings. For domestic transport, spatial segregation is achieved by delivering the different batches in separate vehicles or freight cars or by packing them separately in such vehicles or cars. Virtually the same principle applies for container cargoes. The following requirements must be met when segregating cargo batches of the same kind or with the same appearance using segregation materials: the materials used must not cause any damage to the goods or their packaging, either by depreciation of their sales value or as a result of loss or shrinkage; marking of the goods must not be rendered illegible by segregation; and segregation must be clearly recognisable. With the exception of talcum, all materials described in the section "Dunnage" are used as segregation materials, e.g., wooden boards, wooden panels and walking boards, squared lumber, planks, boards and wooden dunnage, sailcloth, canvas and tarpaulins, reed, rattan or bamboo mats, films, cardboard and paper etc. The properties of these materials and their benefits and disadvantages are also described in the section "Dunnage" and will therefore not be discussed further here. Materials which are only used for segregation are:

- Thin synthetic plastic nets or cargo nets.
- Ropes, cloth strips or jute fabric strips; and
- Paints, marking pens, etc.

Thin synthetic plastic nets are relatively inexpensive. Segregation with these nets is easily recognised and does not cause any loss of stowage space. In addition, they are easy to transport and store, clean and reuse. Another benefit is that they do not hinder any necessary ventilation measures. A disadvantage is that they pose a hazard if anyone becomes entangled in them. To a certain extent, they are a trip hazard. They cannot be used to segregate bulk cargo. Much the same applies to **cargo nets** as to thin nets. However, they are comparatively more expensive, more bulky and very difficult to free from odour infection. Under certain circumstances, they can cause pressure marks on the cargo. **Ropes** are suitable for segregating long cargo items, such as structural welded steel mesh, round wood and cut lumber, TOR steel, etc., provided there is no risk of pressure marks. Fibre ropes are preferable to wire ropes, since the latter can cause rust damage. Impregnated ropes could cause a cargo to be infected by odours. **Strips made of various materials or adhesive strips** are preferable to ropes, as they do not cause pressure marks and lead to smaller cavities and spaces. They are also generally cheaper. Cloth strips or jute fabric strips should never be used for fatty organic cargoes, as they could encourage or cause the cargo to ignite spontaneously. Marking pens and paints can be used to indicate packages which belong together or the same batches, by means of clear signatures, colour codes or stripes. They are only suitable for segregating goods if they do not restrict the further processing and actual marking of these, and if value of the goods does not depreciate as a result of any segregation indications. For instance, descaled metal sheets must not be marked with wax crayon, as otherwise it will subsequently be impossible to coat them without additional work. In the case of soft or absorbent woods, partial or total damage could be caused by using e.g., water-based paints. Sufficiently heavy-duty and dense materials should be used to segregate bulk goods. Otherwise, parts of the various batches could become mixed if the segregation materials were to tear or develop leaks. Covers made of sailcloth, tarpaulin, canvas or strong plastic sheeting are suitable. The materials used to segregate grains and foodstuffs must be non-toxic and neutral in odour.

15.2 Dunnage

Dunnage materials or, more simply, dunnage, is the name used for all materials which are not firmly attached to the cargo transport unit, the cargo or its packaging, and are used to protect the load. This includes wooden dunnage, beams, planks, boards, wedges, plywood and hardboard, walking boards, mats, paper, sailcloth, canvas and tarpaulins; plastic and metal sheets, spray covers; cardboard and paper board, packing paper, oiled paper and fabric paper, talcum powder etc. Whether a material is to be classed as dunnage, segregation or load securing material does not depend on the material itself but on its application. The main function of dunnage is:

- Protection against sweat and condensation water.
- Protection against moisture and liquids.
- Protection against soiling and contamination; and
- Protection against mechanical damage.

Sweat water can occur as container sweat or cargo sweat. It is primarily the roof surfaces that are susceptible to container sweat with the sides and the floor of the container only being a secondary source of problems. Top dunnage and side dunnage can absorb sweat dripping from the container roof until they reach saturation point. This can sometimes significantly delay, or even completely prevent, moisture damage. Dunnage cannot protect against cargo sweat, although this is rarer when goods are transported in containers. To ensure protection against moisture and liquids, cargoes which are at risk and sensitive parts of the means of transport must be shielded. A check must be made for any possible leakage in the means of transport or other cargo items. Potential causes of this could be leaky walls, doors and roofs on the cargo transport units, or damaged drums, cans and other similar packaging receptacles for “wet cargoes”. The effects of wet or very moist load securing materials must also be considered. Protection against soiling and contamination must be ensured. This can be caused by the presence of contaminating goods or contaminated components on the cargo transport units. Contaminating goods must also be enclosed by dunnage to prevent them from damaging other goods or parts of the means of transport. Preventive measures are always needed to protect against mechanical damage, if contact with components on the cargo transport unit itself or other loads or load items could cause mechanical damage. For instance, it can be necessary to adequately protect the load against contact with corrugations, corner posts, doors, lashing points, stanchions, edges and corners of other cargo items, load securing materials, protruding nails, screws, etc. Appropriate dunnage must also be used to reduce the effects of harmful pressure forces. The nature of the materials used must ensure that the packages or the cargo transport unit are adequately protected. The materials themselves must not be wet, moist, odour-tainted or contaminated. The quality of the material must determine for each individual case by assessing the value and susceptibility to damage of the load, the means of transport, the intended protection and other similar factors. The special properties of the particular type of dunnage used must be considered.

Wooden dunnage made of squared lumber, planks, boards, slats and battens, are very suitable for distributing pressure, bridging or lining cavities and gaps, creating air channels and enhancing friction. If wooden dunnage is laid too tightly, it impedes air circulation (this is a particular consideration for ventilated containers), causes a significant loss of stowage space and thus increases costs. Large spaces promote air circulation but can cause mechanical damage to the packages if the load-bearing area is too small. Wood easily absorbs moisture and releases it into the environment at a low vapour pressure. If it has a strong intrinsic odour or has taken up a foreign odour, it must not be used with odour-sensitive loads. Since wood can be infested by pests, only wood which has been impregnated against insects or fumigated may be used on certain trade routes. To distinguish such treated wood from normal wood, it must be appropriately stained or marked in some other way. Wooden boards, wooden panels, walking boards, chipboard and hard boards are very suitable for distributing pressure and have a high loading capacity provided they are sufficiently thick. Since water-repellent, pest-proof or fire-retarding wood is odour-tainted as a result of such treatment or could have other harmful effects, it must virtually never be used with foodstuffs or similar items. Mats made of rattan, bamboo, reed etc. only protect against surface dirt, not against dust. They absorb moisture and under certain circumstances release it back into the atmosphere. This can protect adjacent loads. Mats are easy to put in place and roll back up again, and do not cause any significant loss of stowage space. Provided a container is actively ventilated, mats can release back into the atmosphere any sweat or drops of water they have absorbed. Canvas has a water-repellent impregnation and therefore not suitable for odour-sensitive goods. They are however dust-tight and impede air circulation. Tarpaulins are watertight, dust-tight and airtight, and very expensive. They must be handled with care if a long life is to be expected. It is very rare that tarpaulins will not be needed as a cover to protect against moisture and similar influences. Jute coverings protect against surface dirt but not dust. They are breathable and allow a minimal amount of ventilation. They pass on absorbed moisture to the wrapped or surrounding cargo items. When used with fatty organic goods, jute coverings can spontaneously ignite or promote the spontaneous combustion of other goods. Sailcloth has similar properties to canvas.

Plastic sheeting is available in a number of different thicknesses. It is neutral in odour, inexpensive and generally acid-proof and alkali-proof. It is watertight, dust-tight and airtight but only impervious to water vapour to a limited degree. It has a low loading capacity if it is not sufficiently thick. Composite sheeting is often very heavy-duty but comparatively expensive. Paper, e.g., normal craft paper, is inexpensive and dust-tight, but very sensitive to moisture. Fabric paper has a higher strength whereas oiled paper is watertight and tissue paper is acid-free. Dispersible material such as talcum powder and similar materials can be used to prevent cargoes from caking together.

Depending on their *use and application*, a distinction should be made between floor dunnage, interlayer dunnage, top dunnage, and side dunnage. **Floor dunnage** is used for the following tasks:

- Dissipating and localising sweat and moisture.
- Protecting the load from moisture and dampness on moist or wet stowage surfaces or stowage spaces.
- Dissipating or restraining moisture leaking out of the load, to protect other packages or the cargo transport unit.
- Protecting the load from soiling, contamination or mechanical damage from the stowage surface or stowage spaces; and
- Protecting the cargo transport unit against negative influences from the goods.

There are no generally binding regulations for laying dunnage. The condition, properties and intended protective action of the available materials and similar considerations must be considered. In many cases, only compromises can be considered, since some fundamental requirements can contradict each other; especially if materials are to assume a number of protective functions simultaneously. The following guidelines are intended to help make practical decisions. Since not all variants can be shown, the action and decisions taken will need to reflect the given circumstances. How to lay dunnage most appropriately and what materials are to be used depends on the stowage space, the structural design of the cargo transport unit and the restrictions anticipated during transport. These include the climatic zones through which the cargo will travel, changes in weather, storage duration and voyage duration, anticipated mechanical, climatic and biotic shipping stresses etc. Wooden dunnage is differentiated according to the quantity and thickness of the wood used and the method used to lay it. For example, simple wooden dunnage comprises boards laid out side by side whereas double dunnage or criss-cross dunnage comprises boards laid at right angles to or obliquely over each other and raised dunnage can comprise planks or squared lumber laid in single layers, or a combination of boards, planks or beams laid at right angles to or obliquely over each other. Another good option is to use pallets as stowage surfaces.

The orientation of and space between the wood used depends on the intended protective effect and the nature of the means of transport or the stowage space. If the main aim is to dissipate moisture or damp, the wood at the bottom must be laid in such a way that water and other fluids are conducted along the quickest route outside or to any available discharge outlets and do not produce any accumulation of damp. It is not practicable to dissipate liquid to the outside in standard containers which have not been structurally modified, since these containers are spray-tight and thus generally prevent water from flowing out freely. If actively or passively ventilated containers can be used, criss-cross dunnage can enhance the ventilation effect. When laying out dunnage, it should be remembered that ships are not constantly on an even keel. They can be trimmed by the stern or trimmed by the bow. Rolling and pitching can cause the deck to slope. These slopes, as well as any special factors in the ship's construction, must be taken into consideration when laying out wooden dunnage. With open containers, dunnage can be laid out transversely. Liquids can immediately flow out when the ship rolls. If the ends of the wooden boards are taken right up to the side rails, load forces are distributed appropriately at the same time. It should be remembered that closed containers are spray-tight and that dissipation of moisture to the outside only happens to a limited degree and then only slowly. If a good distribution of load forces is the primary consideration, the wood at the bottom must be laid out in such a way that as many load-bearing components of the cargo transport unit are covered as possible. If good moisture dissipation is as important a consideration as a good distribution of load forces, the best possible compromise must be reached, for instance by laying out the base wood diagonally. If the most important consideration is to protect against mechanical damage, it is essential that potential shifting or movement resulting from shipping stresses and typical for the particular means of transport are considered. If goods which are sensitive to dirt, moisture or mechanical damage are loaded, additional materials such as mats, jute cloth, paper, sheeting etc. are to be used as required.

If there is a danger of dampness or contamination of the packaging spreading to the contents, dunnage must be laid with special care. If plastic sheeting and similar materials are used to prevent dampness from penetrating the cargo, their orientation—i.e., the way they are arranged—plays an important role. In these cases, consideration must be given to where damp could enter and where it can be dissipated. Under no circumstances must "water pockets" be allowed to form, which would enable collected damp to enter the cargo. Dunnage must first be laid out at the point where most water is able to penetrate. Laying out floor dunnage to protect from damp in containers will always be necessary when "wet goods" are stowed together with moisture-sensitive goods. Occasionally, criss-cross dunnage and double dunnage will be necessary. Damp must be prevented from infiltrating sensitive cargoes by laying dunnage at appropriate heights and using additional material, e.g., plastic sheeting. In box containers, it is generally more appropriate to lay out wooden dunnage in a longitudinal direction, because of the distribution of load forces and the use of ground conveyors. On open containers, it should be laid out transversely. In both cases, the load-bearing components must be bridged for heavy packages so that the permissible line load is not exceeded. On open chassis and trailers, wood must be laid out to distribute load forces in such a way that the load is taken by the vehicle's chassis. Walking boards made of glued-laminated plywood at a size of 1 m $\times$ 2.5 m, for instance, or wooden boards of a suitable length and adequate dimensions are very suitable for this. Floor dunnage made of wood or plastic sheeting is used to prevent contamination. Since it is possible that cargo transport units can be transported on deck, sufficient protection against damp must be ensured. When distributing load forces, the vehicle chassis and cross members must be covered.

Floor dunnage is very often not needed for freight carriages with a wooden floor. If there is a risk of damp infiltration with moisture-sensitive cargoes, appropriate measures must be taken. On covered freight carriages, the preferred orientation of the wood is longitudinal whereas in the door area it is transverse. On open freight cars and freight cars with sliding walls, wood is preferably to be laid out transversely for loading and unloading with ground conveyors. Exceptions to this must be made when end-loading platforms are used. The design of the chassis and of the side rails and cross members must be considered with regard to the distribution of load forces. Here, inexpensive materials such as paper, plastic sheeting and so forth are often sufficient for protection against contamination. What has been said about freight carriages applies in principle to cargo transport units for road transport as well. On box bodies, however, wood should always be laid longitudinally where possible, for handling reasons. To be able to effectively protect cargo before packing and after unpacking cargo transport units, dunnage can also be required in stores, sheds, warehouses, hangars and open areas. This can be for levelling out storage areas, protecting against damp and dirt or for distributing load forces. The orientation should be suited to the transport paths and the equipment used. The materials used must be able to provide the necessary protection. Especially on open areas, weather changes such as rain, snow, and ice, etc., must be taken into full account. Other factors to consider are bacteria on the ground and rot pathogens which could damage the cargo.

Side dunnage has similar functions to floor dunnage in that it is intended to afford protection against damp or moisture penetrating from the side as a result of sweat or leakage from other cargo, but also to provide protection of other cargo or of the means of transport against liquids escaping from the side of a load. Side dunnage also offers protection against soiling and contamination on the sides, both of a single load against other loads or from parts of the means of transport or protecting the latter against a specific load and protection against mechanical damage or other damage, such as chafing, tearing, chemical reactions, etc., both with regard to a specific cargo and to other cargoes or parts of the means of transport. The same materials are used for lateral dunnage as for floor dunnage. The use of wooden dunnage, however, is less significant. To protect against damp or moisture or harmful dust penetrating from the sides, watertight and dust-tight materials such as plastic sheeting, oiled paper and tarpaulins are to be used. Under certain conditions, canvas and mats should also be used, but never wrappers or paper. Sweat formation always occurs when temperature differentials are significant, either with a positive temperature difference against the load or a negative temperature difference against components of the means of transport. For further information on this, see the section on "Sources of sweat". The risk of sweat is higher on cargoes with a high-water content. If there is a risk of cargo sweat and the cargo is sensitive to moisture, it is urgent that a corrosion protection method is used. Wrapping cargo in plastic sheeting can have the opposite effect to that which is desired. Since hot air rises, water vapour can condense under the sheeting. Practical experience shows that wooden dunnage boards, wooden boards, mats, etc., provide good protection against mechanical damage. Whether or not it is necessary to use additional materials depends on the particular circumstances.

Top dunnage is laid to prevent the consequences of leakage, sweat formation, bad weather, vertical shipping loads, etc., which could damage a particular cargo or the cargo transport unit and its components. The main functions of top dunnage are to protect against damp or moisture penetrating from the top or escaping upward; to protect against dust or substances which threaten contamination or chemical reactions, either penetrating from the top or escaping upward; and to protect against mechanical damage anticipated to arise from above or acting in an upward direction. All previously described materials can be used as top dunnage materials. To protect against damp or moisture penetrating from the top, or to prevent moisture from being allowed to permeate upward, plastic sheeting, oiled paper, tarpaulins and other watertight materials have proved themselves in practice. If only small quantities of moisture are anticipated, the best materials to use are ones which can absorb small quantities of water. Since the weight increases when these properties are used, they must be attached carefully. When using watertight materials, it is essential that liquid which has collected in the depressions (water pockets) is prevented from flowing into the cargo when the vehicle moves. For this reason, plastic sheets and the like must overlap each other properly otherwise wood, cargo items or the stowage method must be used to create an appropriate slope. Respiring organic goods, such as tobacco, coffee, cocoa and fish meal, which are transported in passively ventilated containers, must not be covered with airtight materials or materials impervious to water vapour. If top dunnage is used to protect against contamination or to prevent harmful chemical reactions, the same principles apply as for protecting against damp, but the materials must be dust-tight and chemically compatible. A certain degree of air circulation must also be ensured here occasionally.

Interlayer dunnage variously has the functions of floor dunnage, side dunnage or top dunnage, depending on how it is arranged. It is laid between two different batches or individual goods in a batch or load type. In some cases, it simultaneously helps secure the load by increasing friction, improving the distribution of load forces or improving the stability of the load. All materials which have been described can be used as well as the addition of powdery substances such as talcum powder. These are introduced in the form of an intermediate layer or are sprinkled between layers, to prevent loads from sticking together, for example, bales of raw rubber. Interlayer dunnage comprising of mats provides adequate firmness and protects against mechanical damage. Further measures should be taken as well (Fig. 15.1).

15.3 Basic Stowage Methods

Stowage instructions and basic stowage methods. To be able to provide and follow appropriate instructions, it is necessary to be familiar with some of the fundamental concepts and principles which apply in this field. To ensure that they are well understood, some of them will be explained here. "*Stowage for ready access*" or "*Packing*

Fig. 15.1 Photograph of interlayer dunnage on a load of pails

for ready access" means that a load must be immediately accessible. Ease of access may apply to a specific (intermediate) destination, i.e., that a given load must be readily accessible in, for instance, Birmingham and can be unloaded without the need to move other loads in Southampton. To this end, Sect. 4.3.7 of the container guidelines states (Fig. 15.2):

> Dangerous cargoes consignments which form only part of the load of a [container] should, whenever possible, be packed adjacent to the doors with markings and labels visible. Particular attention is drawn to 3.3.1 concerning the securing of cargo by the doors of a unit.

Common Agricultural Policy (CAP) products from within the European Union which are subject to customs checks should also be stowed for ready access, so that customs officers have immediate access to the goods. This avoids the need to re-stow. "***Top stowage***" (occasionally referred to as "***on-top-stowage")*** means no further items are stowed on top of the load. This instruction is not to ensure that a batch should be readily accessible but is issued because packages are particularly sensitive or fragile. If stowage of this type is required, appropriate indications must be attached to the

Fig. 15.2 Example of cases with class 1.4 G air bags, stowed for ready access

Fig. 15.3 Example of correct stowage suitable for handling with forklift trucks in a box container

packages. Written instructions such as "Do not over-stow", "Please load on top" etc. are not necessarily universally understood. Sensible, comprehensible markings can be more effective. The original ISO symbol should be known universally. It must therefore be assumed that this symbol is the most likely to be understood. "***Stowing or packing correctly for a fork-lift truck or a crane***" means that appropriate arrangements must be made to ensure that the goods can be lifted by a ground conveyor or lifting gear without requiring special preparation and without any time delay. The dimensions, strength and loading capacity of the aids used, such as squared lumber, wooden dunnage boards, etc., must ensure that the goods do not suffer any damage during shipping. If necessary, stowage surfaces must be prepared with walking boards. Cargo items with a high loading capacity or other usable aids may have boards placed on top so that work can be carried out safely and the batches that have already been loaded can be over stowed with other loads (Fig. 15.3).

Bottom and intermediate dunnage must be arranged so that forks, strops, chains, claws and similar cargo handling equipment and slinging equipment can be used without any problems. ***Tier*** means a layer or stack. It can refer to either horizontal layers of cargo or cargo items stowed vertically one above the other. If circumstances do not make it clear from the outset what type of layer is meant, it makes sense to distinguish between a ***vertical tier*** and a ***horizontal tier***. The stowage position on board container ships is generally documented according to the bay-row-tier system or the bay-tier-row system. In this specific case, "tier" designates the horizontal layers of containers. Layers are counted from the bottom to the top (Fig. 15.4).

Loaded upright or loaded on its side relates to the shape of the consignment items. In most cases, the meaning is clear:

The same applies for wheel rims, pipes, steel bars, narrow wire rod coils and other similar loads. It is also rare for these to be confused. With cylindrical, or roller-shaped goods, the expression "upright" should really be used uniformly. Such cargo items are generally understood to be "upright" if.

- The axis is vertical, and the length of the axis is greater than the diameter; or
- The axis is horizontal, and the length of the axis is less than the diameter.

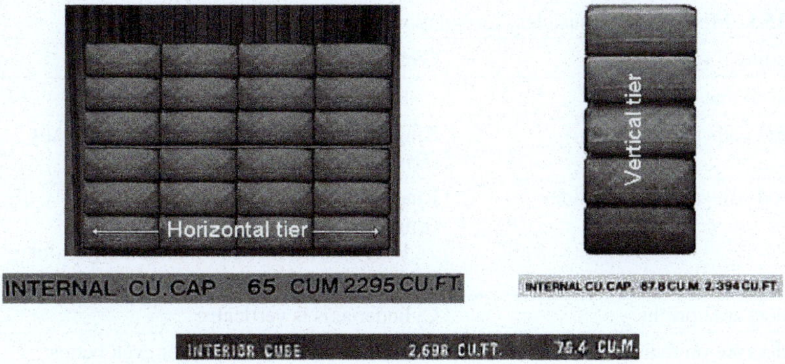

Fig. 15.4 Example of horizontal and vertical tiers

The expression "on its side" also has a standard meaning for the same goods. By definition, cargo is "on its side" if:

- The axis is horizontal, and the length of the axis is greater than the diameter; or
- The axis is vertical, and the length of the axis is less than the diameter.

Differences in understanding of these terms can lead to errors and misinterpretations and thus incorrect stowage. Another problem arises if the length of the axis and the diameter are almost the same size. This can easily lead to misinterpretations of the terms in practice, for example:

- Is the coil upright or on its side?
- Is the roll upright or on its side?

Incorrect expressions in the loading instructions or an incorrect interpretation of correct instructions frequently leads to damaged cargo. Such damage can be avoided if a reference is made in the stowage instructions to the orientation of the axis. Instead of instructions which read "load rolls upright" or "load rolls on their sides", the expressions "roll axis vertical" ("eye to the sky") or "roll axis horizontal and transverse" or "roll axis horizontal and longitudinal" should be used instead (Table 15.1).

As already mentioned, the terms "upright" and "on its side" are clear for most cargo items such as billets, pipes, profiles, bars, etc., but it can be necessary in certain cases to provide or request more precise stowage instructions, such as "lying on its side". To prevent subsequent claims, it is important to always ask if anything is unclear. It is more cost-effective to obtain precise instructions from a superior or to make a telephone call to the customer than to cause reputation loss by having to answer for a claim for damages. In most cases, it is sufficient to use two arrows pointing upward to indicate that goods are to be transported in a particular direction (i.e., with the top of the cargo facing upwards and the base at the bottom). Stacking means loading, packing or stowing packages in layers so that the edges of the packages lie flush above each other. Pressure-sensitive cartons with homogeneous dimensions must always

Table 15.1 Misleading versus clear formulation

Misleading formulation	Clear formulation
... load so that	... load so that
Wire rod coils are upright	Ring axis/winding axis positioned vertically (English: "eye to the sky")
Wire rod coils are on their sides	Ring axis/winding axis horizontal (English: "eye to lie"); or even better: ... Ring axis horizontal longitudinally ... or horizontal transversely
Cylinders are upright	Cylinder axis is vertical
Cylinders are on their sides	Cylinder axis is horizontal; or even better: Cylinder axis is horizontal and longitudinal Cylinder axis is horizontal and transverse

be stacked carefully ensuring that the sides and end faces lie flush. This ensures that only the load-carrying vertical sections are subjected to loads. When using cartons made of corrugated board, it is essential to check, where possible, that the flutes of the boards are vertical. If it could be disadvantageous that individual packages are not firmly connected to each other, anchors made of materials with a large surface area, such as hardboard, paper board, and plastic sheeting, are to be introduced during stowage. Wall boards and ceiling boards, plastic profiles and similar loads generally have single or double belt battens as "built-in packaging". To prevent any harmful loads from being applied to the boards themselves, the belt battens must be stacked flush on top of each other, and not for instance staggered to save on stowage space. Load securing material must also be attached to these points. For particular loads, stacked stowage is an essential requirement, e.g., for pressure-sensitive cartons. In this case, any loads which arise must always be directed onto the load-bearing parts of the package. Even the smallest deviation from the vertical position can cause damage to the cargo. The risk of packing errors can be avoided by using adequate interlayer dunnage. The only intermediate layers which are permitted are plastic sheeting, craft paper and other similar materials which can be used to "anchor" the layers to each other. Despite not being bound tightly, the stack as a whole should be comparatively stable. If gaps need to be left for reasons of ventilation or small gaps are required for stowage considerations, they can be achieved by using rolled up mats or equivalent materials. If only narrow lattices, laths, boards or other strip-shaped materials can be used to ensure adequate air circulation, it must be ensured that they are placed on parts of the load which have sufficient load-bearing capacity. With cardboard cartons, the load-bearing points should be vertically below the side walls. Tightly filled paper bags or sacks should be stacked in the same way. If the bags do not completely fill a hold or do not form a complete transverse tier, a sufficiently stable stack or interconnection of bags or layers can be achieved by inserting wooden dunnage or paper anchors or similar materials. In most cases, it is sufficient to just insert the anchors in the top layers and or on the sides; Generally, segregation material, dunnage and load securing material should be used sparingly, both for reasons of expense and to prevent stowage loss. If dunnage has to be laid because "wet cargo" has to be packed together with dry cargo in a container, the dunnage must be laid without large gaps, otherwise, tightly filled paper bags and similar loads could burst or tear, leading to "bleed".

High lead levels are produced when stowing cargoes in layers with intermediate layers made of squared lumber or wooden dunnage, since only linear bearing surfaces are produced in the area of the bottom and intermediate layers. Only cargoes insensitive to load forces may be stowed in this way. Harmful pressure can be reduced to an acceptable level by using relatively soft wood or by using intermediate layers made of rubber. The pressure-sensitivity of fibre cement pipes or the susceptibility of ceramic pipes to breakage must be taken into consideration when selecting the type of dunnage. All strong and resilient goods of various shapes or sizes can be packed in a ***cross-tie stow*** formation. Remaining gaps must be filled. Wooden dunnage and similar intermediate layers should also be used as anchors. The need to bridge or pad out stowage gaps can result from a difference in the dimensions of cargo items. In principle, this involves the cargo items ***interlocking*** with each other. It can generally be used to minimise the effort required for securing cargo. Methods like this improve the stability of the cargo. But, as we have said, this method may only be used with resilient goods. Alternatively, a ***complete transverse cross-tie*** refers to stowage in which goods fill the stowage space from side wall to side wall. Anchors do not need to be inserted if equivalent measures are taken. This stowage method is suitable for all means of transport having load-bearing side walls or with low lateral stresses, e.g., railway transport. If the walls of a cargo container are not sufficiently strong, the cargo items must be stowed as a cross-tie, interconnected with anchors or secured using other special measures. It should always be considered that the speed of unloading can be increased by selecting the right anchor materials. The use of mechanical aids should be encouraged and not hindered in any way. Some examples of a complete transverse cross-tie have already been provided under the other stowage methods. That said, the following warning must be taken seriously: complete transverse cross-tie stowage is only safe if the sides of the loading space can withstand the load forces! Special care is required when loading curtain siders, lorries with tarpaulins and roof bows, open-sided containers, etc. Vertical lateral tiers can be secured well on the "open" side, e.g., in the direction of a container door, by staggering the layers backward as the tier increases in height. The lateral tiers are then no longer vertical but at a slight angle to the vertical. A gradient of around 10° is recommended. At a stowage height of 2.5 m, the top layer is offset from the bottom layer by about 0.40 m. ***Diagonal stowage*** can be practicable under certain circumstances, from the point of view of loading and unloading, on flat racks, trailers, container bolsters, lorries, certain types of freight carriages and similar vehicles, so that a number of teams can work at the same time. With this method, cargo of the same type or goods destined for the same recipient are loaded diagonally, to enable the cargo items to be accessed as quickly as possible. In contrast to conventional ship's cargo, all goods must be destined for a single place of unloading. Onward transport with remaining diagonal load blocks is not possible.

Fig. 15.5 The lower a package lies in the gap, the greater the opening angle and therefore the greater the forces transmitted

The terms ***bilge*** and ***cant-line*** refer to a loading method in which packages are stowed or fitted into indentations produced by the round or rounded shape of other packages. Genuine bilge and cant-line stowage "on the quarter" is produced when stowing wooden barrels. Today, this stowage method is only used very rarely. There are some modified methods for stowing drum and barrel loads in containers, which are to be found in the section dealing with the loading and securing of drum loads. The terms bilge and cant-line are usually also used to mean "bilge and cant-line on the quarter", although this is not strictly correct. If sensitive cylindrical cargo is stowed using the bilge and cant-line method, the load forces are transferred over two load-bearing lines onto the cylinders stowed below. The cylinders in the bottom layer only rest on one load-bearing line. If wooden dunnage or squared lumber is laid as floor dunnage, only load-bearing "dashes" are present for each cylinder, which can cause harmful load forces. Items stowed using the bilge and cant-line method exert considerable forces on the layers below them as a result of their opening angle. Only precise stowage can minimise these forces (Fig. 15.5).

It should be remembered that at an opening angle of 120°, a package already exerts a force of double its weight on the underlying packages and at 151°, this force is even four times its weight. If packages of this type are in turn bearing heavy loads, they will exert extreme downward forces. On welded pipes, the welded joints should not lie on the contact lines between the pipes. If cargo items do not fit exactly into the vehicles, improper stowage can result in gaps being left between the goods, which results in large opening angles and resultant forces. If gaps cannot be avoided, they must be left on the outer sides and padded out accordingly. If cylindrical goods are loaded with stowage gaps and stowed with the bilge and cant-line method, a cargo item which protrudes low into the cant-lines of the underlying layer could exert considerable lateral forces on the stanchions or other edges of the loading area. This would result in damage to the cargo and/or damage to the means of transport (Fig. 15.6).

The middle pipe exerts lateral forces (a) as a result of its opening angle, which are greater than its own weight (b). With resilient cargo items and load-bearing walls

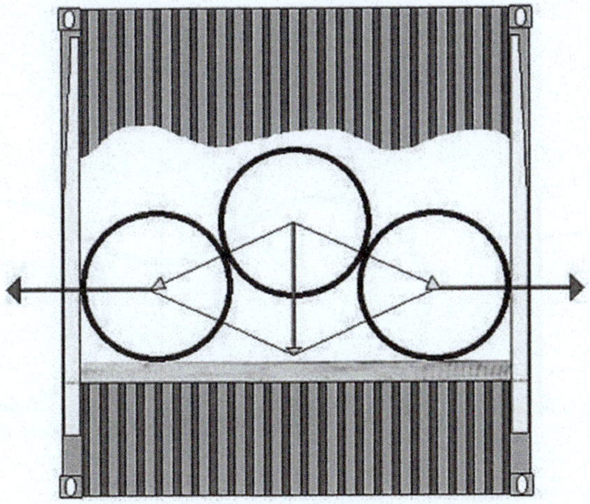

Fig. 15.6 Load forces on other cargo and components by the means of transport caused by the opening angle

of the means of transport, the forces generated by the opening angles can be used to secure the cargo. The best-known example is the locking coil. This method can be used on any type of cylindrical cargo item. This is a good securing method, as long as other cargo items are prevented from riding up elsewhere. If necessary, additional lashing must be used (Fig. 15.7).

If the forces exerted by a cargo item are too great, they can damage other goods or components of the means of transport. The gaps should therefore be no greater than half the diameter of the cargo items. If all the items are of equal diameter, the locking element then protrudes by around a third of its diameter into the cant-lines of the layer below (Fig. 15.8).

Other goods can also be stowed using the bilge and cant-line method such as bagged cargo, more robust paper board cartons, and so on. Using this stowage method, the seams or edges in all layers will lie in parallel. The packages of the vertical tiers are connected to each other, but not to the packages of adjacent tiers. In a ***cross-tie stow***, packages are stowed alternately longitudinally and transversely. In this way, they are firmly interconnected. ***Full cross-tie stowage*** enables goods to be completely connected to adjacent tiers. This means cargo items in one tier extend half-way into

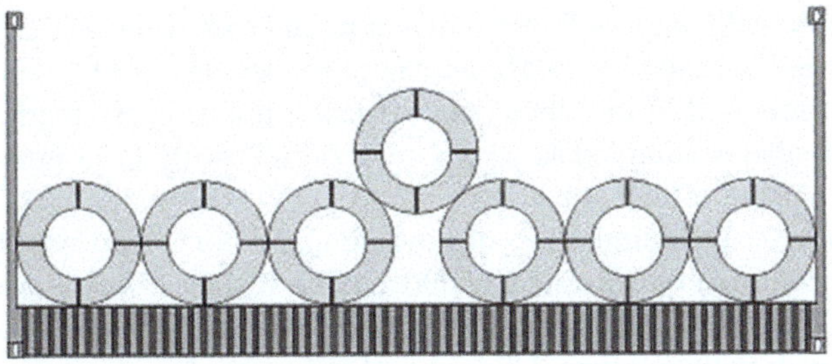

Fig. 15.7 Principle of the locking coil

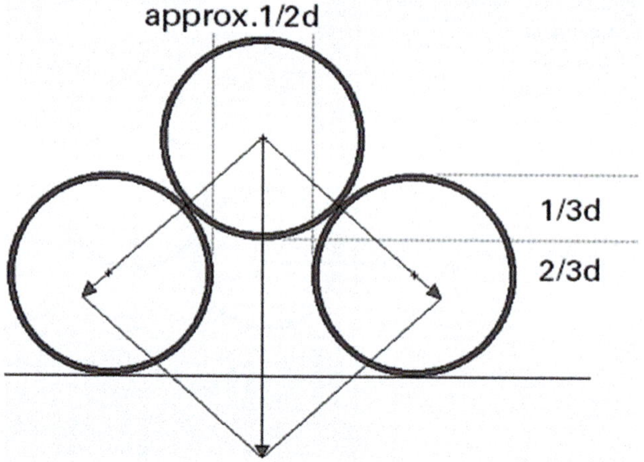

Fig. 15.8 Maximum protrusion for goods stowed with the bilge and cant-line method

the adjacent layer. Full cross-tie stowage is recognisable by the fact that no continuous lines can be seen either longitudinally or transversely. If cargo items have different dimensions, the stow must be arranged in such a way that cargo items which are as flat, long and stable as possible loaded in the outermost tier with the long side pointing into the stack. If bagged cargo is loaded in such a way that free-standing tiers remain at the edges, the seams or closures should point into the stacks, as they are weak points of the packaging. If this is not observed, bags can "bleed", which in turn endangers the stability of the stack. ***Chimney-style stowage*** is extremely well suited for forming pre-slung cargo. With palletised bagged cargo, it is suitable for certain bag sizes. The bags are arranged layer by layer so that they overlap around a cavity. The bags are thus connected to each other. This produces a square footprint, which is very suitable for packing containers if it has the correct dimensions. Nearly all goods can be stacked using the bilge and cant-line method or can be stacked and offset, if sufficient interlayer dunnage is laid, allowing the load forces to evenly distributed, and thus preventing point loading. Soldier stowing or soldier packing is used to describe a method of stowage where individual goods (such as barrels and drums) are positioned in an upright tight formation. Assuming the 20 ft container used has an internal length of 5,895 mm and an internal width of 2,350 mm drums packed using soldier stow will have an external diameter of 585 mm. This means exactly 40 barrels will fit into the container using this stowage method. The same container will also accommodate a total of 160 pails with an external diameter of 290 mm when using soldier stow.

Offset stowage can minimise loss of stowage space, but not always. Offset stowage of the pails with an external diameter of 290 mm, for instance, will allow the container to accommodate a total of 172 pails. The same container will only however accommodate 39 barrels with an external diameter of 585 mm using offset stow—compared with 40 when using soldier stow. Vertical bilge and ***cant-line stowage*** would also be a way of describing offset stowage. Whether stowage space can be saved depends on the dimensions of the goods and the dimensions of the means of transport. More goods can be stowed if the loading areas are large and or the load diameters are fairly small. This is only sometimes the case for fairly small spaces and large load diameters. Alternatively, raising loads can prevent damage, improve the usage of stowage space, and save on load securing materials. Drums, pails and similar cargo items have construction-related problems. Corrugated drums ride up on the corrugations as a result of transport forces. If closures protrude over the edge of quick lever-closure drums, the drums will snag on the closures. This results in unwanted opening of the drums and or damage to the drums. Resulting gaps can also lead to further damage. The closures on fibre drums, the handles on drums, pails, cans and other similar containers can cause damage to cargo stored on top of or next to them as a result of high point loading. Alternately raising individual cargo items or groups can reduce the risk of damage. For further information, see the section on drum loads and mixed loads.

The manual ***interleaving*** of I beams, and U profiles is rarely practiced. "Interleaved" loads are often delivered as a bundle, however, and then stacked using normal stowage anchors and squared lumber anchors. ***Nested stowage*** reduces loss of stowage space or can be used to create level cargo surfaces for access with forklift trucks. Some variants have their own names. Manually stowing on site is also becoming increasingly rare for this method, as it is growing much more common for cargo to be stowed in bundles beforehand at the factory. ***Face stow*** refers to the loading of coils, rolls etc., with their axis oriented parallel to the longitudinal axis of the skid or means of transport. Braiding involves placing ring-shaped loads, such as tires, wire rod coils with a short winding axis etc. in the core of other ring-shaped loads. ***Braiding*** has lost its importance through the use of special vehicles and the now common practice of transporting smaller batches in containers. The method is still used, however, on complete loads or fairly large batches (Fig. 15.9).

In this chapter we have discussed the segregation of cargoes and the use of dunnage and stowage techniques to ensure safe and efficient packing of containers ahead of shipping by road, rail and sea. It should be noted that alternative principles apply to cargoes subject to air transport, as the provisions contained in the container packing guidelines do not cover air transport. In the next chapter we will look at some of the main packing safety procedures as per the IMO container packing guidelines.

Fig. 15.9 Photograph of nested I beams

Chapter 16
Packing Safety and Prohibited Practices

This chapter introduces some basic packing rules and discusses prohibited practices that should be avoided wherever possible.

DO NOT LOAD INCOMPATIBLE GOODS TOGETHER! THE STOWAGE OF INCOMPATIBLE GOODS TOGETHER IS ASSOCIATED WITH RISKS!

If it is essential for incompatible goods to be stowed in a single container, The goods must be segregated from each other so that they can no longer cause any damage

SEGREGATE INCOMPATIBLE GOODS IF THEY HAVE TO BE LOADED TOGETHER!

Where such segregation is necessary, the precise method used needs to be chosen on a case-by-case basis. Covering goods with plastic sheeting, wrapping them, introducing barrier layers and similar measures are possibilities.

DO NOT PACK ODOUR RELEASING GOODS WITH ODOUR ABSORBING GOODS

If it is not possible to avoid packing odour-emitting and odour-absorbent cargoes together in a container, they must be isolated from the atmosphere of the hold so that they cannot influence each other:

A. A. Olsen and F. Karkori, *Containerized Cargo Handling and Stowage*, Springer Series on Naval Architecture, Marine Engineering, Shipbuilding and Shipping 22,
https://doi.org/10.1007/978-3-031-63329-4_16

ODOUR CONTAMINATION MUST BE PREVENTED!

Time is the factor that determines how extensive these measures should be and what segregation materials are to be selected. The longer the duration of the journey or of storage, the more resistant the screening materials have to be against molecular penetration/diffusion.

DO NOT PACK WETTING GOODS TOGETHER WITH MOISTURE SENSITIVE GOODS

If, contrary to this rule, liquid cargoes or moisture-emitting cargoes have to be packed into a common cargo transport unit, the moisture-emitting cargoes must be loaded at the bottom:

DO NOT PACK 'LIQUID CARGOES' ON TOP OF SOLID CARGOES ... BUT DO PACK 'SOLID CARGOES' ON TOP OF 'LIQUID CARGOES'

These basic packing rules are also defined in the container packing guidelines. Section 3.2 demands:

> Heavy cargoes should not be placed on top of lighter cargoes and liquids should not be placed on top of solids. When it is intended that packages are to be stacked on top of each other, attention should be paid to the strength of pallets and the shape and condition of the packages. Attention is drawn to annex 1 of the IMDG code on stacking tests. It may be necessary in some cases to ensure stability of such a stack by introducing dunnage or solid flooring between tiers of the stow. In cases of doubt, especially with heavier packages such as IBCs for liquids, it should be ascertained from the shipper or manufacturer of such packages whether or not they are designed and strong enough to be stacked on top of one another, especially where part of the transport will involve a sea voyage. The centre of gravity should be below the half-height of the cargo space.

Particular care is needed with IBCs! The last sentence in Paragraph 3.2.6 of the container packing guidelines contains a further rule:

THE CENTRE OF GRAVITY SHOULD BE POSITIONED BELOW THE HALF-HEIGHT OF THE CARGO SPACE

The formulation "heavy cargoes should not be placed on top of lighter cargoes and liquids should not be placed on top of solids" does not mean that liquid containers may not be packed next to or behind other cargo items. This will often be the case.

Then, depending on the quantity of liquids involved, floor dunnage, raised dunnage or criss-cross dunnage should always be laid below the loads endangered by leaking liquids. For further information, see also the section on "Dunnage." Another basic packing rule is:

DO NOT PACK GOODS WHICH PRODUCE DUST TOGETHER WITH DUST SENSITIVE GOODS

If this packing rule cannot be observed, either the dust-producing goods must be prevented from passing dust particles into the air in the stowage space or the dust-sensitive goods must be protected against thc dust particles. How this kind of protection is implemented in practice will depend on the outlay of materials and work. The most cost-effective method should be used. This, however, only applies when the result is neutral—otherwise, the better method must be preferred.

DO NOT PACK DIRTY GOODS WITH DIRT SENSITIVE GOODS

The following also applies here: If this packing rule is unable to be observed, the sources of the contamination must be neutralised or the goods at risk of contamination must be protected by appropriate measures. The following packing rule is also not always practicable:

DO NOT PACK PRESSURE-SENSITIVE LOADS TOGETHER WITH LOADS WHICH EXERT PRESSURE!

If the rule cannot be observed, steps must be taken to ensure that no pressure can be exerted on the pressure-sensitive goods, or that the pressure can be sufficiently reduced so that it cannot exert any damaging influence on the goods. Appropriate measures must be taken if the following packing rule is not complied with:

AVOID PACKING POINTED AND SHARP-EDGED GOODS TOGETHER WITH GOODS WHICH MIGHT BE DAMAGED

BY THEM!

Either pad the pointed and sharp-edged parts or protect the sensitive goods against damage by using appropriate packaging or dunnage. There is no alternative to this packing rule:

DO NOT PACK FOODSTUFFS TOGETHER WITH TOXIC SUBSTANCES!

Failure to comply with this rule puts human life at risk or at least poses a real risk of (serious) damage to health. It may also represent a breach of the law which could lead to criminal proceedings. The situation with this packing rule is different:

DO NOT PACK HEAVY GOODS WITH FRAGILE GOODS!

In practice, this rule will have to be broken more frequently. The risk to fragile goods can be eliminated by taking appropriate load securing measures for the heavy goods. An important packing rule is as follows:

CHEMICALLY INCOMPATIBLE GOODS MUST NOT BE PACKED TOGETHER!

This rule relates primarily to the entire range of dangerous goods—but it also affects perfectly normal goods which can be damaged or have their marketability reduced by chemical reactions. Chemical incompatibility does not always represent a danger to humans, animals or the environment. The reader must understand that this handbook cannot offer detailed information on loading dangerous goods. Special packing rules are dealt with in other sections of this book which do not have to be mentioned here. To take an example: cargoes that can be damaged by magnetism, e.g., compact discs, hard drives, etc. must not be packed together with (in immediate proximity to) permanent magnets or similar. For advice on this, see for instance the section on "Packaging and marking." The following stowage rules are contained in annex 5 of the container packing guidelines, although with different illustrations, in the form of "Dos" and "Do not's":

DO NOT CONCENTRATE HEAVY LOADS ON SMALL AREAS OF THE FLOOR - DO DISTRIBUTE HEAVY LOADS OVER A LARGE FLOOR AREA

Further information on distributing loads is dealt with in the section on load securing aids under "Wood" and under the line loads of containers.

DO NOT SECURE LOADS IN A WAY THAT PRODUCES HEAVY FORCES ON SMALL AREAS OF THE INSIDE STRUCTURE OF A UNIT!

DO SECURE LOADS IN A WAY THAT FORCES ARE DISTRIBUTED OVER A LARGE AREA OF A UNIT!

This demand can be met by using appropriate squared lumber or other sufficiently strong materials with a large surface area.

DO LOAD WITH THE CENTRE OF GRAVITY IN THE CENTRE OF THE CONTAINER

Modern container gantries with a high load-bearing capacity and multiple hoisting points can cope with a centre of gravity which is off-centre. Containers may, however, need to be handled by cranes or gantries with lower pulley blocks which are close together or cranes or gantries with a single hoisting point. This will hinder handling, or even render it entirely impossible. Such slanting loads must also be forbidden for security reasons. In addition, traffic regulations regarding uniform load forces also play a role. The container packing guidelines also provide the following additional rules:

NOT BUILD UP IRREGULAR LAYERS OF PACKAGES

DO USE A BLOCK STOW WHEN PACKING CARGO

Another recommendation relates to packaging: Material with a non-slip surface should be used to prevent packages from sliding. It has already been mentioned and explained a number of times that the construction of goods and packages plays a key role in load securing. Finally, an important stowage rule:

IT IS ACCEPTABLE TO PUT LIGHT ON HEAVY… BUT NEVER HEAVY ON LIGHT!

Fig. 16.1 Avoid concealing CAP products

This important basic rule is very often broken! If in an exceptional case, heavier goods must be packed on top of light goods, either construct "false decks," or support the heavy goods from beneath or take similar measures. CAP products should be loaded in the door area. CAP products are of special interest to the Customs authorities. Since a physical check is also made for the existence of goods of this kind, it is sensible to stow them in the door area of box containers. At the very least, they should be packed in such a way that the goods can be inspected by Customs without the need to unpack and re-pack the container. This is not the case with the container in Fig. 16.1 where CAP products were placed in the middle of other packages. Customs staff could only check and inspect the container after it had been partially unloaded. The area in which containers are packed should be spacious enough to prevent staff from obstructing each other. If too many cargo items are made available in a tight space and several fork-lift trucks are used, this can easily lead to a reduction in efficiency and damage to the goods.

AVOID CONFINED SPACES

Quick changes from brightness to darkness are extremely demanding for fork-lift truck drivers' eyes. They can cause damage by bumping etc. before their eyes have adjusted to the darkness.

ALWAYS USE THE FORKLIFT TRUCK LIGHTS!

When loading under tight space conditions, there is a risk of damage. In the case of bagged cargo, "bleeding" paper bags can endanger the stability of the entire block. To minimise the risk of damage, a helper should hold a sufficiently wide strip of hardboard vertically between the cargoes when the pallet is inserted. Work on homogeneous cargo items should be carried out using attachments suited to the goods in question. With heterogeneous loads, attachments should only be used if the devices can be attached and detached very quickly. The drum clamp shown above is picked up and put down by inserting or removing the fork. It is quickly secured by two pins and released again by removing the pins. It is always more efficient to work with fork-lift trucks equipped with a fork adjustment device and whose forks can be moved laterally in parallel.

HIGH RISK WORKING METHODS SHOULD BE AVOIDED

Packages are often mechanically ruptured by forks on loading and unloading as a result of carelessness, inexperience etc. If the forks are used to push a package into place, the forks can slip down and cause damage. This risk is particularly high if the equipment is in poor condition and the tips of the forks are not exactly parallel horizontally. If the forks are too long, goods stowed further behind are often also affected. Should damage of this kind remain unnoticed, it can pose risks to the environment and in the case of dangerous cargoes, cause substantial consequential losses and signify a violation of the law. As Sect. 4.2.7 of the container packing guidelines says:

> Packages of dangerous cargoes should be examined, and any found to be damaged, leaking or sifting should not be packed into a [container].

Damage caused during stacking or when pushing a load into place can effectively be prevented by using pieces of squared lumber of a suitable length and format, laid on and/or fastened to the rear of the forks.

HIGH RISK WORKING METHODS SHOULD BE AVOIDED

Packages should always be picked up resting fully on the fork and set down level. Loads should not be pushed with slightly raised fork tips. This could otherwise cause damage to the bottom of certain loads. When handling long goods, the forks should be set as wide apart as possible from each other. Here, however, the packaging is the main concern. If long goods have to be pushed into containers, skids running in a longitudinal direction must be provided. Strapping must not run along the belt battens at the bottom. They must run along a milled groove. If over stowage of the cargo items is an issue, the steel strapping should run in a groove at the top in the same way. An all-round groove is ideal. To prevent the strapping from tearing off, suitable hardboard strips can be placed under the belt battens with the smooth side facing down. This reduces friction, which in turn makes it easier to push loads in (Fig. 16.2).

CORRECT ARRANGEMENT OF SKIDS

Fig. 16.2 Stackreacher

Cases built like this are easy to push into containers provided they have been set down with the front end in the door area. This has not been done here. Loads should not be pushed over a ramp, since usually the metal sheets are textured or made of deck plate.

INADEQUATELY PACKAGED GOODS

Even the best packing company in the world cannot be expected to work well if the goods they are delivered are atrociously packed. Here, the load unit was destroyed by a braking manoeuvre of the fork-lift truck. Only professionally prepared units should be loaded. Matters may be improved in the long term by inspection of incoming goods, good customer contact and regular quality control at the consignor end. Pushing gently with a pallet as padding can be tolerated for certain goods in exceptional cases. It cannot be tolerated with goods which could be damaged by such actions. There is a considerable risk of damage on unloading at the destination, if the packers do not have the stripping of the container in mind when loading it. Cargo must always be packed in such a way that the cargo transport units can be unloaded as easily as possible. This can also mean attaching belts, slings or similar to individual cargo items which can then be used to pull out the items. If, for instance, it would be difficult for recipients to attach slinging equipment to loads in open-top containers, the shipper should leave it in place and tie it together in such a way that it is immediately accessible from the top. When working with lifting gear and single hoisting points, auxiliary ropes should be attached to the load or the hoisting equipment to be able to safely guide the cargo items.

16.1 On Completion of Packing

The container packing guidelines provide detailed instructions on what action is to be taken after packing has been completed. Special instructions are given for cargo transport units with dangerous cargoes. Section 4.4.2 deals with the documentation. These states:

4.4.2.1 For transport by sea, regulation 5 of chapter VII of SOLAS 1974, as amended, requires that the person responsible for the packing of dangerous cargoes into a container or road vehicle shall provide a signed Container Packing Certificate or Vehicle Declaration stating that the cargo in the unit has been properly packed and secured and that all applicable transport requirements are met.

4.4.2.2 The IMDG Code recommends the following declaration:

- The container/vehicle was clean, dry and fit to receive the cargoes.
- If the consignments include cargoes of class 1 other than division 1.4, the vehicle or freight container is structurally serviceable in conformity with Section 12 of the introduction to class 1 of the IMDG code.
- No incompatible cargoes have been packed into the vehicle or freight container (unless authorised by the competent authority concerned in accordance with 12.2.1 or 17.6.3.1 of the General Introduction to the IMDG code).
- All packages have been externally inspected for damage, leakage or sifting, and only sound packages have been packed.
- All packages have been properly packed into the vehicle or freight container and secured.
- Drums have been stowed in an upright position, unless otherwise authorised by the competent authority.
- The vehicle or freight container and the packages therein are properly marked, labelled and placarded (sic).
- The dangerous goods declaration required in subsection 9.4 of the General Introduction to the IMDG code has been received for each dangerous cargoes consignment packed in the vehicle or freight container.

4.4.2.3 A Container Packing Certificate/Vehicle Declaration is not required under the RID, ADR, ADN or ADNR regimes, even though they may be required for inland domestic transport in certain countries. However, such certificates will be needed if the transport operation includes sea voyages. They will then need to be provided prior to loading, as port authorities, berth operators and ship masters may wish to sight them (or a copy) before accepting containers or vehicles packed with dangerous cargoes into their premises or aboard their ship. In accordance with section 8 of the GGVSee, manufacturers and distributors have special duties. Usually, they are referred to as the "shippers."

4.2.2 The shipper should also ensure that dangerous cargoes are packaged, packed, marked, labelled, placarded (sic) and provided with the required signs, in accordance with the applicable regulations. A declaration that this has been carried out is normally required. Such a declaration may be incorporated into or attached to the transport documents.

4.4.2.5 The functions of the dangerous goods declaration (see 4.2.2) and of the Container Packing Certificate / Vehicle Declaration may be incorporated into a single document; if not, these documents should be attached one to the other. If these functions are incorporated into a single document, e.g., a Dangerous Goods Declaration, shipping note, etc., the inclusion of a phrase such as "it is declared that the packing of the cargoes into the container/vehicle has been carried out in accordance with the provisions of section 17 of the General Introduction to the IMDG code" is sufficient. Where both declarations are included in a single document, separate signatures are required for the two declarations.

CONTAINER SET DOWN, CLOSED AND SEALED PROPERLY

THE CENTRE OF GRAVITY SHOULD BE POSITIONED BELOW THE HALF-HEIGHT OF THE CARGO SPACE

3.3.3 After closing the doors, it should be ensured that all closures are properly engaged and secure. Usually, a seal is applied to a container. Care should be taken that sealing procedures are carried out properly.

UNIQUELY NUMBER CONTAINER SEAL

When using flat racks with a load covered with a tarpaulin, it must always be taken into consideration that the flats are transported on deck. The tarpaulins must therefore be secured against storms.

> 3.3.4 Where [containers] have hinged or detachable fittings, a check should be made that they are properly secured, with no loose equipment likely to cause a hazard during transport.

THE CENTRE OF GRAVITY SHOULD BE POSITIONED BELOW THE HALF-HEIGHT OF THE CARGO SPACE

HINGE MECHANISM ON A COLLAPSIBLE FLATRACK

A check should additionally be performed on refrigerated containers to ensure that the correct refrigeration temperature has been set. Containers carrying dangerous cargoes must be provided with the relevant placards. For further information, see the sections on refrigerated containers and marking of containers with dangerous cargoes. If a container is shipped to a country in which there are quarantine regulations regarding the treatment of wood, e.g., Australia, New Zealand and China, it must be ensured that all wood carried in the container conforms to these regulations. Either all parts must have previously been fumigated or heat-treated and marked appropriately, or the entire container must be subjected to special treatment. A copy of the wood treatment certificate must always accompany the container. This should be stored directly inside the right-hand container door or in a weather-resistant pouch on the outside of the door.

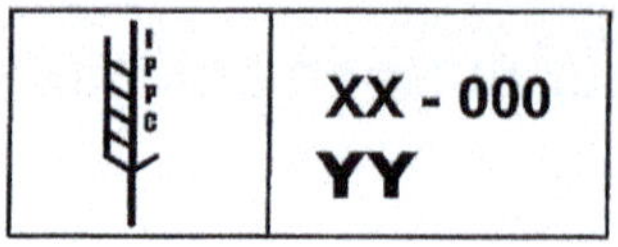

NOTICES ON PLYWOOD CASES AND PALLETS

16.2 Final Work in the Door Area

The requirements laid down in the container packing guidelines with respect to a secure load face in the door area can be met in a number of separate ways. One option is to place a wooden lattice or pallets against the last portion of the load and to brace these against the rear corner posts. Lattices can also made by nailing inexpensive, untrimmed wooden dunnage boards together. How lattices are to be placed in front of the load depends on the forces which have to be absorbed and what type of load it is. In the case of stacked, palletised goods which are firmly secured to the pallet, the variation shown on the right is suitable. Then, the horizontal boards in contact with the load should cover the lower area, the seams of the stack and the upper area. In the case of goods stacked in three layers, lattices must therefore be made of four horizontal boards. If the mass to be braced is not too large, the bracing can be placed directly against the lattice (left figure). If forces from larger masses are to be absorbed, an additional cross piece must be fitted. If pallets suitable for use with containers are available, for instance with the dimensions 1,150 mm × 1,550 mm or 750 mm × 1,150 mm, they can also be used instead of a lattice. If three pallets are arranged with the narrow edge running transversely, squared lumber can be inserted between the pallet boards. If double deck pallets are not used, but four-way pallets are used instead, for instance, the completely boarded side should face the load. Unsatisfactory packaging results in errors in load securing which could only have been prevented by additional outlay of material and work. The visible stowage gaps must be filled in any case. The container packing guidelines require that when during the final stages of packing a container, care should be taken as far as is practicable, to build a secure face of the cargo. This is to prevent cargo from falling out when the doors are opened. A board fitted between the corrugations is not capable of ensuring this. In order to come anywhere near fulfilling this requirement, it would have been necessary to have fitted a number of transverse boards or—better still—a lattice made of several boards. It can be assumed that if the loads are otherwise correctly stowed, the boards or lattices will be held in place by the container doors, preventing any longitudinal shifting of the load. Since dynamic forces are no longer acting on the load when the container is opened, it is not to be expected that cargo items will be able to fall out when the doors are opened. In this way, the requirements of the container packing guidelines for a secure load face can be met. It is important that boards are placed and nailed to each other in such a way that their position in relation to each other and to the container cannot change. If the load is not braced against the corner posts, the wood fitted horizontally between the corrugations cannot hold the face of the load securely. This can result in damage to the container corrugations if the forces from the load become too high and the squared lumber is pressed toward the door. Fitting squared lumber and boards in the container corrugations is a quite easy method of securing, but this method should be used on its own only for light loads. In all other cases, boards and squared lumber must be fitted horizontally, supported against slipping down and also braced along the length of the container. With the previous method, the transversely fitted squared lumber is subjected to bending stress across its whole width. The wood inserted at the container wall only has little to "get its teeth into" on the corner posts. Securing material must therefore cover a large area and where possible be placed obliquely against the corner posts to ensure a high load-bearing capacity. If the total weight of the load in the container is heavy or extremely heavy, additional squared lumber bracing against the doors must be provided. It is constantly forgotten that during transport, the vehicle is subject to changes in speed and direction. The inertia forces resulting from these changes set the load in motion relative to the vehicle.

If we refer to Fig. 16.3, we can see that dangerous cargoes have been loaded in this container but not—as the regulations demand—in the door area; instead, they have been "concealed" within the rest of the load. Section 4.3.7 of the container packing guidelines states that:

> Dangerous cargo consignments which form only a part of the load of a [container] should, wherever possible, be packed adjacent to the doors with markings and labels visible.

Attention has already been drawn to the specific measures required for securing cargo by the doors of a unit several times. It is dubious whether two crossed boards can actually prevent the load from falling out if the container has travelled for some distance. This is particularly relevant if the boards are not bevelled. The simple reason being that without a bevel the fitted board will not have enough grip "to get its teeth into" in the groove of the container corrugations to withstand the potential loads. Importantly, bevelled boards can withstand loads acting toward the door much better. A weak point of all securing measures of this kind, however, is that the boards slip away toward the top and then no longer provide any protection. Theoretically, one viable solution to this would be to brace the boards toward the top, but it would

Fig. 16.3 Incorrectly loaded container with dangerous goods

be difficult or awkward to achieve this in practice. It is not always possible to secure diagonally inserted boards by nailing. This type of securing method should therefore be avoided entirely. Horizontally fitted boards or lattices are suitable for light loads. Squared lumber or other lumber measuring 5 cm × 8 cm is a better solution for medium heavy goods. Professional bracing must be built for heavier loads.

Part V
Container Load Securing

Chapter 17
General Load Securing Methods

A significant volume of material has been published about load securing over the last few years. The necessity for correct load securing has been unanimously confirmed by industry experts and academics. However, the most effective methods are subject to fierce discussion. New regulations are constantly introduced, and old regulations reworked or revoked. A range of recommendations and handbooks have appeared on the market designed to help the user, but these books do not always live up to expectations. When selecting load securing methods and using load securing materials, economic considerations should obviously play a significant role, but safety should never be compromised as a result. This part of the handbook is designed to provide the user with information, particularly with regard to the practical techniques of load securing. But at the same time, as far as this is possible, the background and context will also be discussed from the theoretical point of view. It is impossible to do full justice to this topic given the limited number of pages available however, the author hopes that this work will offer a suitable introduction to this topic.

17.1 General Load Securing Methods

Because containers are carried on all types of transport vehicle we shall introduce some specific terms for loading and securing methods as used by the global shipping industry. All fields of shipping refer to load securing in one way or another, yet each field has its own specialist terms. For example, Sect. 17.2 of the German rail traffic regulations (EVO) defines a general obligation to secure loads. Accordingly, these states:

> The objects to be loaded must be fundamentally safe and secure and should not be able to move as a result of jolts or vibrations that can be expected under normal operating conditions. The possibility of longitudinal movement is permitted as far as the magnitude of this movement is limited by the use of appropriate means and there is no risk posed to safety.

A. A. Olsen and F. Karkori, *Containerized Cargo Handling and Stowage*, Springer Series on Naval Architecture, Marine Engineering, Shipbuilding and Shipping 22,
https://doi.org/10.1007/978-3-031-63329-4_17

This paragraph provides an important piece of information, namely that rail transport permits two fundamentally different securing methods; these being:

SLIDE LOADING ***RIGID LOADING***

The slide or sliding loading method allows cargo to move in the direction of impact as a result of the jolting forces that occur during rail transportation. With this method, jolting energy is absorbed by friction work. The magnitude of deceleration in g (1 g = 9.81 m/s^2) corresponds to the size of the friction coefficient between the package and the loading surface. The lower the friction coefficient, the lower the forces acting on the cargo, but the greater the travel and vice versa. As a result of the limited travel, which is possible, this method is geared toward the friction between wood and wood with travel of at least 1.5 m on each side of the object. If the possible travel is less than 1.5 m, then the use of non-slip materials placed between the load and the floor of the freight carriage is required. The German railroad authorities recommend the use of this method for sensitive loads, for example, valuable machinery and equipment, if the properties of the load and the length of the loading area permits the method and if the goods are not in danger of tipping. The lower end edges of the load should be bevelled in order to ensure that they do not snag on uneven areas of the freight car floor. If the bottom surface of the goods does not allow them to slide, they should be rigidly attached wooden skids or sleds. The leading edges of the skids should be shaped like runners (Figs. 17.1 and 17.2).

The loads must be secured against sideways movement. Squared wooden guide rails attached next to the loaded goods or sleds in such a way that the load not prevented from sliding have proved themselves in practice. Any movement by the

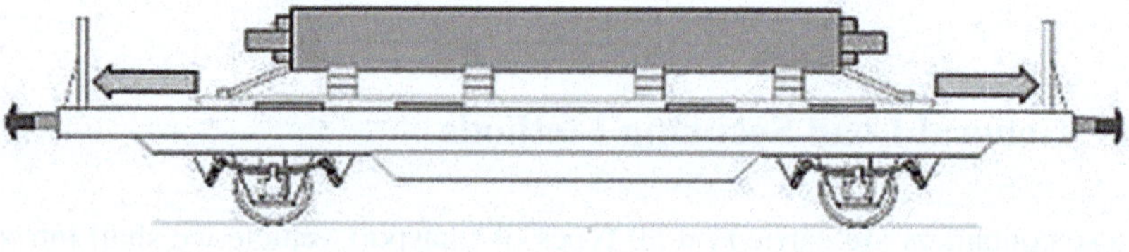

Fig. 17.1 The sliding loading method—side view

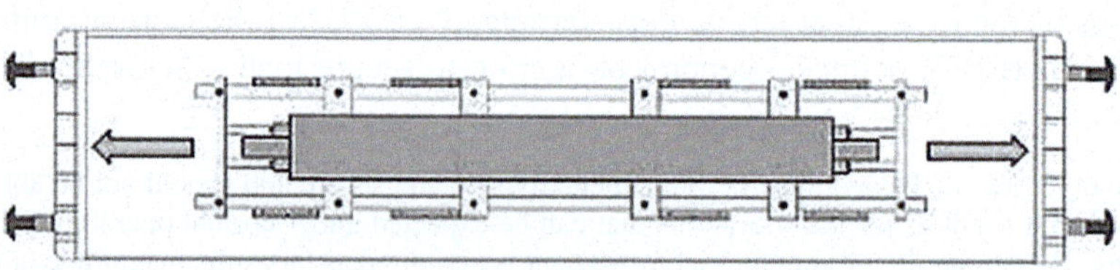

Fig. 17.2 The sliding loading method—plan view

load must not cause permitted axle loads to be exceeded. Thus, the maximum permissible load of the freight carriage cannot be fully exploited. The axle load must not exceed the value that results when the sum of the tare weight plus the permitted maximum load is divided by the number of axles or bogies. The ratio of the axle loads of cars with two axles must not exceed 2:1; that of bogie cars must not exceed 3:1. If the floor surface is of the same nature, goods will slide the same distance, regardless of their mass. Thus, a number of partial loads or load groups consolidated as blocks can be loaded on a car using the sliding loading method. The loading guidelines require a gap of 50 cm between the individual goods/groups. When using this type of loading, several consecutive jolts during switching should not occur in the same direction. Cars loaded in this manner are specially checked at the switching yards.

> **!** WARNING!
>
> THE SLIDING LOAD METHOD MAY ONLY BE USED IN A CARGO TRANSPORT UNIT IF YOU CAN BE ABSOLUTELY SURE THAT THE UNIT WILL BE TRANSPORTED EXCLUSIVELY BY RAIL. THE SLIDING LOAD METHOD SHOULD NEVER BE USED WITH ANY OTHER MEANS OF TRANSPORT

The term rigid loading comes from the loading guidelines published by the International Rail Union (UIC) and is used as a blanket term to cover both compact loading and individual securing.

Rigid loading method

⇩ ⇩

Compact loading | Individual securing

The term "rigid loading" means exactly what it says: the load must be secured on or in a cargo transport unit in such a way that it is unable to move. If this demand is met, the load is unable to topple, fall down or become damaged/cause damage in any way. However, the term does not specify the method of securing that is to be used. The term ***compact loading*** also clearly indicates the concept. A compact load is one which is stowed without gaps or, as far as gaps remain, where these gaps have been filled by resilient material. The term also immediately suggests that the containers must have borders of sufficient load bearing capacity around the loading area in order to allow a compact unit to be formed in which nothing is able to move. Even this simple description makes it clear that a container that does not have fixed side or end walls is not suitable/only partially suitable for securing loads in this way. ***Individual loading*** refers to individually secured loads which are attached to the transport unit using special load securing aids. The term does not, however, indicate exactly how securing is to take place. It does, however, imply that certain components must be available or certain requirements must be in met in order to allow loads to be secured individually: These components may be lashing points or other securing mechanisms on either the transport unit itself or the load.

Aim of load securing mechanisms

⇩ ⇩

Securing against sliding	Securing against tilting

If these load securing measures are carried out correctly, both of the important requirements of load securing will have been met: The load is unable to slide or tip. If we go into more detail, tipping can fall into two categories:

Securing against tilting

⇩ ⇩

Securing against tilting	Securing against tipping over

17.2 Achieving a Tight Fit Using Container Components and Special Components

This type of tight-fit is designed to transfer the forces from an item either directly to components of the transport unit, such as the walls or stanchions or directly via parts of the load which are capable of withstanding these forces. This cost-effective method of load securing has already been referred to as compact loading. If the packages have the right shape, size, strength and compatibility they can be packed together in tight contact. This forms a coherent loading block. If this block is unable to move in any direction, then no further securing measures will be required. This method of packing may also be described as a complete longitudinal and transverse block. All the inertia forces acting on the load are transferred directly by the load to the components of the means of transport. The prerequisite is that the components of the transport unit are capable of withstanding such forces long-term. Goods can often be packed successfully against the end walls and the sides. If the packages do not fill the entire length of the container, appropriate securing mechanisms should be used in the door area. If there is a risk of the packages working loose from the block as a result of the movement of the ship, then additional lumber bracing or other similar securing mechanisms should also be applied on top of the load. With flat racks, they are specially designed to accommodate stacked receptacles. They are used to deliver securing elements such as stacking cones, twist locks, bridge fittings and other equivalent items used for securing containers on board ships. Solutions of this type are also suitable for transporting other small items. Considering the costs of packing and securing items, the use of a more expensive special container may, overall, actually pay for itself, particularly as the half-height saves on transport volume. Platform containers of normal construction can usually be loaded in a compact/tight-fit fashion if the dimensions of the load are identical to the dimensions of the container loading area. That said, this constellation is very rarely seen in practice. Shortcuts are not permitted in the field of load securing, but if the intended dimensions of a case are close to those of the loading area of a flat rack, then it is possible to make the case fit the flat rack by building it slightly larger than originally planned. The additional costs invested in building the case are recouped by the costs saved on securing the case, and the risks of carriage are also lower.

As an example, the load shown here is a substation. An over width cargo such as this would normally require a considerable load securing outlay. The adjustable stanchions (a) (and which must be specially manufactured) can be used to ensure an excellent lateral tight fit. In reality, cargo items are not so large that they reach as far as the end walls. In this case, therefore, they would need to be braced with wooden beams or similar material. Securing by lashing would be even more work-intensive. Even if the items to be sent were the same length as the container, the over width would prevent the cargo from being loaded hard up against the end walls since the container would no longer fit into the slots of the ship. A solution would be to use a flat rack with special stanchion pockets that allow the load to be secured as a tight-fit (Figs. 17.3, 17.4, 17.5 and 17.6).

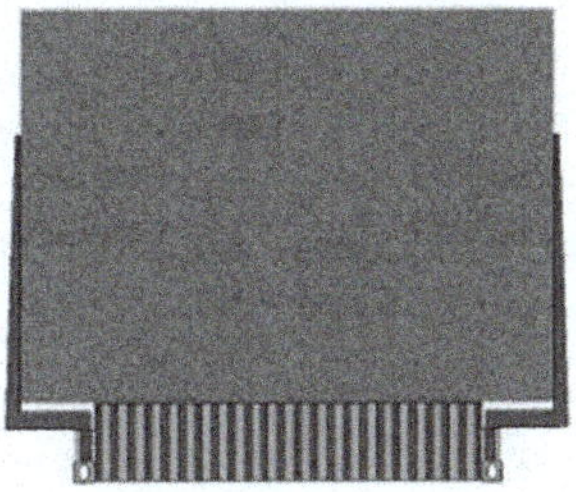

Fig. 17.3 Compact tight-fit securing of a power substation using specially shaped stanchions crosswise on a flat rack

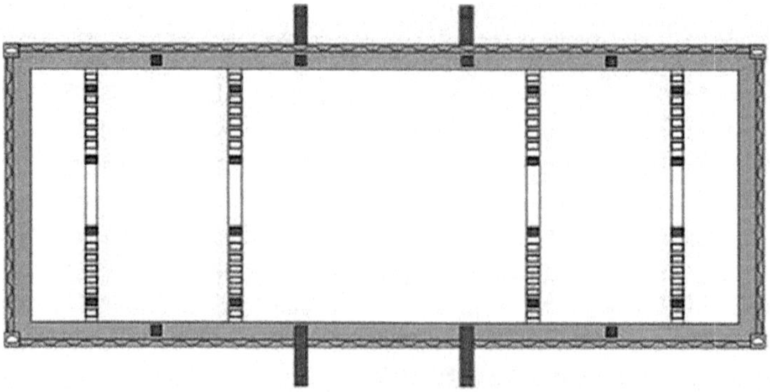

Fig. 17.4 Example of stanchion pockets on a specially designed flat rack

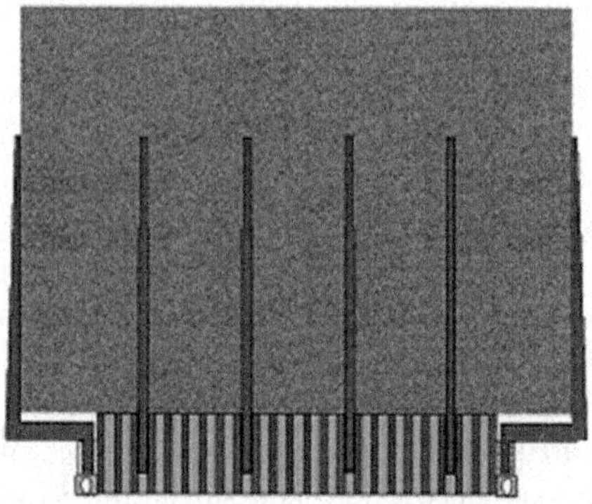

Fig. 17.5 Graphic of a power substation secured using removable and adjustable stanchions

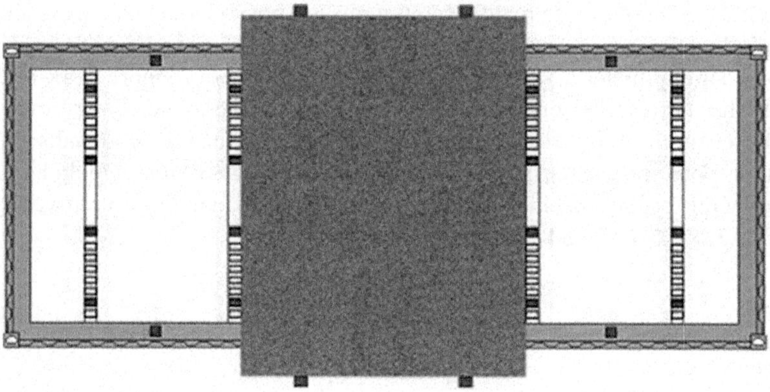

Fig. 17.6 Graphic of power substation secured using removable and adjustable stanchions—top view

Fig. 17.7 Example of incorrect securing by filling

The means of transport are often only partially filled. Gaps may arise as a result of the characteristics of the load or for packing reasons. If so, then the load must be secured using the appropriate methods. Special securing measures may be required if there are edges on the means of transport or the edges available are not able to bear the load, as is the case when using partially open/fully open means of transport. In these cases, other methods will need to be used. These are described in the following sections.

17.3 Achieving a Tight Fit by Filling in Gaps

This method of compact loading involves leaving gaps that must be filled in. When using box containers, these gaps may be longitudinal or transverse and should be filled in appropriately unless there is a better way of securing the cargo. Filling is usually the better and more economical method since lashing inside box containers requires considerable effort. The situation with platform containers is different; here lashing is often more suitable. Objects suitable for filling in gaps are wooden battens, which will be discussed in the next section, paper board which can be used for small gaps, boards and individual pieces of squared lumber for small gaps and pallets for larger gaps. If smooth surfaces are available or measures can be taken to eliminate sharp edges, then air bags may also be used. Compact packing and filling of gaps can only be implemented effectively if all parts of the load are sufficiently robust. If we look at Fig. 17.7, for example, here the movement of the ship and the container while at sea would, eventually, destroy the cartons that are marked fragile. The gaps that occur as a result of this would then put the wooden cases at risk. The fact that the upper of the two cartons has been loaded upside down is not particularly good, but it is also not forbidden. If the shipper wishes to ensure that a packaged item is transported upright, it must be marked accordingly. The gap has been left between the two sturdy wooden cases and filled in using boards and squared lumber. The cartons were loaded into this space. The remaining spaces were filled in using old pallets. The forces acting on the wooden cases are transferred from case to case by the squared lumber and onto the side walls of the container. The pallets are simply transferring the forces from the cartons to the case on the right. When deciding on packaging and cargo-securing material, it should be borne in mind that some countries enforce a garbage- and litter-avoidance policy. This may lead to limitations on the use of certain materials and imply fees for the recovery of packaging at the reception point as well as similar problems for the shipper of the cargo. In such cases, reusable packaging

(a)

(b)

Fig. 17.8 Example of correct securing by filling **a** with pallets and **b** air bags

and securing material should be used. Increasingly, countries are requiring timber dunnage and packaging materials to be disembarked (Fig. 17.8).

The space remaining in the vicinity of the door is filled in using upended pallets placed across the space and squared lumber beams fitted into the corrugations of the container. Care must be taken if the overall mass of the load is large, since the use of this method could cause damage to the corrugations. If possible, the load should be braced against the door posts in the vicinity of the doors. If this is not possible, the door area can be filled in, but in such a way that the load is spread over a sufficient area that no harmful pressure points arise. Leaving spaces in the middle of a container and filling them in requires the packer to calculate the required dimensions exactly. With good planning this is not a problem. There is an advantage to this, since it allows the distribution of weight to be better organised. Since air bags should not be used in the vicinity of the door for safety reasons, the obvious option is to leave space in the middle of the container. The air bags can be affixed and partially inflated before loading the cases in the door area. Once the cases have been put into place, the left-hand door is closed, and the air bags can be inflated fully. Theoretically it is possible to determine the size of the gap and to fully inflate the air bags before loading the boxes. However, this runs the risk of the air bags being crushed without the packers being able to monitor this, since the height of the cases means that the air bags are no longer accessible. This method of securing is practical when used with containers which have side doors and with platform containers which can be accessed from the side.

17.4 Achieving a Tight Fit by Bracing

Bracing elements made of wood are a tried and tested method of load securing when packing containers. Squared lumber, battens and other elements are fitted in such a way that spaces within the load are bridged. Bracing between the cargo and the container, and between the items of cargo are both commonly used methods. The latter is the more cost-effective solution since the required securing forces are lower. Wooden bracing elements used to fill in gaps or empty spaces can often be made quickly and simply. Occasionally, however, they are feats of craftsmanship that require a great deal of experience and ability. The critical issue when bracing is that the forces must be transferred to the load-bearing components of the cargo transport unit or the packaged items that surround the gap. From a technical point of view, the bracing elements in both examples are equally well made. Horizontal elements are nailed onto vertical elements. The bracing elements are formed by the wooden beams that are cut to fit and placed transversely on top. The majority of this work can be done in advance outside the container. Only the bracing must be put in place inside the container. The advantage of the method shown on the right is that the bracing elements are placed level with the strongest parts of the cases, namely the top and the bottom. This form of bracing assumes that the packages either side of the gap are of almost identical height. Two transverse lengths of squared lumber are placed on the floor at the bottom of the space; depending on the forces that can be expected, at least two or more pieces of lumber are cut to fit and placed horizontally. Outside the transport unit, boards are prepared which can then be nailed to the fitted lumber beams. This prepared bracing element is then lowered into place from above. All elements are stapled in place to prevent them from moving or are cross-braced with wooden boards. Using squared lumber fitted into a space in the way shown above is only possible if the packaged items are able to withstand such point loading. If the cases are not strong enough to allow the ends of the bracing beams to be placed against them directly, boards, planking or squared lumber can be additionally used to distribute the load forces. If this is not sufficient, vertical bracing elements or other methods of bracing should be used. The section on wood contains a number of other useful tips, as well as rules of thumb for calculating strength. There are a number of things to bear in mind with regard to the woodwork and the lashing that have been used here. It is often the case that insufficient attention is paid to the bracing of gaps and or the work is carried out in an unprofessional manner (Fig. 17.9).

Every space in the load represents an increased risk of danger. Spaces should therefore always be filled in. With goods that are liable to topple or palletised goods, which are badly bonded to the pallet, it is not only the floor region which must be filled out, but a greater surface area as well. It is common for containers to experience roll angles of 30°. Such transverse forces occur when a container is transported on a ship in the usual fashion, namely with fore and aft stowage. To put the risks in perspective, it can be assumed that on a long voyage a container will be subjected to several tens of thousands of these roll oscillations. During pre- and post-carriage by rail or by road, the objects in front of or behind the case are the ones most at risk. The

Fig. 17.9 Example of how to fill the space in the vicinity of the door

same would apply if the container were to be stowed athwartships on the vessel. The possibility of the packages being crushed could have been avoided by using squared lumber in the floor region between the case and the sides of the container. If cases are at risk of tipping, they must also be braced close to the lid of the case (Figs. 17.10 and 17.11).

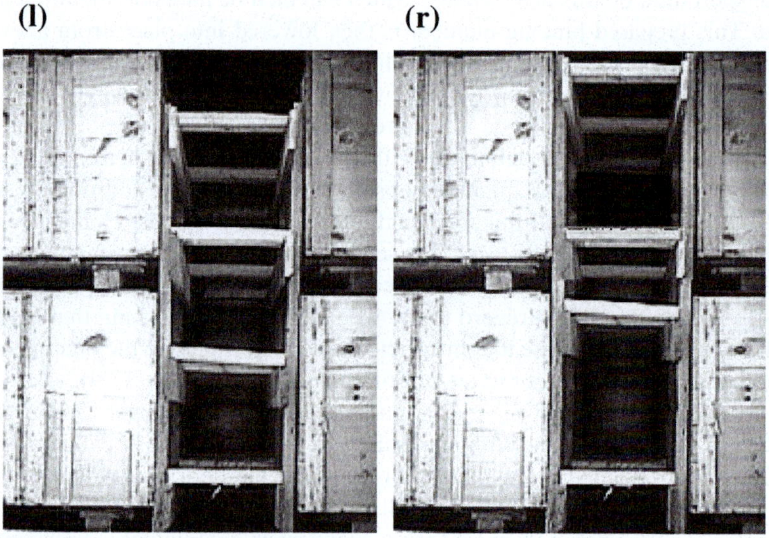

Fig. 17.10 (l) Example of horizontally bracing, poorly placed in part; (r) well placed horizontal bracing elements

Fig. 17.11 Bracing an over height case on a 40′ flat rack

17.5 Achieving a Tight Fit by Lashing

Lashing refers to the use of steel strapping, chains, steel wire, textile straps, ropes and other securing materials which are fixed, on the one hand, to the package and, on the other hand, to the container and then tensioned lightly. The crucial factors that determine the effectiveness of the lashings are the quality of the materials and the fixing points as well as the directions in which the lashings work. This means that lashing angles also play a critical role. Lashing a package is intended to prevent it from moving longitudinally or laterally and to stop it rising from the loading surface or tipping. A lashing is used to create a tight fit. The term direct lashing is also used. The quantity and thickness of the required lashings will depend on the weight of the load that is to be secured, on the anticipated forces, on the maximum securing load of the lashings and the lashing angles. More information can be found in the relevant sections dealing with the materials and in the lashing examples. Lashing has a less key role to play when securing packages in box containers. Since the lashing points in the container usually have a maximum securing load of 1,000 daN (corresponds to a mass of approximately 1t), the use of lashing must be restricted to lighter packages. Another consideration is the fact that lashing points are often not accessible as a result of spatial restrictions and the space occupied by packages. When securing loads in open top containers a higher MSL can be assumed—often in the region of 2,000 daN (corresponds to a mass of approximately 2t). Platform-style containers, on the other hand, offer good lashing facilities, which means that lashing is indispensable on such cargo transport units. Horizontal lashings have no vertical component. Horizontal lashings attached longitudinally, as shown here, have no lateral components. The package is simply held in place longitudinally by this type of load securing. If the lashings are absolutely parallel to the horizontal transverse axis, they will have no longitudinal or vertical components. A package secured in this manner is thus exclusively secured in the transverse direction. The lashing shown here, however, also has a slight vertical component as a result of the position of the lashing points. A purely vertical lashing has no horizontal component at all. These

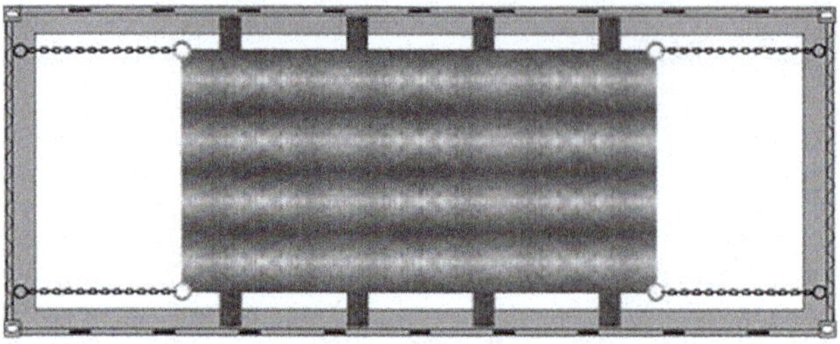

Fig. 17.12 Horizontal lashings with longitudinal components—view from the top

lashings are thus unable to exert either a longitudinal or a transverse securing effect. Longitudinal and transverse forces can cause the load to move (Figs. 17.12, 17.13, 17.14, 17.15, 17.16, 17.17, 17.18, 17.19 and 17.20).

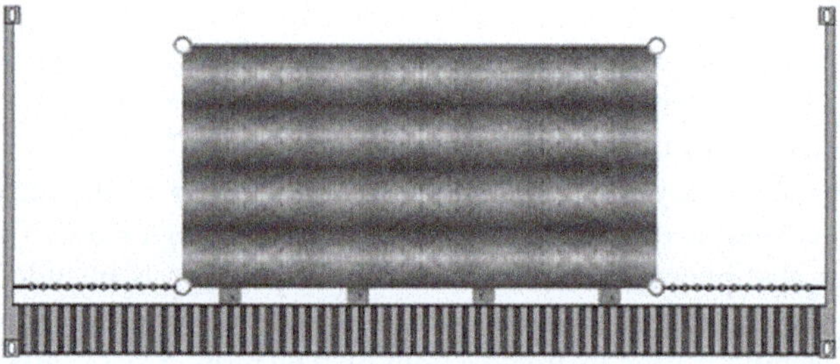

Fig. 17.13 Horizontal lashings with longitudinal components—view from the side

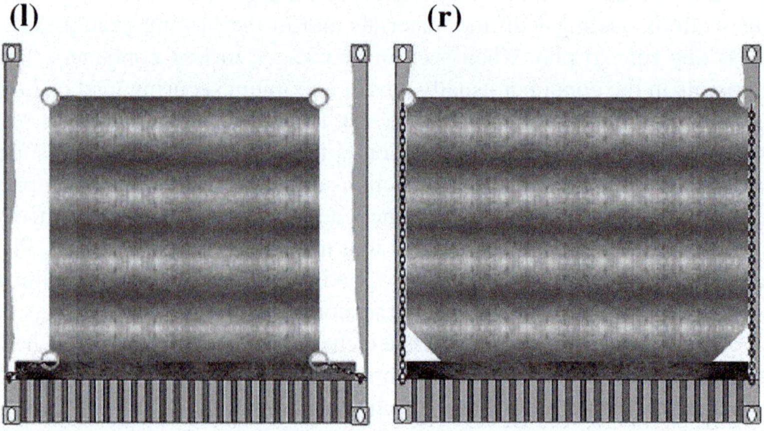

Fig. 17.14 (l) Horizontal lashing with transverse and vertical components—lateral view and (r) vertical lashing without horizontal components—lateral view

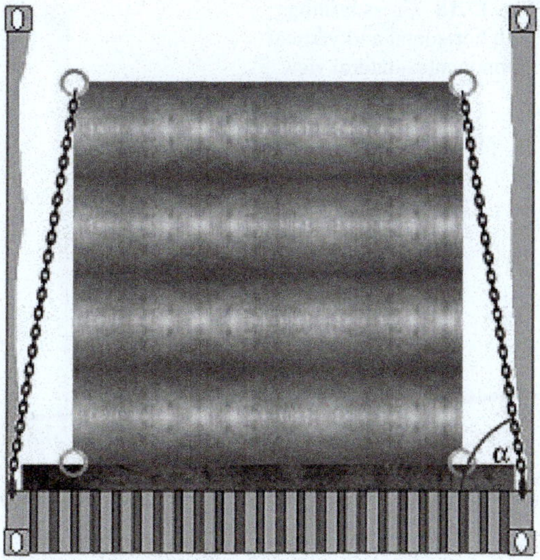

Fig. 17.15 Vertical lashings with slight horizontal transverse components—lateral view

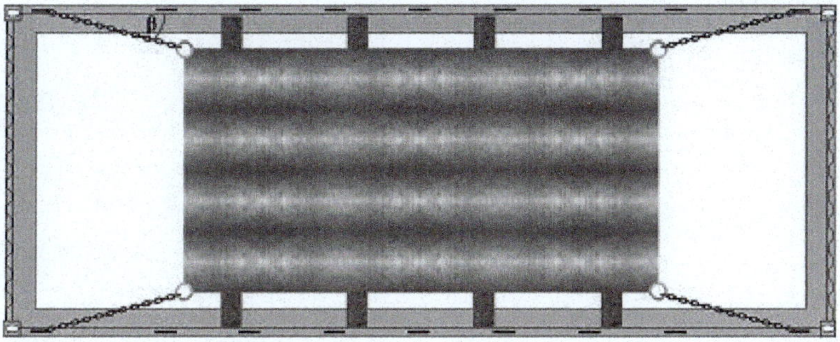

Fig. 17.16 Diagonal lashings with horizontal longitudinal and transverse components and vertical components—view from the top

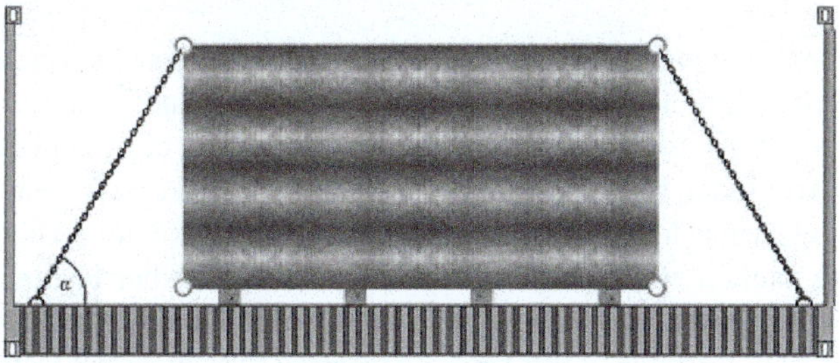

Fig. 17.17 Diagonal lashings with horizontal longitudinal and transverse components and vertical components—view from the side

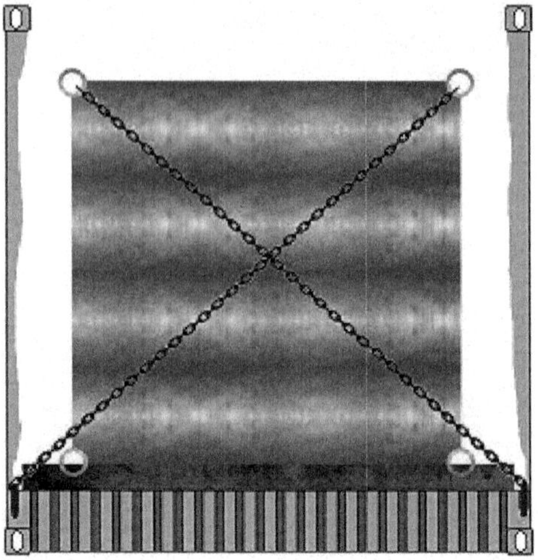

Fig. 17.18 Cross lashings with horizontal and vertical components—lateral view

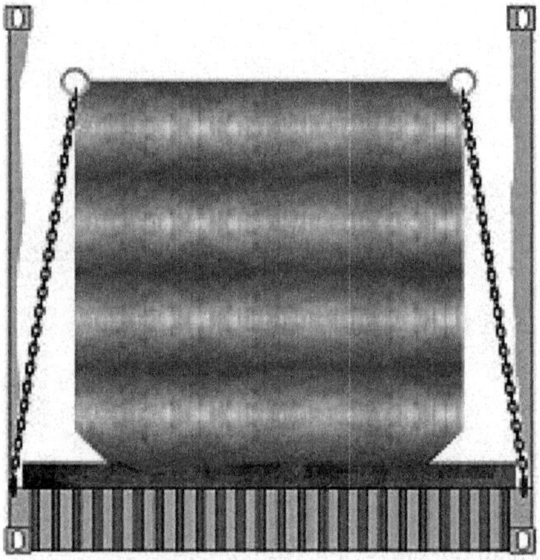

Fig. 17.19 Diagonal lashings with small horizontal transverse components

If this happens, the purely vertical lashings are displaced sideways and sometimes placed under considerable tension. The tension that is created as a result of the load moving exerts a force on the load pushing it down on the container floor. This increases the frictional forces on the floor. Frictional forces for each lashing are calculated using the product of the MSL and the friction coefficient. Diagonal lashings form a lashing angle α with the horizontal plane and a further lashing angle b with the horizontal longitudinal direction of the transport unit. According to the terms defined by the VDI, lashings of this type are referred to as diagonal lashings. With the diagonal lashings shown here, the vertical components are the greatest. The horizontal longitudinal components are smaller, and the horizontal transverse components even smaller still. Cross lashings are simply a special form of diagonal lashing. These are advantageous if the lack of space requires greater horizontal components. The

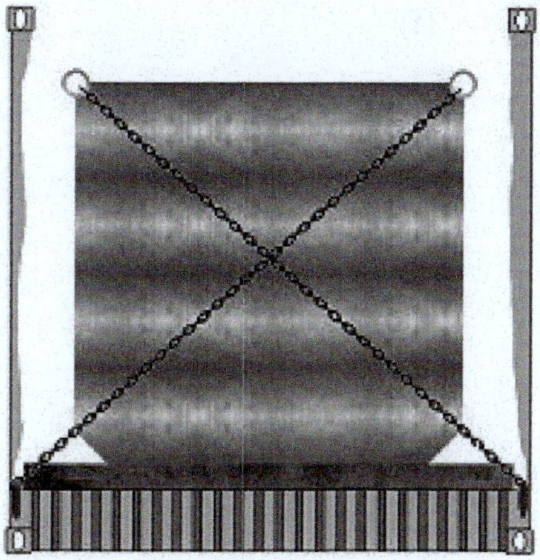

Fig. 17.20 Cross lashings with better horizontal transverse components

package on the right Fig. 17.21 (r) is better secured laterally than the one on the left Fig. 17.21 (l). Under such conditions cross lashings are often suitable. But beware: This no longer applies if the package is to be secured against tipping (Fig. 17.22).

The optimum angle for securing against tipping is to imagine a line to the outside at right angles to the line connecting the lashing point and the tipping edge. A clear warning must be issued against the so-called round turn lashings where the lashing material is fixed at one side, wrapped around the object and then fixed at the other side Fig. 17.23 (l) whereas round turn lashings allow the loads to move freely Fig. 17.21 (r).

Loop lashings are unfortunately rarely used as a form of tight-fit securing although they are extremely effective. This method can be used very effectively with all long, cylindrical loads (Figs. 17.24, 17.25 and 17.26).

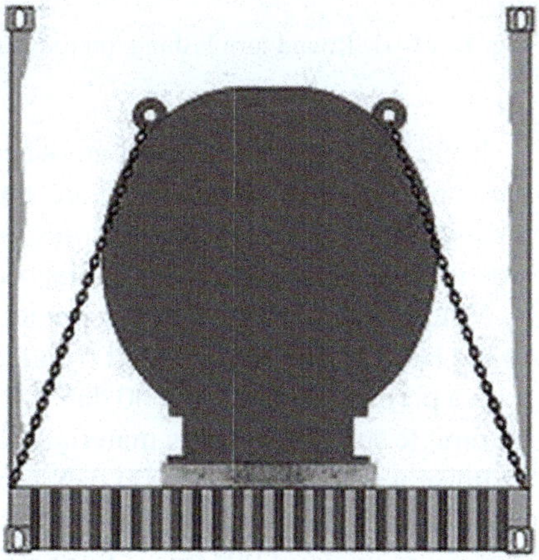

Fig. 17.21 (l) Ineffective components for securing against tipping; (r) effective components for securing against tipping

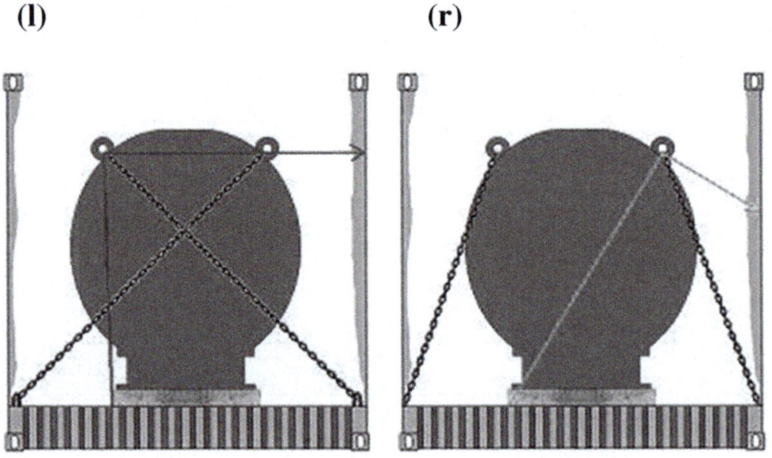

Fig. 17.22 (l) Optimum lashing angle to protect against tipping, but impossible to implement; (r) the real lashing angle is close to the optimum to protect against tipping

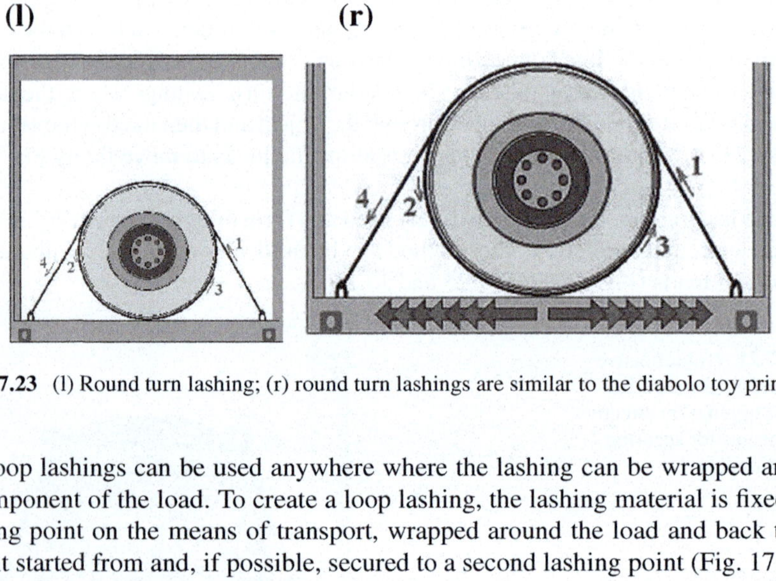

Fig. 17.23 (l) Round turn lashing; (r) round turn lashings are similar to the diabolo toy principle

Loop lashings can be used anywhere where the lashing can be wrapped around a component of the load. To create a loop lashing, the lashing material is fixed to a lashing point on the means of transport, wrapped around the load and back to the side it started from and, if possible, secured to a second lashing point (Fig. 17.27).

When using separate lashing points, the permitted lashing force of a single strand of the lashing can be increased by a factor of up to two at best. With a lashing point with a permitted load of 2,000 daN (approximately 2t) and a permitted lashing force of a single strand of lashing material of 2,000 daN (approximately 2t) as well, this will result in a maximum of 4,000 daN (approximately 4t) securing force. The opening angle between the ends of the lashing material leads to a certain reduction: with a maximum angle of 90° this is thirty percent. The pre-tensioning force in the lashing will also reduce the securing force. However, depending on the circumstances, the

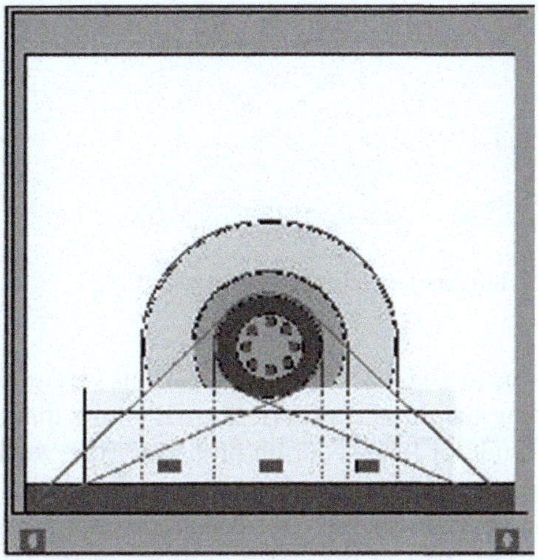

Fig. 17.24 Loop lashing used to secure a piece of machinery—lateral view

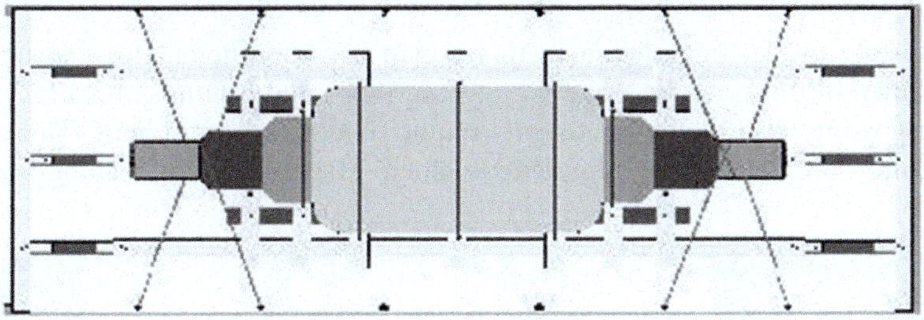

Fig. 17.25 Loop lashing used to secure a piece of machinery—view from the top

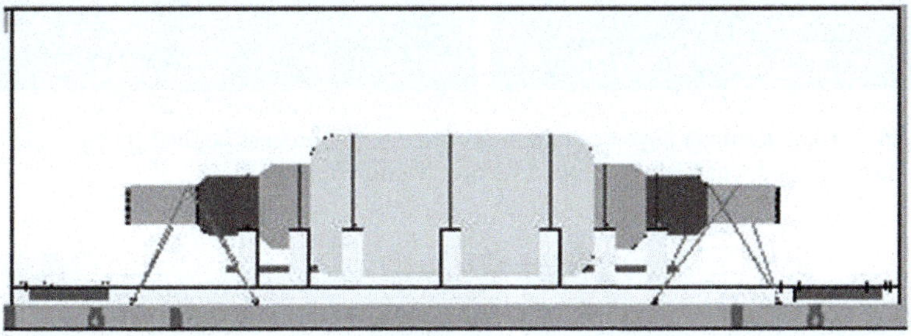

Fig. 17.26 Loop lashing used to secure a piece of machinery—side view

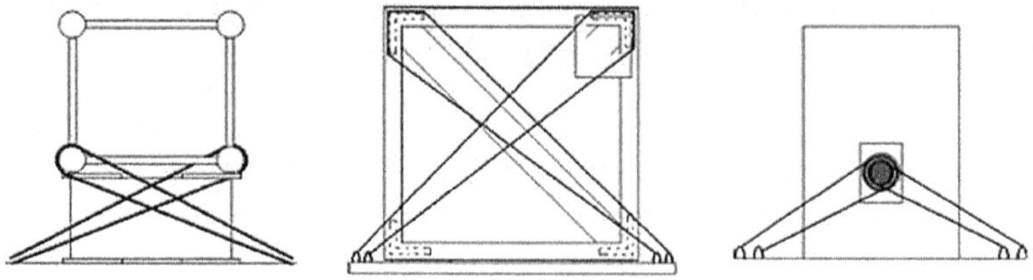

Fig. 17.27 Examples of different loop lashing techniques

frictional forces may be increased. There are numerous variations on the loop lashing principle. A little imagination and experience will show that this principle can be used to solve a multitude of load securing problems. The weaknesses of the round turn lashing, and the strengths of the loop lashing have been investigated by the *Fortbildungszentrum Hafen Hamburg* (the training centre for the Port in Hamburg) in a highly effective test. Part of the series of photographs taken from the test are shown below (Figs. 17.28 and 17.29):

If there are no securing points, lashings can be attached with "head loops" (Fig. 17.30)

Note that the lashing shown in this sketch is not sufficient for carriage. It is only shown to illustrate the principle of head loops and lashings attached to them. Securing using head loops will usually require two members of staff. This method is much too time-intensive to implement alone. The term spring lashing is used to

Fig. 17.28 (l) The lashings have been applied—the tipping process begins; (r) an angle less than 30° has been reached—part of the friction forces are still effective

Fig. 17.29 (l) The round turn lashing fails, the loop lashing remains firm; (r) impact: the round turn lashing fails even more, the loop lashing still remains firm!

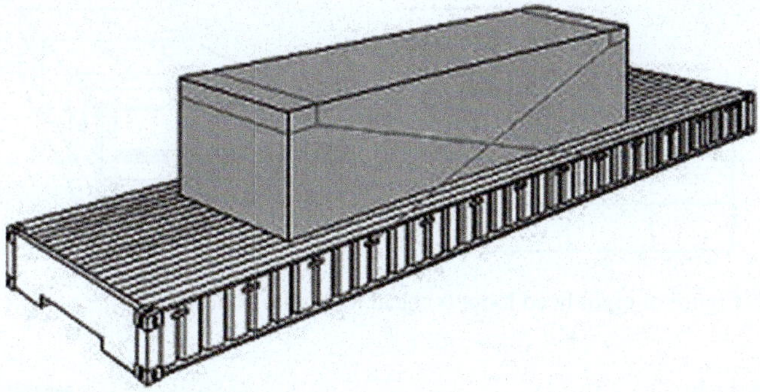

Fig. 17.30 Head loops with spring lashings in the form of a cross lashing

described lashings that run from the outside to the inside, regardless of whether they cross. Head loops can take a number of different forms. This handbook and the GDV Cargo Securing Manual contain a number of samples in the numerous examples.

If layered goods run the risk of the head loop cutting in between the individual packages, then a figure of eight head loop can be used to achieve the same result. The use of the head loops as shown above, requires the packages to be stable enough to take the full MSL of the lashing. If this is not the case, boards, lattices, walking boards, square lumber dunnage and other similar aids can be used to cushion them in order to meet the prerequisites for the head loops. This is usually needed when securing bundles, IBCs, units, bagged cargo pallets, or other similar packages. The majority of securing errors are made when making and securing the lashings. This picture shows the use of a hook which has already been bent open, and which has also been inserted in the wrong direction. The chain tensioner shown has not been used correctly. It must be tightened down by at least 45° and must not protrude at right angles, otherwise there is a danger that the link of the chain could escape from the lug of the tensioner. The lashing angles are not suitable. High vertical components have been achieved, but the longitudinal and transverse components are too low. Figure 17.31 shows the use of a load securing tarpaulin with textile straps that have been secured to the ends of the roll trailer, thus ensuring a tight-fit. This type of securing is only suitable for ocean-going transport if the roll trailers are stowed in a "tight-fit" constellation next to each other and either along the length of the ship or athwartships. Since the load has simply been strapped down at the sides of the roll trailer, the securing is sufficient for terminal transport, but not for ocean-going transport (Fig. 17.32).

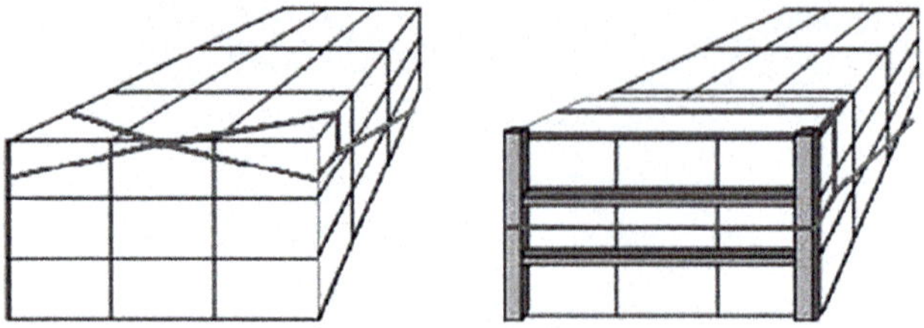

Fig. 17.31 (l) Figure of eight head loop; (r) head loop with padding

Fig. 17.32 Tight-fit securing of rolls of paper using a WisaFix lashing cover

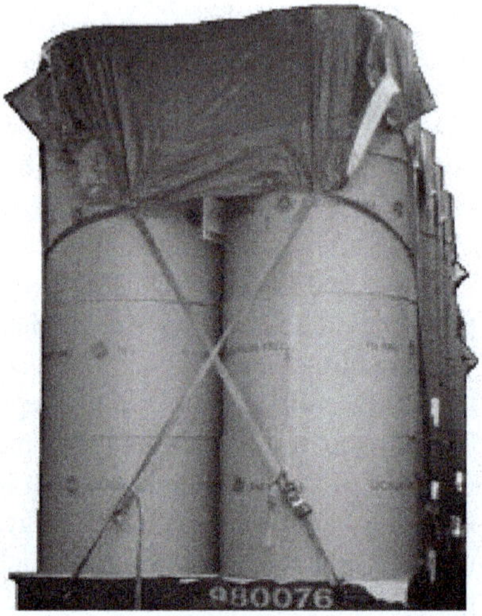

17.6 Friction Securing

Friction securing or friction loop securing means that the frictional forces between the packages and the intermediate layers or underlying surface are greater than the forces caused by transportation, and which are working on the load. Only if this requirement is fulfilled when the load is packed and can be maintained during the entire transportation process, can the load be considered "secure." A simple example taken from everyday life is the passenger traveling on a bus who places their foot on top of their briefcase in order to prevent it sliding away from them in a corner or during acceleration and deceleration (Fig. 17.33).

The best-known and most frequently used form of friction securing is the round turn or friction loop. The VDI guidelines and other road traffic publications refer to this method as tying down, and to the lashing as a tie-down. The rail transport industry uses the term tie-down. Whatever term is used, the guidelines are referring to the same method: With the aid of tensioning elements, a certain degree of pre-tensioning is introduced into the lashings which effectively puts additional artificial weighting on the secured package. This causes additional frictional forces on the

Fig. 17.33 Friction loop made of steel strapping

bottom of the package. In conjunction with the "normal" frictional forces this will give the overall securing force of a given lashing. The effectiveness of this method of securing increases with the friction and the effective pre-tensioning. If tie-down lashings can be referred to as "advantageous" at all, this is as a result of fact that the frictional forces act in all directions. Thus, it is not significant from which side of the object the tie-downs are applied (Fig. 17.34).

The general securing principle is that the vertical components of the pre-tensioning force introduced to the lashing aid increase the downward pressure between the load and the surface it is resting on, thus generating additional frictional forces. The smaller the vertical components and the frictional values, the lower the securing forces that can be achieved. Where tensioning aids are applied to just one side of an object, the pre-tensioning force decreases towards the opposite side. The rougher the contact surfaces over which the material passes, the lower the pre-tensioning force on the opposite side. The use of smooth edge protectors helps to ensure that as much of the pre-tensioning force created on the one side is distributed to the other as possible. Jolting may cause the pre-tensioning force on the tensioned side to decrease and to increase on the side that is not tensioned. If it is not possible to use smooth edge protector corners with tensioners used on one side only, it is sensible to tension the lashings on alternate sides. The lashing capstans would have been more useful if three of them had been installed on each side of the flat rack. Here, it is not ideal

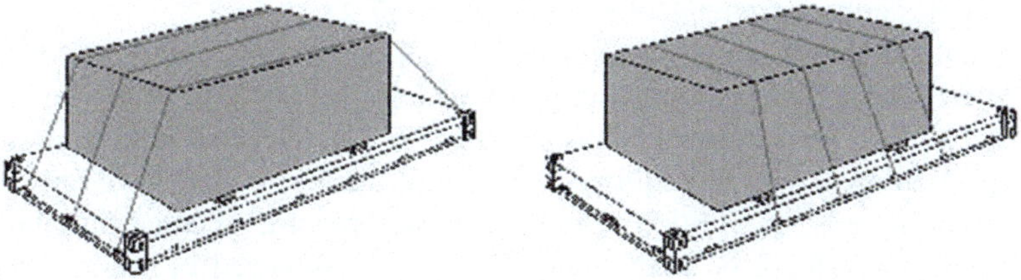

Fig. 17.34 (l) and (r) frictional forces act in all directions

that all the steel straps are tensioned on the same side. What is important is that the pre-tensioning force always remains constant. Thus, it is only sensible to use tie-downs with elastic materials. Wire rope, chains and even steel straps are unsuitable for use as tie-downs unless they can be rendered elastic by the use of special padding materials. When transporting materials that give way easily, a padding with a high degree of recovery, such as tyre rubber, is not going to be highly effective. The loss of pre-tensioning force is quite dramatic. Unfortunately, steel wire ropes tend to lose tension. To combat this, we can use tyre rubber as an ideal method of reducing damage to the tarpaulins and to the load. The magnitude of the pre-tensioning forces that can be introduced into the lashing material is dependent on the tensioning forces used, the leverage on the tensioning element and the elasticity of the lashing material. Experience shows that the following lashing materials are able to attain the following pre-tensioning forces in the lashing material when tensioned with normal physical effort:

- Normal ratchet tensioners approximately 300 daN per side.
- Tensioner for single-use webbing approximately 500 daN.
- Long-lever ratchet tensioners 500–750 daN per side.
- Commonly available spindle tensioners and capstans 500–1,000 daN.
- Manual steel strap tensioners 1,000–1,500 daN; and
- Lashing turnbuckles approximately 3,000 daN.

These values are not definitive; they are based on practical experience gained in a seaport. Practitioners should calculate the appropriate values for their field depending on the materials used. The use of special tensioning aids may make it possible to apply greater pre-tensioning force. The use of pretension indicators is recommended. These allow both accurate measurement of the pre-tensioning force during lashing, as well as allowing the residual pre-tensioning force to be checked during transportation. However, it must be said that this is of little use if the containers cannot be checked during the transportation process. For reasons that will not be explained here, the pre-tensioning force is only permitted to be, at the most, half of the MSL or the permitted lashing force of a single lashing. The vertical component of the pre-tensioning forces decreases as the lashing angle of the tie-down decreases. From a mathematical point of view, the rate of decrease is by the sine of the lashing angle. The following values can be used as guidelines (Table 17.1).

The table below shows examples of the magnitude of the overall securing forces of tie-downs with a total pre-tensioning force of 600 daN when used with various lashing angles and coefficients of friction (Table 17.2).

The table shows that with a total pre-tensioning force of 600 daN, a lashing angle of 90° and a coefficient of friction of 30% it is possible to achieve a securing force of 180 daN per tie-down. Thus, in order to secure a load with a mass of 9t, 50 (fifty!) tie-downs will be required. This shows that tie-downs have only limited use as securing aids. In addition to this, tie-downs require an equal distribution of the pre-tensioning force on both sides. This can be achieved by using tensioning elements in pairs, centrally or alternately. The effective securing forces can be very accurately calculated on site with little effort and some basic mathematics, as long

Table 17.1 Influence of the lashing angle on the vertical component

Lashing angle	Magnitude of the vertical component			Graphical component
90	1.00	or	100%	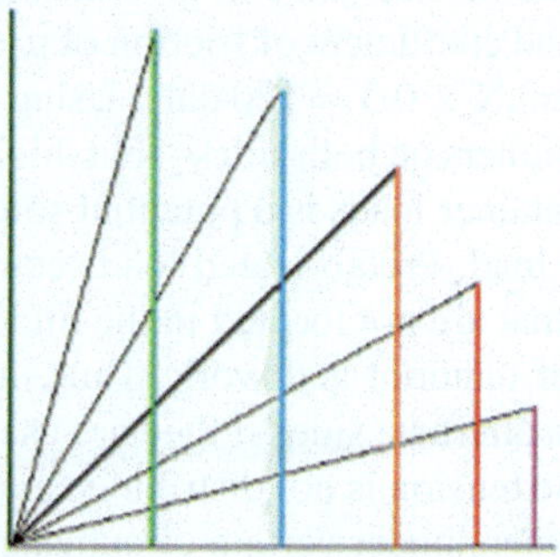
75	0.97	or	97%	
60	0.87	or	87%	
45	0.71	or	71%	
30	0.50	or	50%	
15	0.26	or	26%	
0	0.00	or	0%	

Table 17.2 Securing forces depending on pre-tensioning force, lashing angle and friction

Total pre-tensioning force in daN	Lashing angle	Vertical component	Friction						
			0.10	0.20	0.25	0.30	0.40	0.50	0.60
600	15	0.26	16	31	39	47	62	78	94
	30	0.50	30	60	75	90	120	150	180
	45	0.71	43	85	107	128	170	213	256
	60	0.87	52	104	131	157	209	261	313
	90	1.00	60	120	150	180	240	300	360

as the pre-tensioning and frictional forces are known or can be estimated with some accuracy (Fig. 17.35).

The effective length of the lashing aid in the example is measured at 1.00 m (dark-green line). The vertical component of the lashing in the example is 0.80 m

Fig. 17.35 Determining the vertical component of the pre-tensioning force of a lashing

(light-green line). The vertical component of the pre-tensioning force is thus 80% or 0.8 (always divide the smaller value by the larger value). Multiplying the value with the pre-tensioning force of the lashing aid and the coefficient of friction between the package and the surface will give the securing force. With a pre-tensioning force of 500 daN and coefficient of friction of $\mu = 0.3$, the lashing applies a securing force of 0.8×500 daN $\times 0.3 = 120$ daN. Using conventional materials, it would be possible to use tensioners on both sides—but this is not practical. The central use of tensioning aids in container loads is a potential source of accidents since staff need to climb on top of the load. Prefabricated load securing materials usually only have tensioning elements that are not located in the middle and would thus require additional outlay in terms of material and work. Thus, it is recommended that tie-downs should be secured on alternate sides if they are used at all. Shifting of the load is more likely to occur if the tension is not distributed equally. Tie-down lashings can only be used in box containers under certain circumstances. They are easy to use with all forms of platform container but are not to be recommended. As has already been mentioned, in order to ensure adequate securing of the load, it is necessary to maintain the pre-tensioning during the entire transportation process. Even the slightest of gaps in the load, the settling of the load, yielding or compression of the load, or the cutting of lashing aids into the load as a result of a lack of edge protectors can all reduce the pre-tensioning force to zero. Loads which have an overhang are unsuitable for tie-down lashings for another reason (Fig. 17.36).

As a result of the unsuitable lashing angle in the lower area, it will be impossible to attain sufficient pre-tensioning force regardless of the type of lashing material used. Thus, the risk is particularly high that the package will shift during transport (Fig. 17.37).

To summarise, friction loops or tie-downs can only really be used as an effective form of securing if there are no gaps in the stow, or if any gaps are filled. With small gaps, the use of wedges or tapered blocks is recommended. Larger gaps in the load should be braced using wood or another equally durable material. It must also be possible to create sufficiently high pressure. If this is not possible, loads should be secured using other methods. Lashing materials without sufficient elasticity are not suitable for use as tie-downs. Those with a suitable degree of elasticity, for example

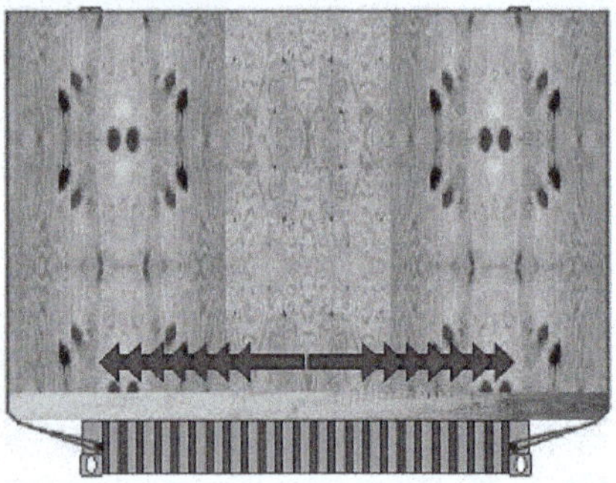

Fig. 17.36 Potential movement of the load caused by too little securing force applied to a wide load

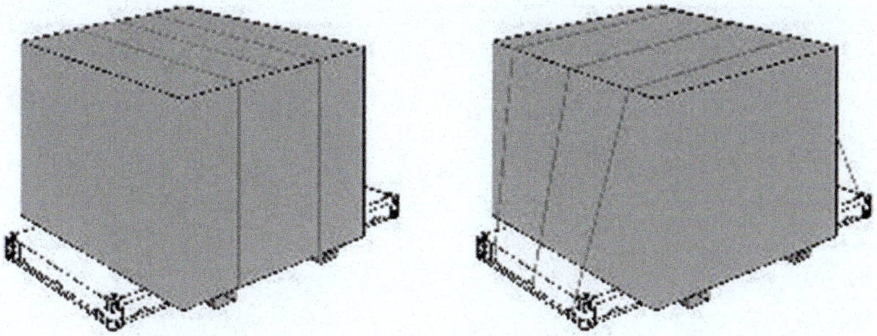

Fig. 17.37 (l) Poor strapping; (r) still poor strapping, but slightly improved

lashing straps, will gradually lose their pre-tensioning force during use. This process will be accelerated if the load is exposed to vibrations. Tie-downs must thus be checked and, if necessary, tightened at regular intervals. Since this is not possible when a load is stowed in a container, tie-downs are not suitable for securing loads inside containers. Again: If another method of load securing can be used, tie-downs should be avoided! One thing tie-downs always do is to create a direct lashing acting against vertical movement. Friction securing can also be achieved using wedges. Another variation is to put pressure on the load forcing it down. In the event of an absence of available/accessible lashing points, this method of securing may be used in box containers. The prerequisite is the accessibility of components of the means of transport that are able to withstand this load. Depending on the magnitude of the securing forces to be created, the sensitivity of the packages and the resilience of the components of the means of transport, squared lumber, planking and boards can be used. Caution: at very sharp angles, driving the diagonal stays in laterally can create extreme vertical forces.

17.7 Securing Against Tipping and Other Hazards

The stability of packages will depend on the position of the centre of gravity and the load bearing area. Gravitational forces act on the centre of gravity (Fig. 17.38).

These act vertically and downwards. On a steep incline, the gravitational force is divided into two components: one acting vertically to the plane and one in the direction of the slope, the shear force. Each of these forces acts on the centre of gravity. Inertia forces arising as a result of a change in speed or direction within a given time period will also act on the centre of gravity. They may cause packages to slide or topple. When an object tips, the centre of gravity is raised. If the centre of gravity remains within the tilting edges, the package will return to its normal position, as soon as external forces cease to act on it. If the centre of gravity of an object is located directly over the tilting edge, the package is unstable. The degree to which the centre of gravity must be raised before it reaches that position is a measure

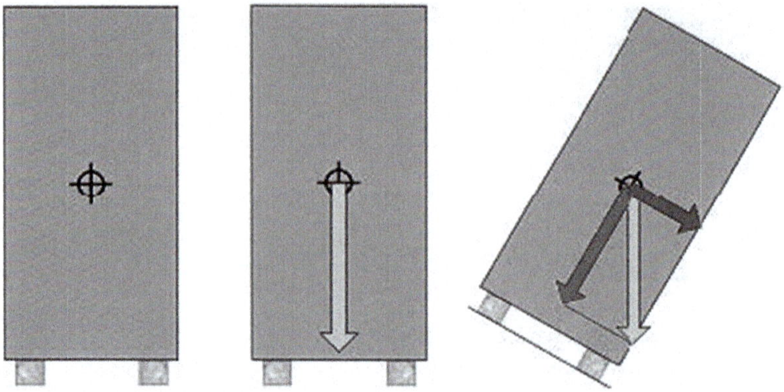

Fig. 17.38 Examples of gravitational forces

of the stability of the object. The product of the weight of the object and this distance gives the stability moment. The lower the centre of gravity, or the greater the lower surface area of the package, the greater the stability moment. If the centre of gravity moves beyond the tilting edge, then the package will topple. A number of different methods can be used to prevent tipping, such as:

- Combining a number of units that are liable to tip to form a single block.
- Bundling.
- Increasing the lower surface area by using blocks or lashings; and
- Lashing with head loops in the absence of lashing points.

Caution: when using this method, the opposing forces created are not as high as those sketched previously.

- Using bracing elements to prevent tipping; and
- Using diagonal stays to prevent tipping.

The packing and formation of load units is responsible for protecting against shifting or racking. Improvements must be made if packages or unit loads are clearly too weak. This may be done using walking boards, wooden lattices etc. and additional strapping. The collapse of an entire unit load as a result of packaging that is not strong enough and/or excessive stack loads can be avoided by selecting out or rejecting packaging that is not strong enough, by improving the packaging, employing sufficient interlayer dunnage and by building "dummy decks."

17.8 Securing by Nailing

A popular method of securing loads is to use wooden constructions that are fixed by nailing. Squared lumber and wooden wedges are used, but also planks and battens. These solid wooden constructions can only really be effective if the nails used are strong and long and can penetrate deep into the wood and into a surface which will take nails. Container floors have a thickness of just 25–30 mm. Also, the majority of floor types will be damaged by the use of nails. Thus, these methods cannot really be recommended for securing in containers. They could be used on softwood planking, as is sometimes used on platform containers, but are not suitable as the only form of securing for heavy goods. Another variation is to create an additional, strong, specially-designed base made of planks or squared lumber beams and driving nails into this. If the method is not used for light packages that are at risk of tipping, as it usually is, then it must be ensured that only top-quality wood is used, with a thickness of at least 50 mm. Top-quality in this case means, healthy, free from cracks and dry. Shear strength and extraction resistance are influenced by the diameter of the nails, the penetration through the wood, the penetration into the underlying surface, the angle at which the nail is driven in, and the properties of the nail surface. Nails with grooves or with coatings will have higher resistance than ordinary wire nails. Experience suggests that a nail, 0.5 mm thick, that is driven in vertically and which penetrates through 50 mm of the grain side of a piece of wood and is driven into a depth of 40 mm can be expected to offer 400 daN of resistance. Testing suggests that this value can be halved for wet wood. If the nail is driven in along the grain the values can also be halved. The resistance values are also influenced by the pattern of nails in the wood and the distance between the nails. Nails should always be driven in vertically or at a slight angle to the direction of loading. They should not be too close to the edge of the piece of wood and should not run in a line along the grain. Nails can be driven in efficiently using a compressed air nail gun. Magazines of nails, either rolled or in strips provide a source of nails that can be used with little physical effort. When using wedges to secure loads, badly cut wedges are often selected. Wedges should be cut and positioned in such a way that the long side of the wood, i.e., the grain side, is the side that is nailed, and not into the grain. Wedges of this type are also called chocks, roll or pipe wedges. The critical factor is that the grain side of the wood is available to take the nails. Specially designed wedges of a variety of shapes can be used to secure special types of loads. The securing effect is achieved by the shape of the wedge itself. The nails serve purely to hold the wedge in place.

Chapter 18
Load Securing, Filling Materials and Use of Friction

In this chapter, we will look at how to determine the maximum securing load of materials used for load securing using simple rules of thumb and explains some basic principles for practical work (Figs. 18.1 and 18.2)

Before we do though, it is worth explaining some of the key terms used in load securing. Some of these terms we have already seen and used in earlier chapters, and some will be entirely new to us. ***Kilonewton (kN)*** is a unit of force useful for describing for instance the breaking strength or breaking load of load securing material. It has replaced the previously common metric ton which is the unit reserved for describing mass according to the SI standard. The conversion is easily learned: 1 kN ≡ 0.1 t or 100 kg. Anyone who was used to calculating mass in kilograms can use the unit decanewton (daN) as a unit of force. The following English terms are in common use in maritime transport across the world. ***Securing element*** is an individual item of equipment on board ship which is used for load securing, e.g., a shackle, a deck ring, a turnbuckle, chain or wire rope. ***Securing device*** is a suitable combination of elements which together form a means of load securing, e.g., lashing or bracing. ***Securing Arrangement*** is a reasonable arrangement of load securing means with the aim of securing a cargo item or a cargo block. ***Breaking Load (BL)*** is the nominal breaking load, specified by the manufacturer. However, it can also be estimated using rules of thumb. ***Maximum Securing Load (MSL)*** in kN, is the greatest permissible force which can be applied to a load securing element or device. Calculation Strength (CS) in kN, is an arithmetic force determined by reducing the MSL by the formula: CS = MSL/1.5. CS values are only used to assess the efficiency of securing arrangements as per annex 13 of the CSS code. The relation between Breaking Load and Maximum Securing Load is shown in Annex 13 and in accordance with Table 18.1. ***Lashing elements and lashing materials***. There are no international standards on tie down lashings. It is to be expected, however, that manufacturers or dealers will provide information on or certification of the nominal breaking load on purchase. It is, however, unclear how this value was determined and under what conditions it is valid. No reference is made to any other properties, such as

A. A. Olsen and F. Karkori, *Containerized Cargo Handling and Stowage*, Springer Series on Naval Architecture, Marine Engineering, Shipbuilding and Shipping 22,
https://doi.org/10.1007/978-3-031-63329-4_18

Fig. 18.1 (l) Steel wire rope; (m) lashing point; (r) steel strap

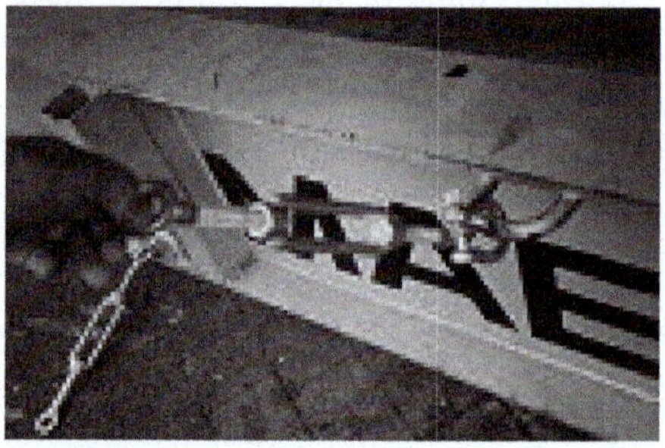

Fig. 18.2 Inspecting lashing equipment

elasticity and fatigue strength. The table below provides a list of the most important materials and elements with the usual characteristic values. An accepted rule of thumb is used for the breaking load. If millimetres are chosen instead of centimetres for the dimensions, the breaking load values will be in decanewtons [daN] instead of kilonewtons. For economic reasons, it is advisable to try to homogenise load securing equipment and load securing arrangements:

- ***Homogeneous load securing equipment*** comprises elements which where possible have the same MSL values.
- ***A homogeneous load securing arrangement*** comprises load securing equipment which are arranged in such a way that, when subjected to extreme loads, they bear the part of the load appropriate to their strength.

Table 18.1 Determining the maximum securing load from the breaking load

Material	Maximum securing load
Shackles, Rings, Deck Eyes, Turnbuckles of mild steel	50% of breaking strength
Fibre ropes	33% of breaking strength
Web lashing	70% of breaking strength
Wire rope (Single use)	80% of breaking strength
Wire rope (Reusable)	30% of breaking strength
Steel band (Single use)	70% of breaking strength
Chains	50% of breaking strength
Lumber	0.3 kN per cm^2 normal to the grain

To summarise the problems of load securing, some examples are provided below as a sort of "guide" on how to calculate the number of securing devices required, what such a device can withstand and what can be expected of it (Table 18.2).

18.1 Tie Down Lashing

Worked example. Let us assume that the wooden case on a 'standard' flat rack has a weight of 12,000 daN. Without considering the risk of this over height case tipping, the package must be secured for overseas shipment. Lateral acceleration forces of 0.8 g can be expected. This means that lateral forces of 12,000 daN × 0.8 or 120 kN × 0.8 i.e., 9,600 daN or 96 kN are anticipated. The shipper has decided to use single-use webbing belts which have a breaking load of 3,433 daN. This equates to 2,403 daN at an MSL no greater than 70% of the breaking load. No more than half of this, i.e., around 1,200 daN, may be used as the pre-tensioning force. It should be noted that in practice this value can neither be achieved nor maintained throughout the entire voyage. The effective length of the belt from its attachment to the lashing point to the edge of the case is 3.0 m. The effective height is 2.93 m, an extremely high vertical component (97.6%). This component can be determined for any load by dividing the effective height by the effective length. Multiplying this by the pre-tensioning force gives the force with which the tensioned side of the load is pulled onto the flat. In the example, this is 97.6% of 1,200 daN, or 1,171 daN. If we assume ideal conditions and this force were completely transmitted to the other side, a total pre-tensioning force of 2,342 daN per lashing can be assumed. Assuming a friction coefficient of 0.3, a single tie-down lashing can achieve a securing force of around 703 daN. 13.65 belts are theoretically required to secure the case (9,600 daN/703 daN). In reality, the webbing belts used would be able to maintain a maximum pre-tensioning force of around 100 daN through 200 daN during the voyage. This means that a single belt is able to maintain a long-term securing force of 30 daN through 60 daN. To have

Table 18.2 Common securing materials and accepted breaking loads

Material/Element	Breaking load	Remarks
(a)		
Natural fibre ropes (Manila, Sisal, Hemp)	$6 \times d^2$	d = diameter of rope in cm. Natural fibre ropes are sensitive to decay, acids and alkalis. All fibre ropes are sensitive to chafing from sharp edges. Knots on synthetic fibre ropes can slip open. Heavers of sufficient thickness should be used to tighten them and these in turn should be secured to prevent them from unwinding
Polypropylene	$12 \times d^2$	
Polyester	$15 \times d^2$	
Polyamide	$20 \times d^2$	
Hercules (Sisal)	$6 \times d^2$	
Hercules (Polypropylene)	$12 \times d^2$	
(b)		
Wire rope 6 × 9 + 1 FC Wire rope 6 × 19 + 1 FC Wire rope 6 × 37 + 1 FC	$50 \times d^2$	d = diameter of rope in cm. Producing conventional wire rope lashings with turnbuckles and rope clips is technically demanding and can give rise to a number of potential problems
Wire rope 6 × 9 + 7 FC Wire rope 6 × 12 + 7 FC Wire rope 6 × 15 + 7 FC	$25 \times d^2$	
(c)		
Shackles	$20 \times d^2$	d = diameter of bolt in cm. The breaking load formula only applies to shackles made of standard strength steel
Turnbuckles	$20 \times d^2$	d = diameter of thread in cm. The breaking load formula only applies to turnbuckles made of standard strength steel
(d)		
Untreated steel strap Blued steel strap	$70 \times w \times t$ $85 \times w \times t$	w = width of strap in cm t = thickness of strap in cm
Long- and short-link chains with different tensioners	See manufacturer's instructions.	Tie down lashing chains are always made of higher strength steel to save weight. Calculation of the breaking load is therefore dependent on the manufacturer's specifications
(e)		
Deck eyes and eye plates	$20 \times d^2$	d = diameter of eye material in cm. The breaking load formula only applies to material made of standard strength steel

(continued)

Table 18.2 (continued)

Material/Element	Breaking load	Remarks
(f)		
Synthetic fibre lashing belts	$20 \times d^2$	Lashing belts are produced in a number of different grades. They are highly elastic but can become permanently deformed when subjected to threshold stresses greater than 50% of the breaking load and therefore quickly become loose. They must not be knotted. They are sensitive to external influences in the same way as synthetic fibre ropes
(g)		
Weld joints subjected to shear loads	MSL = 10 kN per cm	Single-layer weld, 4 mm thick
	MSL = 10 kN per m	Three-layer weld, 10 mm thick
(h)		
Softwood used for bracing	MSL = 0.3 kN per cm^2	Compressive load perpendicular to the grain
Softwood used for bracing	MSL = 1.0 kN per cm^2	Compressive load parallel to the grain
(i)		
Special equipment for RoRo ships	–	Trailer horses, trailer jacks, wheel chocks; breaking loads usually unknown
Special equipment for container ships	Refer to the manufacturer's instructions	Lashing rods, Turnbuckles, Twist Locks, D rings, Sockets, Bridge Fittings, Tie Plates, etc. Strength and material properties as per the requirements of the relevant classification society

really "secured" the case, somewhere between 160 and 320 belts would have to be provided!

This guide is simpler and more precise than a calculation which uses the lashing angle a, since in practice distances are easier to measure than angles. If the vertical component of a lashing is to be calculated using the lashing angle, the permissible lashing force must be multiplied by the sine of the lashing angle: vertical component = MSL $\times$ sin α. The smaller the lashing angle, the smaller the vertical component will be. At a lashing angle of 90° it will be 100% (sin 90° = 1), at 75° 97% (sin 75 = 0.9659), at 60° 87 % (sin 60° = 0.866), at 45° 71% (sin 45° = 0.7071), at 30° 50% (sin 30° = 0.5), at 15° 26% (sin 15° = 0.2588) and at 0° 0% (sin 0° = 0) (Fig. 18.3).

TIE DOWN LASHINGS ONLY PROVIDE SECURING FORCES OF THE VERTICAL COMPONENT OF THE PRE-TENSIONING FORCE MULTIPLIED BY THE FRICTION COEFFICIENT

THE PRE-TENSIONING FORCE MUST NEVER BE GREATER THAN 50% OF THE MSL OF THE WEAKEST SECURING ELEMENT

Fig. 18.3 Example of tie-down lashing

18.2 Direct Lashing

The main difference between direct lashings and tie-down lashings is that with direct lashings, the pre-tensioning force can and should be kept as low as possible.

THE PRE-TENSIONING FORCE SHOULD BE AS LOW AS POSSIBLE ON DIRECT LASHINGS. HOWEVER, SLACK MUST NEVER BE ABLE TO DEVELOP IN LASHINGS

The pre-tensioning force must, however, be sufficiently high to prevent a lashing from becoming slack. The reason for this is that the lashing may be loaded up to its MSL under stress and the vertical components resulting from this produce additional frictional forces (Fig. 18.4).

For those who enjoy mathematics, the relevant lashing forces can be calculated by first measuring the lashing angle α (47.5°) and then the sine and cosine of this angle to determine the vertical and horizontal components and using these in conjunction with the permissible lashing force of the chain. Using another guide, it is unnecessary to determine this angle or its sine and cosine. Basic arithmetic will suffice. The following lengths are determined by using a tape measure or metre rule: the effective length of the lashing chain (3.61 m), the effective vertical component (2.66 m) and the effective horizontal component (2.44 m). Using rules of thumb or the manufacturer's specifications, the breaking load and the MSL of the 13 mm link material diameter high-tensile chain. This corresponds to 10,000 daN. Checking the size of the lashing point gives a steel diameter of 28.3 mm. This gives a breaking strength of 16,000 daN and an MSL of 8,000 daN. This value represents an upper threshold if the chain components have a higher MSL. Vertical securing force: 2.66 m: 3.61 m ×

Fig. 18.4 Example of direct lashing with chains

10,000 daN = 7,368 daN. A lashing chain secures the package against vertical movement with a force of 7,368 daN. But this is not so important. This force becomes effective when the package is moved horizontally causing the chain to be tightened; the package is then pulled toward the floor with this force. Assuming a sliding friction coefficient of 30% ($\mu = 0.3$), the package is secured by a lashing in all directions with a force of 7,368 daN × 0.3 = 2,210 daN. Shortfall in securing force: 2.44 m/ 3.61 m × 10,000 daN = 6,759 daN. A lashing chain directly secures the package laterally with 6,759 daN. To this are added the frictional securing forces of 2,210 daN as previously determined. The calculated chain lashing secures the machine component against movement laterally towards the right with a force of 8,969 daN. Since no longitudinal components exist, a chain only secures the machine component longitudinally with the frictional forces produced by the vertical component of 2,210 daN. To calculate the longitudinal, transverse and vertical securing forces using the lashing angles α and β, use the following method (Table 18.3):

Table 18.3 Calculation for the longitudinal, transverse and vertical securing forces using the lashing angles α and β

Component	Calculation
Vertical component	MSL × sin α
Horizontal component	MSL × cos α
Additional frictional forces	Vertical component × μ or $MSL \times \sin\alpha \times \mu$
Pure lateral component	Horizontal component × sin α or $MSL \times \cos\alpha \times \sin\alpha$
Pure longitudinal component	Horizontal component × cos α or $MSL \times \cos\alpha \times \cos\alpha$

Since the additional frictional forces produced by the vertical component may be added to the forces produced by the lateral and longitudinal components, the securing forces produced are (Table 18.4):

Again, a guide is quicker, simpler and more precise. Four distances must be measured and the MSL of the weakest load securing element must be determined. The friction coefficient must be determined or estimated (Table 18.5).

Calculated example. A doubled, single-use webbing belt is used for securing as previously. For reasons of caution, the MSL is not assumed to be 70% but only 30% of the breaking load. The sketch below shows all the data needed (Table 18.6):

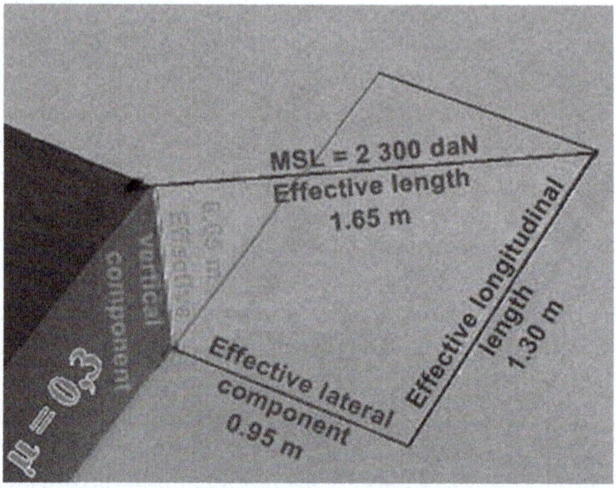

Table 18.4 Additional frictional forces produced by the vertical component

Securing forces	Calculation
Vertical securing	MSL × sin α
Horizontal securing	MSL × cos α × cos α + MSL × sin α × μ
Longitudinal securing	MSL × cos α × cos α + MSL × sin α × μ

Table 18.5 Additional frictional forces produced by securing forces

Securing forces	Calculation
Vertical securing	Effective vertical component/effective length × MSL
Additional frictional forces	Effective vertical component/effective length × MSL × μ
Lateral securing including additional frictional forces	Effective lateral component/effective length × MSL + effective vertical component/effective length × MSL × μ
Longitudinal securing including additional frictional forces	Effective longitudinal component/effective length × MSL + effective vertical component/effective length × MSL × μ

Table 18.6 Additional frictional forces produced by securing forces

Securing forces	Calculation
Vertical securing	0.65 m/1.65 m × 2,300 daN = 906 daN
Additional frictional forces	0.65 m/1.65 m × 2,300 daN = 906 daN × 0.3 = 271.8 daN
Lateral securing including additional frictional forces	0.95 m/1.65 m × 2,300 daN = 1,324 daN + 271.8 daN = 1,595.8 daN
Longitudinal securing including additional frictional forces	1.30 m/1.65 m × 2,300 daN = 1,812 daN + 271.8 daN = 2,083.8 daN

As far as the overall securing is concerned, it should always be borne in mind what lashings are subjected to loads under what circumstances (Fig. 18.5):

It can be concluded that the lashing arrangement would not be homogeneous if the load in the example were loaded on a flat rack transported using fore and aft stowage on a container ship, where acceleration forces of 0.4 g occur longitudinally and acceleration forces of 0.8 g occur laterally. In this instance, the lashings would need to have been arranged so that the lateral component was twice the size of the longitudinal component (Fig. 18.6):

If it is assumed that the flat rack is transported using fore and aft stowage on a container ship and transported during pre-carriage and or onward carriage on a road

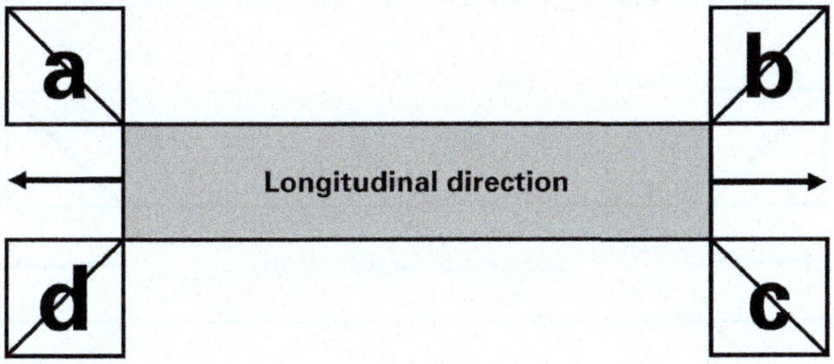

Fig. 18.5 Package secured with four diagonal direct lashings

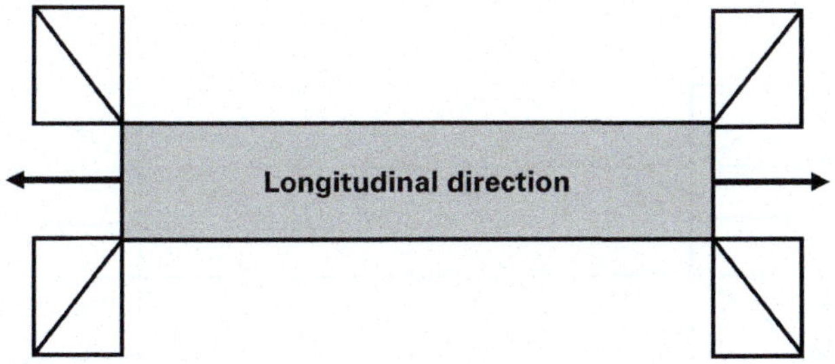

Fig. 18.6 Lateral components twice as large as longitudinal components

vehicle, longitudinal and transverse acceleration forces of 0.8 g should be assumed for the lashings. The decision in this instance is simple: both components must be the same size (Fig. 18.7).

If the flat rack is transported during pre-carriage or onward transport on a freight car in multi modal operations, longitudinal acceleration forces of 1 g are assumed. If lateral acceleration forces of 0.8 g are assumed, the longitudinal components must be 25% larger than the lateral components or the lateral components can be 20% smaller than the longitudinal components (Fig. 18.8):

Using tie-down lashings also produces angles which cause a reduction in the MSL. The effects are not overly dramatic because they are the cosine of half the opening angle between the ends of the lashing. It does not really matter whether angles are measured, and forces are calculated, or lengths are measured. It is easier for people familiar with mathematics to calculate using the trigonometric function than to measure distances—provided a pocket calculator with trigonometric functions or a relevant table is available. Both methods are feasible. Anyone who is familiar with the shape of a cosine wave or has certain values in their head, can also make a good estimate (Fig. 18.9).

The angle γ is 121.552°, but who can ever determine an angle as precisely as this? Half this angle is 60.776°. The cosine of this angle is 0.488. This makes it obvious that this type of securing is ineffectual. The mathematically calculated part only generates securing forces which are less than half. It would have been better to have secured the cargo with a single-strand direct lashing. The cosine of 60° is 0.5. It is therefore clear that lashings are only worthwhile if their opening angle—i.e.,

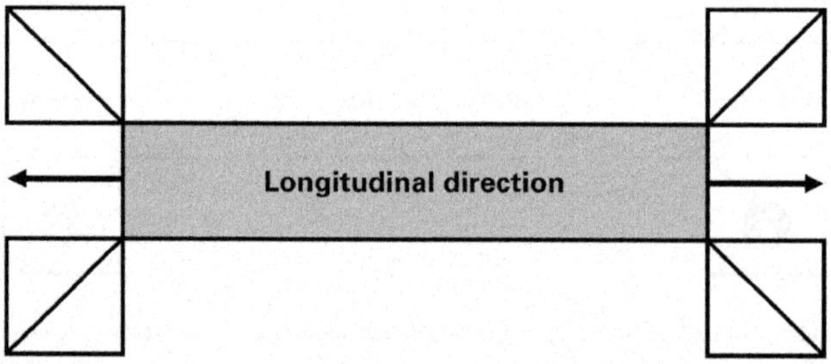

Fig. 18.7 Transverse and longitudinal components same size

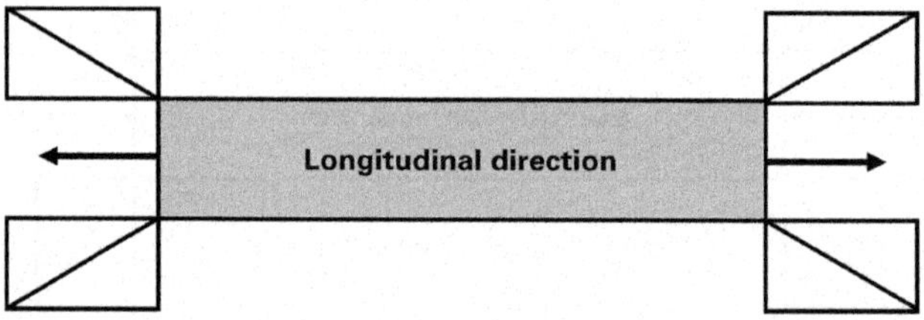

Fig. 18.8 Longitudinal components 25% larger than lateral components

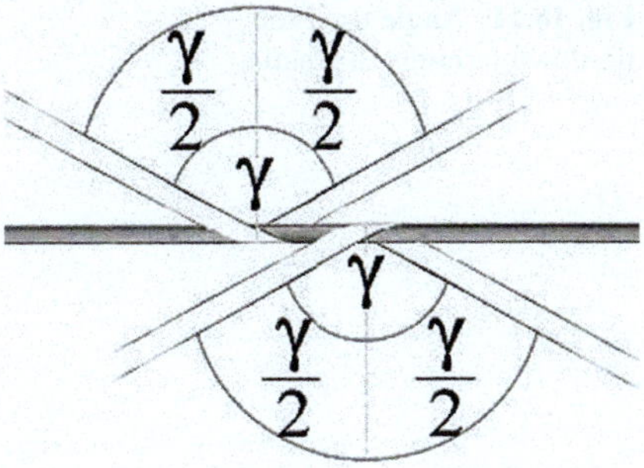

Fig. 18.9 Angle which is too large to be permitted

the complete γ—is significantly less than 120° and $\gamma/2$, significantly less than 60°. The same could have been determined—more quickly—by taking measurements (Fig. 18.10):

The effective length of one end of the lashing is 197.5 cm, the effective lateral dimension is 96.5cm. This is less than half. Both ends of the lashing together are thus less than 1. It can also be seen and proved from this that this lashing is not worthwhile. This does not apply here. The opening angle here is γ 27° ($\gamma/2$ is thus 13.5°). The cosine of 13.5° is 0.973—i.e., close to 1. This means that this securing aid achieves nearly double the MSL, or to be precise 1 time the MSL of a single strand (Fig. 18.11).

Measuring would have produced the same results. If—using our guide—the smaller measured value is divided by the larger measured value, the result is 0.973. That means, the lashing secures 2 × 0.973 times, i.e., 1.946 times the MSL of a single strand (Fig. 18.12).

To be precise, the angle between the ends must be measured in three-dimensional space, since lashings which are attached to separate lashing points form an "oblique" angle to each other. These problems should not occur during measuring, even if one end comes from the top and the other end comes from the bottom. Here are the rough guideline values for the MSL at different estimated angles (Table 18.7):

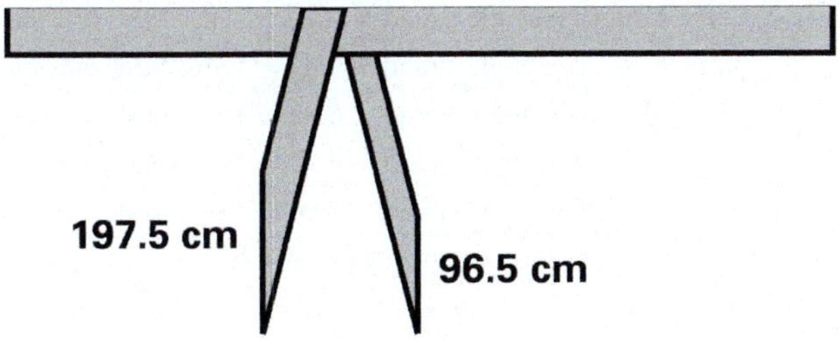

Fig. 18.10 As previously, the angle which is too large to be permitted

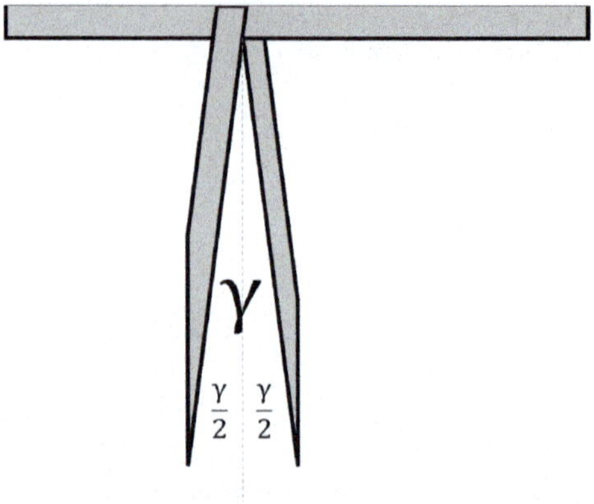

Fig. 18.11 Angle used for tie-down lashings in practice

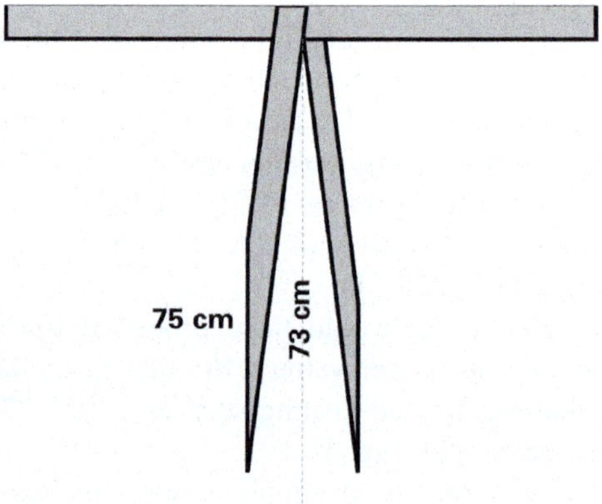

Fig. 18.12 As previously, angle used for tie-down lashings in practice

Table 18.7 Guideline values for the MSL at different estimated angles

Estimated angle	MLS of the lashing
120	2 × MSL of a single strand × 0.50 = 1.0 × MSL of a single strand
90	2 × MSL of a single strand × 0.70 = 1.4 × MSL of a single strand
60	2 × MSL of a single strand × 0.87 = 1.7 × MSL of a single strand
40	2 × MSL of a single strand × 0.92 = 1.8 × MSL of a single strand
30	2 × MSL of a single strand × 0.97 = 1.9 × MSL of a single strand
<30	2 × MSL of a single strand × 1.00 = 2.0 × MSL of a single strand

18.3 Wood

In this section, we will explore the basic properties, the various formats, technical terms, designations and methods of using wood to aid in load securing. For clarity, rules of thumb for calculating the strength of wood are also provided. Wood is

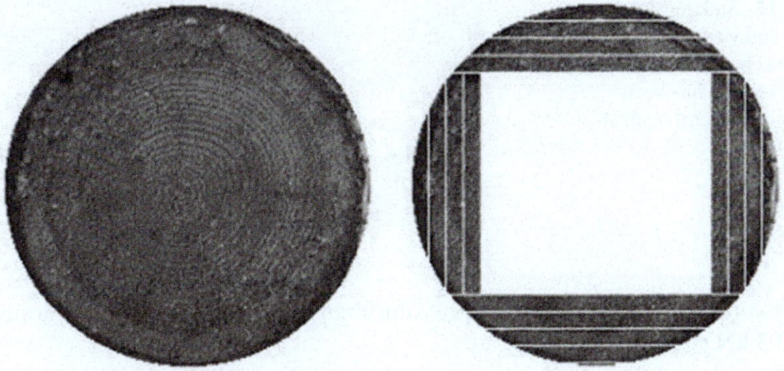

Fig. 18.13 (l) untrimmed wood log; (r) trimmed wood log in direction of the grain

indispensable as a load securing material. A distinction is always made between the long and the short side of the wood (across the grain and with the grain). Wood is less able to withstand loads on its long edge, i.e., when the load is perpendicular to the grain. The extraction resistance of nails and screws, however, is higher when they are nailed or screwed perpendicular to the grain. In contrast, wood is far better able to withstand loads applied on the end grain, i.e., parallel to the direction of the grain (Fig. 18.13).

Untrimmed edge boards or waney edge lumber are often cut from the outer sections of a log for use as wooden dunnage. Depending on the thickness of the trunk, balks, squared lumber, battens, laths, planks and boards are cut from the trunks (Fig. 18.14).

Balks have large cross-sections, with edges usually over 15cm long, squared lumber starts as of about 8 cm × 8 cm. On solid wood, the heart, medulla or pith is located in the centre, on halved lumber on the side and on quartered lumber on one corner. 1:2 can be a size such as 16 cm × 8 cm. Dimension stock could, for instance be 6.3 cm × 7.5 cm (2 1/2″ × 3″). Thicker battens might be 3.2 cm × 5 cm (1 1/4″ × 2″) or 3.8 cm × 6.3 cm (1 1/2 × 2 1/2″). Thin laths are used as an intermediate layer between lumber so ensure that it does not develop mould or rot. Wood with an extremely high frequency of knots should not be used for bracing. It must be rejected beforehand (Fig. 18.15).

The following rules of thumb can be applied to sound lumber to calculate the maximum securing load:

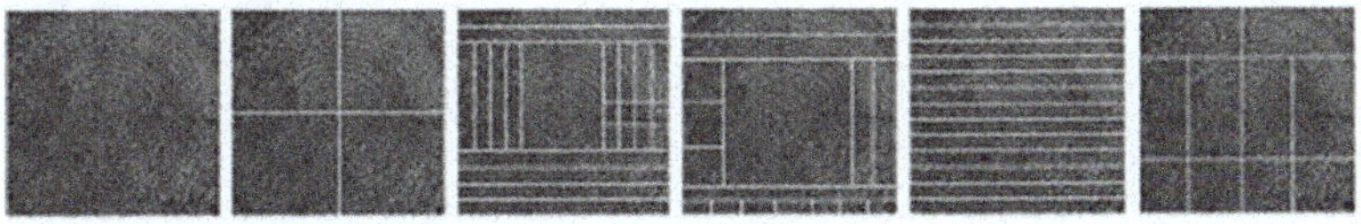

Fig. 18.14 (l–r) trunk, balks, squared lumber, battens, laths, planks, boards

Fig. 18.15 Squared lumber, battens and wedges—wood which has been stained indicates that it has been treated to protect it against the sirex wasp

- Softwood used for bracing, compression strength perpendicular to the grain, MSL = 0.3 kN per cm^2
- Softwood used for bracing, compression strength perpendicular to the grain, MSL = 30 daN per cm^2.

Assuming that the contact surface between a and b is 10 cm × 10 cm = 100 cm^2, a securing force of 3,000 daN (30 kN) can be introduced via a and transmitted to b. The following rule of thumb applies for the maximum securing load parallel to the grain (Figs. 18.16 and 18.17):

- Softwood used for bracing, compression strength parallel to the grain, MSL = 1 kN per cm^2
- Softwood used for bracing, compression strength parallel to the grain, MSL = 100 daN per cm^2

Note: For road and railway transportation only, 200 daN/cm^2 can be assumed for the load parallel to the grain, since it is assumed that only isolated or few load events occur during each transport operation. For maritime transport, the IMO halve

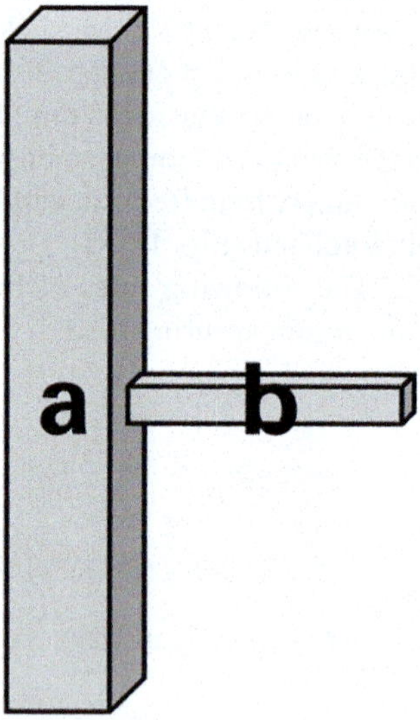

Fig. 18.16 Applying the rule of thumb to compressive strength parallel to the grain of the wood

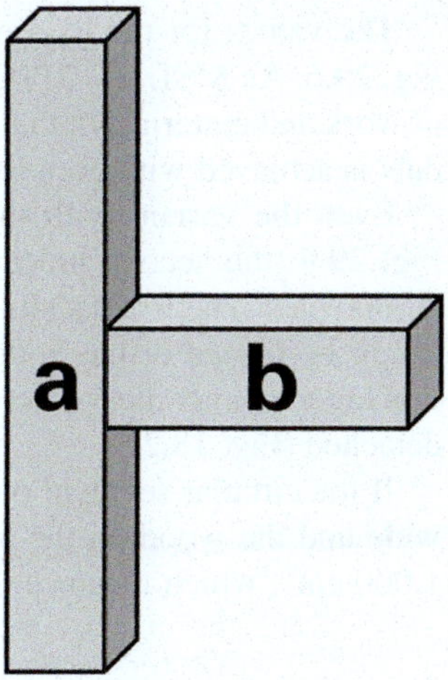

Fig. 18.17 Applying the rule of thumb to compressive strength parallel to the grain

these values to 100 daN/cm^2 due to the constantly changing stresses in rough seas. Assuming that the contact surface between the steel component a and the wood b is 10 cm × 10 cm = 100 cm^2, a securing force of 10,000 daN (100 kN) can be transmitted between a and b. Since the grain side is crucial for the bracing shown in Fig. 29.6, four horizontal bracing elements measuring 10 cm × 10 cm will achieve an MSL of 12,000 daN (Figs. 18.18 and 18.19).

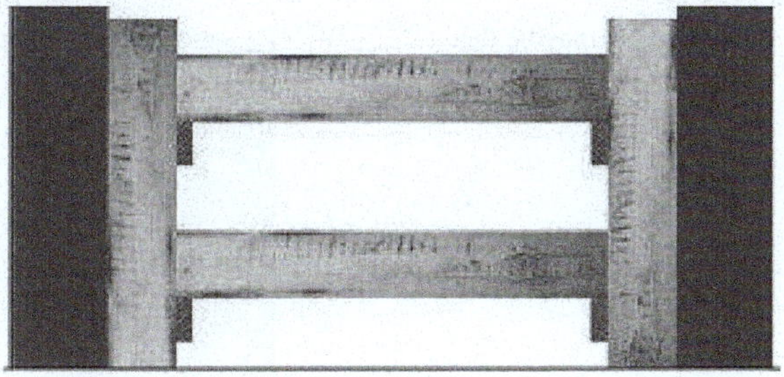

Fig. 18.18 High outlay—low strength

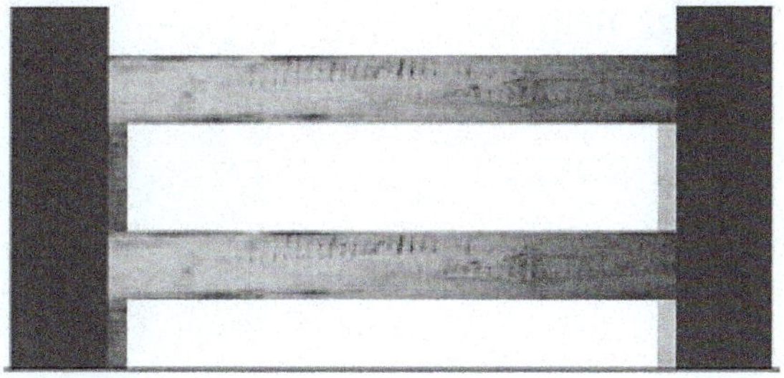

Fig. 18.19 Little material—high strength

The values for the load on the end grain can be used for the bracing shown in Fig. 29.6. An MSL of 40,000 daN is achieved with significantly less outlay in terms of work and materials In this instance, Fig. 29.7 offers the best solution since 40,000 daN is achieved with even less material (Figs. 18.20 and 18.21).

Even the variant with only two bracing elements would suffice as shown in Fig. 29.9 (the second bracing element is hidden by the first). Even this achieves 20,000 daN. The wood should be placed at the top for goods that are likely to topple. It can be placed at the bottom for flat goods which are not likely to topple. The bracing elements must be cross-braced against shifting so that they do not become detached (Fig. 18.22).

If the circular segment on which the wooden-clad drive shaft is placed is 50 cm wide and the wood on the trestle is 20 cm wide, there will be a projected area of 1,000 cm^2, which means a total bearing surface on the trestle of 2,000 cm^2. Since

Fig. 18.20 Even less material—high strength with four horizontal bracing elements used in pairs

Fig. 18.21 It may be that one pair of bracing elements may be sufficient

Fig. 18.22 Heavy drive shaft on trestles

the compression load is perpendicular to the grain, the trestle can absorb 2,000 cm^2 × 30 daN/cm^2 at this point, i.e., 60,000 daN. Since the shaft has been placed on several trestles, they have obviously been over dimensioned here. The workmanship also leaves something to be desired. There is still some potential for rationalisation here (Fig. 18.23).

Lumber measuring 10 cm × 12 cm has been used here. This would produce an area of 120 cm^2 at the contact points. Since two pieces of wood were used as bracing on each side, the total area is 240 cm^2 per side. The stays would therefore theoretically absorb 240 cm^2 × 100 daN/cm^2, i.e., thus 24,000 daN. There is however a slight problem here: The stanchions are not as wide as the lumber (Fig. 18.24).

Since the stanchions are only 8 cm wide, only an area of 8 cm × 12 cm = 96 cm^2 per wooden element can be used for calculation, i.e., 192 cm^2 per pair of stays. In reality, therefore, the stays can only be loaded with 19,200 daN per pair. Since the total length of the stay is 1.47 m and the transverse component is calculated as 1.05 m, a pair of stays secures the beam against tilting with 1.05 m/1,47 m × 19,200 daN = 13,714 daN. The following rules apply:

- If it is feasible, always brace horizontally; and
- If it is feasible, brace so that the end grain loading capacity is used.

Always brace against the parts with the greatest load-bearing capacity, e.g., on cases, use the plate effect provided by the base and lid or guide the forces into the vertical sides. Securing measures of this kind can only be created using wooden aids

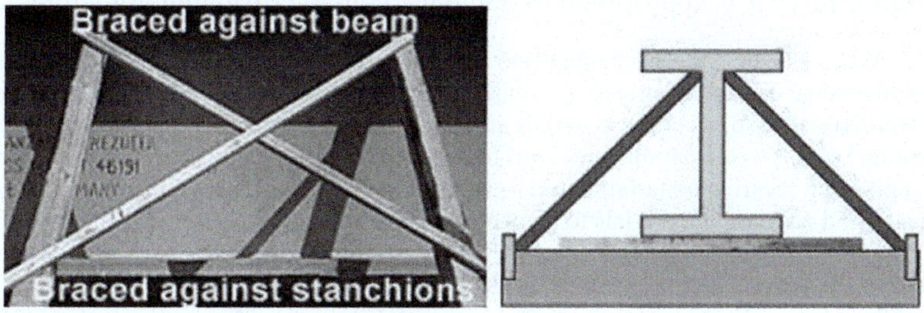

Fig. 18.23 Securing with diagonal stays

Fig. 18.24 Always take note of the smallest contact surfaces!

and driving wedges, which are removed again after securing has been completed. Driving wedges can be used individually or in pairs. They should be made of hardwood. The very shallow angle here produces large leverage effects which can be used to achieve extremely high forces. Caution is therefore demanded. Only drive wedges in gently and then fix them into place. Sometimes only quite simple structures are required to fill in a gap in the container. But one last note: The less material has to be used for load securing the better. The best and most cost-effective load securing measure is a compact stow using well-protected and sufficiently strong packages. In this section we have discussed the nature of using wood as a form of dunnage and securing mechanism. In the next section, we will look at the use of filling materials and air bags as an alternative to solid wood.

18.4 Filling Materials and Airbags

Expanded synthetic plastics have extremely poor recovery properties and no elasticity. They deform very quickly under load, i.e., they become thinner under compression leads. They can therefore only be used for filling gaps with light loads. In all other cases, they should not be used as a filling material. The same applies for cardboard and corrugated board. Used tyres may to a certain extent be suitable for filling in gaps. To enable them to fill in a large area, they could be arranged so that they are pressed between two walking boards which each cover a large surface area of the cargo. Once more it is appropriate to quote the container packing guidelines:

> 3.2.12 When deciding on packaging and cargo-securing material, it should be borne in mind that some countries enforce a rubbish- and litter-avoidance policy. This may lead to limitations on the use of certain materials and imply fees for the recovery of packaging at the reception point as well as similar problems for the shipper of the cargo. In such cases, reusable packaging and securing material should be used. Increasingly, countries are requiring timber dunnage and packaging materials to be debarked.

One particularly useful and inexpensive method for filling in spare space is to use inflatable air bags. The container packing guidelines provides us with the following guidance:

> 3.2.4 If air bags are used, the manufacturer's instructions on filling pressure should be scrupulously observed. Allowance should be made for the possibility of a considerable rise in the internal temperature of the [container] above the temperature at the time of packing. This may cause the bags to expand and burst, thereby making them ineffectual as a means of securing the cargo. Air bags should not be used as a means of filling space at the doorway unless precautions are taken to ensure that they cannot cause the door to open violently when the locking bars are released.

The most commonly used form of air bag comprises a multilayer paper bag available in a wide variety of varied sizes. This type should only be used if the areas to the left and right of the gap are as smooth as possible and have a high load-bearing capacity. If necessary, walking boards are to be provided to ensure correct distribution of pressure. Specific measures are required for working with air bags, in particular with white and brown goods, since the parts of the cartons which are not padded would be pushed in. The minimum inflation pressure should be 200 hPa. For many air bags, this is around 300 hPa. This can still achieve a securing force of 3,000 daN per m^2. Air bags are inflated using compressors. Special inflation fittings with pressure reducers are required. Special fittings enable the air to be evacuated. This is an advantage if a bag is not seated correctly, as it can quickly be deflated and then inflated again. This fitting facilitates deflation of reusable bags. Since it is time-consuming to have to unscrew valves to let out the air, the recipient will deflate single-use bags by puncturing them. When closing gaps, air bags which have already been partially inflated should be placed in the stowage gap and then completely inflated. Suitably large and smooth surfaces are required on the cargo transport units when using them. Only air bags which are capable to a certain extent of withstanding point loadings should be used. One crucial factor with regard to usability is that their side faces should be as near as possible to parallel. Air bags with a pronounced bulge must not be used (Figs. 18.25, 18.26 and 18.27).

Fig. 18.25 Example of filling material

Fig. 18.26 Using car tyres to fill void space

Fig. 18.27 Examples of various types of air bags

18.5 Artificial Tight Fits

18.5.1 Gang Nails

Gang nails, sometimes known as dowels, can be used to provide a tight fit. They are available in a wide variety of different shapes and sizes. Despite the specialist literature referring to these as "shear transfer plates," they are in effect punched pieces of steel sheet with triangular teeth bent upward and downward (Fig. 18.28).

Fig. 18.28 Gang nails—(l) without and (r) with nail holes

Fig. 18.29 Different shapes and sizes of shear transfer plates

To facilitate handling, these devices are also available in "display packaging," where they are shrink-wrapped ready-to-use and can be inserted quite easily. Aids such as these make sense when unitising packages with robust softwood surfaces, for example, single-use pallets, softwood pipe clamps, cases with suitable skids, or belt battens. If gang nails are laid between loads of this kind and then the loads are strapped or bound together in some other way, they will increase cohesion. If it is possible to lay them on platform containers with softwood floors, they can contribute to improving load securing or to reducing the remaining securing work. On their own, they are not sufficient as a means of securing. Under stress, the teeth bend, flatten, and consequently become ineffective. This has been confirmed by tests carried out on unsecured loads. In box containers it is conceivable to use them between individual package layers, but not practical, since accessibility is difficult in most cases unless special aids are used to lay them. Because of the risk they pose, they should not be used on container floors made of plywood or textured coated board (Fig. 18.29).

18.5.2 Chocks

It makes sense to use spike-based chocks on cargo transport units with soft floors which allow this. A further requirement is that cylindrical goods can be placed against the chocks. When securing cylindrical goods with for instance four spike-based chocks, the load would have to be set down, the chocks put in place, the load raised, the chocks secured, and the load then set down again. If the positions of the chocks are measured accurately and they are placed beforehand, the first steps can be omitted, but the product must be put in place from above. The applications for spike-based chocks of this kind are extremely limited. In the field of container traffic, they could only be used on platform containers (Fig. 18.30).

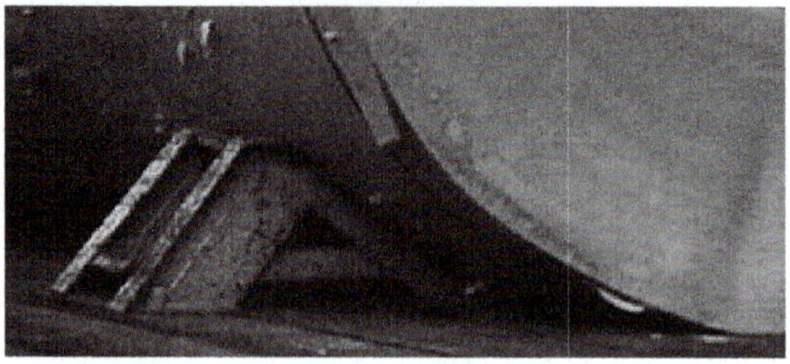

Fig. 18.30 Example of chocks

18.6 Friction and Friction Enhancing Mats

With respect to friction and friction enhancing mats, a fundamental distinction must be made between the two types of friction: external and internal. External friction is used to refer to the resistance that is created as a result of the relative movement as the contact surfaces of two objects rub against each other. External friction is caused, in particular, by forces of adhesion and unevenness on the surfaces of the objects. Friction is independent of the size of the contact surfaces. Friction is often also explained as the meshing of two materials in contact with each other at a microscopic level. This can be seen clearly, if for example, the structure of softwood is examined on the grain edge. It is not difficult to imagine that two surfaces of this nature will become 'meshed' if they are in contact with each other. The frictional force (F_F) is always counter to the current movement of an object. The magnitude of the frictional force is dependent on the magnitude of the normal force (F_N) with which the contact surfaces press against each other and on the friction coefficient μ. Thus, the basic formula for Coulomb's law of friction is:

$$F_F = \mu \cdot F_N$$

In this formula μ stands for the friction coefficient or friction factor. It is dependent on the material, the state of the surfaces (roughness and whether it is lubricated) as well as the sliding speed of the bodies rubbing against each other. The symbol used in the formula is μ, the lowercase Greek letter "m" or "mu." Distinct types of friction are distinguished, depending on whether or not there is relative movement between the bodies, whether the bodies are already moving and if so whether they are sliding or rolling; these being static friction, sliding friction, and rolling friction. If the conditions are otherwise identical, the static friction coefficient is always larger than the sliding friction coefficient and the rolling friction coefficient is usually much lower than the sliding friction coefficient. During frictional processes, mechanical energy is converted into heat, frictional heat. Internal friction is caused by molecular forces. It occurs when individual components of a liquid or a gas, or the various layers that make up a substance, rub against each other. The technical term is viscosity. There are two types of viscosity: dynamic and kinematic.

Contact surfaces between liquids and gases, on the one hand, and the walls of a pipe both exhibit resistance which is referred to as pipe friction. Without friction, our everyday lives would be impossible. All forms of locomotion are reliant on friction. When transferring forces, for example when starting or braking a vehicle or securing a load, elevated levels of frictional resistance are desired. When moving loads the objective is to keep frictional resistance as low possible in order to reduce the loss of energy to a minimum. In order to determine the level of frictional resistance, quite simple tests can be used. The resistance to movement can be determined using measurements of force or comparisons of mass, or measurements can be carried out to show the gradient at which an object on a slope remains in position, starts to move or remains in motion.

Static friction is measured when a measuring instrument, for example, a spring balance, is pulled until the frictional force F_F is almost sufficient to move the object being pressed onto the surface by the normal force F_{N}. The force that is required to keep the object in motion and pull it across the surface corresponds to the sliding friction. When applied to a vehicle, this principle can be used to determine the rolling friction. A more detailed test can distinguish between static rolling friction and sliding rolling friction. In the field of load securing, the values for sliding friction are of most interest. The simple method described here is not admissible for determining friction in genuine scientific studies. There are special standards which detail how friction coefficients should be determined. For example, a block of wood with a normal force or weight—F_N of 90 N is dragged over a cardboard floor. The friction coefficient F_F is measured at 27 N. After transformation of the basic formula $F_F = \mu \times F_{\mathrm{N}}$ this results in:

$$\mu \frac{F_F}{F_N} = \frac{27\,\mathrm{N}}{90\,\mathrm{N}} = 0.3$$

For this example, the friction coefficient of wood on cardboard is calculated as 0.3. This value can also be expressed as thirty percent or three tenths. If the test block is doubled by the addition of an additional block, and increased to 180 N, then the frictional force is increased to 54 N. The frictional coefficient of 0.3 does not actually change, because 180 N divided by 54 N = 0.3. The friction coefficient remains constant for any given pair of materials. If the normal force of an object and the friction coefficient are known, then the frictional force can be calculated mathematically. Thus, the calculation for the above example is as follows:

$$F_F = \mu \cdot F_F = 0.3 \cdot 90\,\mathrm{N} = 27\,\mathrm{N}$$

$$F_F = \mu \cdot F_N = 0.3 \cdot 180\,\mathrm{N} = 54\,\mathrm{N}$$

If the friction coefficient remains the same, the frictional force is always proportional to the normal force or weight. Thus, if the normal force is doubled, the frictional force is doubled. Likewise, if the normal force is halved then the frictional force is also halved. If the friction coefficient and the frictional force are known, the basic formula $F_F = \mu \times F_{\mathrm{N}}$ can be transformed and used to calculate the normal force or weight:

$$F_N = \frac{F_F}{\mu} = \frac{27N}{0.3} = 90N$$

$$F_N = \frac{F_F}{\mu} = \frac{27\text{N}}{0.3} = 180\,\text{N}$$

A test configuration used to calculate the frictional resistance by comparing mass may be as follows:

$$m_N m_N$$

$$m_F \uparrow\uparrow m_F$$

Sufficient weight m_F is added until the mass m_N only just remains motionless. When the masses start to move, weights can be removed. The formula that is used to calculate the friction coefficient and the transformed formulas based on it are:

$$\mu = \frac{m_F}{m_N} \quad m_F = \mu \cdot m_n \quad m_N = \frac{m_F}{\mu}$$

Friction is still being determined using forces, since all masses are forced onto surfaces by the gravitational pull of the earth. For example, a block of concrete with a mass mN of 200 kg is pulled over a wooden surface. The mass m_F required to overcome the friction is determined to be 80 kg. In accordance with the formula μ $= m_F/m_N$ this gives a friction coefficient of μ = 80 kg/200 kg = 0.4. If the friction coefficient and the mass of an object are known, then the required frictional force can be calculated mathematically. Thus, for the example:

$$m_F = \mu \times m_N = 0.4 \times 200\,\text{kg} = 80\,\text{kg}$$

If the friction resistance and the friction coefficient are known, then the mass can be calculated:

$$m_N = \frac{m_F}{\mu} = \frac{80\,\text{kg}}{0.4} = 200\,\text{kg}$$

If no dynamometers or calibrated weights are available to determine the friction coefficient, the values can also be determined using an inclined surface. The surface is raised until the mass just starts to slide or roll. Once it has started to move, a lower angle of slope is sufficient to keep the object in motion. There are two methods of determining the friction coefficient μ using a gradient or a sloped surface. One option is to divide the height of the sloped surface by the length of its base. For example, an aluminium block is placed on a rusty steel plate which is raised at one edge. The height at a is measured at 96 cm. The length b is measured at 3.84 m. Using the formula μ = a/b, this gives a friction coefficient of μ = 0.96 m/3.84 m = 0.25. The friction coefficients determined using the sloped surface always correspond to the gradient of the slope expressed as a percentage. A μ value of 0.25 determined by a test such as this means that the slope has a gradient of 25%. Thus, if the friction coefficient of car tires on an icy surface is calculated at 0.1 this means that the car is able to drive on a maximum gradient of 10% on black ice. If a pocket calculator with trigonometrical functions is available, or a table showing the values of the trigonometrical functions, then μ can be calculated using the angle of the incline. In this case the formula is:

$$\mu = \tan\alpha$$

For example, return to our aluminium block on the rusty steel surface, the angle necessary for the block to begin to slide is 14°. The tangent of this angle is 0.25 and corresponds to the friction coefficient μ. By transforming the formula, two known values can always be used to calculate a third unknown value:

$$\tan\alpha = \frac{a}{b} \quad a = b\tan\alpha \quad b = \frac{a}{\tan\alpha}$$

18.7 Calculating the Friction Coefficient μ For Bulk Cargo

When using a conveyor or grab to move bulk cargo, when shovelling, tipping or pouring bulk cargo, the natural slope angle or angle of repose is formed naturally. Using the formulas for inclined surfaces, the friction coefficients (μ values) can be calculated. The value of the friction coefficient results when the height of the heap (a) is divided by the base length of one of the flanks (b) or the tangent of the angle of repose or slope angle (α) between the slope of the heap and the horizontal plane is calculated. During testing, values are determined by levelling the surface of the contents of a receptacle and then either tipping the receptacle or placing it on a surface which is then raised at one edge until the contents start to slide/flow. For example, a glass of salt with a level surface is tipped until the first grains of salt break from the surface. A tipping angle α of 31° is determined. The tangent of 31° can be used to determine the friction coefficient of this type of salt as μ = 0.6. A receptacle containing levelled peas is standing on a surface which is subsequently raised until the individual peas start to roll/slide. The height a = 1.36 m and the base length of the inclined surface b = 4.00 m are used to calculate a friction coefficient for peas of μ = 0.34. Checking using the measured tipping angle of 18.8° confirms the friction coefficient: tan 18.8° = 0.34. To completely empty bulk containers on vehicles at tipping points, the loading platforms must be raised so that the "tipping angle" α is reached or exceeded. With respect to the safe transportation of bulk cargoes, particular attention should be paid to the fact that loads are levelled off before the vehicle is driven off. If this rule is not followed, the load may begin to slide and overflow at extremely low acceleration. With the cargo loaded in this manner, it requires a lateral acceleration force of 0.6 g to make the cargo slide. The angle α is identical to the slope angle α of 31°. The angle of repose, slope angle or discharge angle as well as the friction coefficients of bulk cargo are all affected to a great degree by the moisture content of the cargo, the size of the grains and other similar factors. Test observations. Running tests and noting the measured data for varied materials and masses or objects with varying weights will allow the following conclusions: bearing area. As far as the objects do not press into the surface, the dimensions of the bearing area have no influence on the magnitude of the friction resistance; weight/mass. Under otherwise identical circumstances, the friction resistance of objects is proportional to their mass/weight. Thus, doubling the mass will double the friction resistance, and halving the mass will halve the friction resistance, etc., material. The materials which are in contact are what determine the values observed. The magnitude of the friction coefficient depends solely on which materials are making contact; and units of measurement when using the basic formulas. For all calculations using the basic formulas $\mu = FF/FN$, $\mu = mF/mN$ and $\mu = a/b$ it is not relevant which unit of measurement is used to measure force, mass or distance. The important thing is that they are used consistently. Since the calculation dimensions in the formula cancel each other out, μ is a dimensionless numerical value.

18.8 Units of Measurement When Using the Transformed Formulas

With these calculations, care must be taken to ensure that the same mathematical dimensions are used, otherwise the results will not be suitable for comparison or will need to be converted. Writing down the mathematical dimensions guarantees the correct results, because errors can be spotted immediately.

18.9 Sliding Friction Coefficients

For load securing, the sliding friction coefficients are of particular importance. The following tables contain sliding friction coefficients for selected general and bulk cargoes as well as a number of materials that are used on a regular basis. The abbreviation FE mat stands for friction-enhancing mat. Materials of these types are ideal for use as interlayer dunnage when layering goods that are unitised with suitable strapping. They are less suitable for use as underlay or interlayer dunnage for unsecured loads or partial loads. As vibrations occur, the friction coefficient drops dramatically, and the items will start to move. These thin materials tend to wear easily. At times, particularly if they are heated, these materials can even have an adhesive effect which can depreciate the value of sensitive goods. These materials can be used to maintain friction coefficients of approximately 60%. The material on the left is available in a number of different thicknesses, but usually 6 and 10 mm thick. The thicker version is preferable for use with heavier goods because it does not wear quite so quickly. Materials of this sort are usually composite materials from rubber scrap and recycled rubber etc. The material on the right is just 2 mm thick and has a ridged surface. This is more suitable for forming unit loads. In some cases, friction-enhancing materials cannot be brought into the containers, because they would hinder the normal work. This method can only really be implemented if the goods can be placed cleanly on prepared pads or strips. This method can also be used when loading open top containers or platform containers where the intended positions are easily accessible to staff and equipment. The use of non-slip materials is impossible if the length or any other characteristic of the load requires it to be pushed into the container. In practice, dry softwood should not be calculated with a friction coefficient greater than 30%. With normal container floors made of plywood or textured coated board, a maximum friction coefficient of 20% can be reckoned with. Even if the floor has been steam cleaned and dried thoroughly, when packers walk/drive across the floor, dust and other contamination is usually deposited thus lowering the friction.

18.10 Factors Which Alter the Friction Coefficients

The values given are just guidelines. Each shipper should themselves determine the values that should be used in their situation. But beware: The values obtained under laboratory conditions are rarely the same as those that occur in practice. External influences, such as humidity, dust and grease can affect the figures. Dirty loading areas with the remains of granular materials, such as sand, etc. reduce the friction coefficient. Some of these values are given in Tables 18.1, 18.2, 18.3, 18.4 and 18.5. The friction coefficient can, in particular, be reduced by vibration, and in a worst-case scenario it may even be reduced to values approaching zero. When using containers with smooth floors, it is recommended not to rely on friction, but instead to pack and secure the load in such a way that it cannot move, even without friction. The VDI 2702 guidelines, published by the Verein Deutscher Ingenieure (VDI), and entitled "Ladungssicherung auf Straßenfahrzeugen—Zurrkräfte" (Load securing on road vehicles—lashing forces) provides the sliding friction coefficients in Table 18.1. The VDI guidelines recommend using the lowest value for the sliding friction coefficient μ if in any doubt. In addition, attention is drawn to the fact that additional material pairs and environmental influences, such as dirt and ice, require that the friction coefficients should be estimated or should be determined specifically by testing. Table 30.6 provides the sliding friction coefficients from the VDI guideline expressed as fractions, as gradients in% and as angles. Rolling friction coefficients are much lower than static or sliding friction coefficients. Values for common materials are shown in Table 18.4. Squared lumber with a square cross-section can act like rollers. This risk is particularly high when using waney edge squared lumber since this often behaves in an equivalent way to round wood. Rectangular formats or, better, planks should be used flat. The thickness should be calculated in order to ensure that lifting gear or ground conveyors are able to handle the goods. A better format is the so-called "sandwich" which combines wood with top-quality friction-enhancing mats glued or screwed to each side. The gluing or screwing can be made more durable if the lumber is shaped to accommodate the mats. The depth should be measured in such a way that even when a heavy load is placed on the mat, the friction enhancing material is not crushed to such an extent that the goods are resting on the wood. If transport jolts are to be expected transversely to the orientation in which they are laid, rectangular formats such as planks are to be used. It would be better to set out the beams parallel to the expected direction of the jolts, but for handling reasons, this is only rarely possible.

As has already been mentioned, frictional forces can be reduced to zero as a result of vibration. A certain minimum amount of load securing should always be implemented to counteract the effects of vibrations. Goods should never be transported unsecured just because the calculated or assumed friction coefficients are greater than the horizontal acceleration that is to be expected. With very few exceptions, for example, when materials are not compatible or the selected handling methods do not permit the use of friction enhancing mats, friction should always be enhanced as much as possible by selecting the appropriate materials. Since the use of tensioning or lashing aids or vertical bracing creates additional frictional forces, the load securing outlay may be reduced considerably if high friction coefficients can be achieved.

18.11 Examples of Different Friction Enhancing Materials

See Figs. 18.31, 18.32, 18.33, 18.34, 18.35 and Tables 18.8, 18.9, 18.10, 18.11, 8.12.

Fig. 18.31 Structure of softwood on the grain edge

Fig. 18.32 Material with a μ value of approximately 0.8

Fig. 18.33 Material with a μ value of approximately 0.7

Fig. 18.34 Material with a μ value of approximately 0.6

Fig. 18.35 Material with a μ value of approximately 0.5

Table 18.8 Sliding friction coefficients for selected materials

Material pairs	Friction coefficient μ as			Discharge angle (°)
	Decimal	Fraction	Gradient (%)	
Stones/loose soil, certain varieties of ore and coal	1	1/1	100	45
Friction-enhancing mats or car tyres/very rough concrete, F_E mat on F_E mat (gum)	0.9	9/10	90	42
F_E mat/rough concrete, rough wood or rough chipboard; tire rubber/macadam; certain types of paper/F_E mats	0.8	4/5	80	38.7
F_E mat/rusty steel or matt varnished wood, tire rubber on rough materials	0.75	3/4	75	36.9

Table 18.9 Sliding friction coefficients for selected materials

Material pairs	Friction coefficient μ as			Discharge angle (°)
	Decimal	Fraction	Gradient (%)	
Rough galvanised sheets, rough ekki (Azobé) or softwood/FE mat; very rough concrete or stone/very rough concrete, F_E mat (gum)/wood or P_E	0.7	7/10	70	35
F_E mats or rough stones or concrete/rough wood, planed oak or smooth plywood/F_E mat; very rusty steel/very rough wood; car tyres/wet road; various cereals as bulk cargo	0.6	3/5	60	31
F_E mat/PA; rough concrete or rough stones/rough wood, very rough concrete/rough concrete, various materials/F_E mats	0.5	1/2	50	26.6

Table 18.10 Sliding friction coefficients for selected materials

Material pairs	Friction coefficient μ as			Discharge angle (°)
	Decimal	Fraction	Gradient (%)	
Aluminium profiles/F_E mats.	0.45	9/20	45	24.2

(continued)

Table 18.10 (continued)

Material pairs	Friction coefficient μ as			Discharge angle (°)
	Decimal	Fraction	Gradient (%)	
Rough wood/matt varnished wood; aluminium/F_E mat; rusty steel/very rough softwood or oak; certain types of paper/PE; very rusty steel/very rusty steel; concrete/wood; concrete/concrete; minimum value for F_E mats.	0.4	2/5	40	21.8
Rough wood/wood; rough wood/PE; aluminium/softwood or PE; rusty steel/rough wood; rough galvanised steel plate/softwood; certain types of paper/cardboard or wood; concrete/wood; wood/wood; wood/special plastic films; car tyres on snow.	0.3	3/10	30	16.7
Matt varnished wood/ PE; rough chipboard/softwood; rough chipboard/PE; planed ekki (Azobé)/softwood; plywood/softwood; paper/wood; steel/aluminium.	0.25	1/4	25	14

Table 18.11 Sliding friction coefficients for selected materials

Material pairs	Friction coefficient μ as			Discharge angle (°)
	Decimal	Fraction	Gradient (%)	
Planed oak, PA/softwood, PE; Bongossi, plywood/PE; F_E mat/sanded wooden loading platform; rusty steel/PE, PA; rough galvanised steel/PA; contaminated steel/aluminium; dusted steel/rough wood; wood, cardboard on textured coated board, PVC, plastic coated wooden floors, etc.	0.2	1/5	20	11.3
PA/PE, rough galvanised steel/PE; rough wood/sanded wooden loading platform.	0.15	3/20	15	8.5
Aluminium, PA, PE, steel, wood and various other materials on heavily sanded or contaminated F_E mats; smooth steel on smooth steel; tyres on ice.	0.1	1/10	10	5.7
Water, oil and other liquids.	0.0	0/∞	0	0

Table 18.12 Sliding friction coefficients in accordance with the VDI Guidelines

Material pairs	Dry	Wet	Greasy
Wood on wood	0.20–0.50	0.20–0.25	0.15–0.05
Metal on wood	0.20–0.50	0.20–0.25	0.10–0.02
Metal on metal	0.10–0.25	0.10–0.20	0.10–0.01

Material pairs	Dry			Wet			Greasy		
Wood on wood	1/5 through 1/2	20% through 50%	11.3° through 26.6°	1/5 through 1/4	20% through 25%	11.3° through 14°	3/10 through 1/20	15% through 5%	8.5° through 2.86°
Metal on wood	1/5 through 1/2	20% through 50%	1.3° through 26.6°	1/5 through 1/4	20% through 25%	11.3° through 14°	1/10 through 1/50	10% through 2%	5.7° through 1.15°
Metal on metal	1/10 through 1/4	10% through 25%	1.3° through 26.6°	1/10 through 1/5	10% through 20%	5.7° through 11.3°	1/10 through 1/100	10% through 1%	5.7° through 0.57

Chapter 19
Lashing Materials and Equipment

In the previous chapter, we covered the principles of load securing; it should be fairly obvious now that simply strapping a piece of cargo down is not sufficiently robust to keep it safe and free from potential damage during transit. There are many factors which need to be considered to ensure the cargo is able to withstand the forces that are likely to act on it both on land and at sea. Now that we have covered the underlying principles of load securing, in this chapter will look at the various means for securing cargo loads.

19.1 Natural Fibre and Manmade Material Ropes

By far, the most common means of load securing is by way of rope. The term "rope" covers a wide range of manufactured ropes made of dissimilar materials, each with quite different purposes. In the field packing and load securing, a distinction is made between fibre ropes, wire ropes and composite or mixed ropes. Fibre ropes are further divided into natural fibre ropes and synthetic or man-made fibre ropes. Ropes are made in a number of diverse ways. Ropes are distinguished between laid rope and braided rope. In many cases, the way a rope is manufactured determines how it is treated and used. The basic process used to manufacture laid fibre ropes is as follows: when they have been obtained from the raw materials, the fibres are combed to separate them. These are then twisted to produce yarns. In the trade, this is known as laying. If the direction of the lay is to the right, it is known as Z-laid, or if it is to the left, it is known as S-laid. In the next stages of the process, the directions of lay are alternated. This is referred to as a 'cross lay' or 'conventional lay.' A cable lay is only used for strong fibre ropes—but this is not relevant for load securing purposes. In the normal direction of lay, three Z-laid ropes are laid to the left to produce a cable; in the opposite direction of lay, three S-laid ropes are laid to the right to produce the finished cable. The normal direction of lay is also known as Z-lay or ZSZS-lay and

A. A. Olsen and F. Karkori, *Containerized Cargo Handling and Stowage*, Springer Series on Naval Architecture, Marine Engineering, Shipbuilding and Shipping 22,
https://doi.org/10.1007/978-3-031-63329-4_19

the opposite direction of lay is known as S-lay or SZSZ-lay. Laid (twisted) ropes are used for load securing purposes. They are typing A hawser-laid, i.e., a 3-strand rope is used. There are both Z-laid and S-laid versions of this type. The normal direction of lay, however, is usually to the right. Natural fibre ropes are no longer used for load securing in most industrialised countries. In some countries, however, natural fibre ropes may be used for securing loads in or on specific cargo transport units. This includes, in particular, ropes made from the following materials:

- Hemp (Ha).
- Manila (Ma).
- Sisal (Si); and
- Manila mixed (Ma/Si).

Natural fibre ropes are extremely sensitive to acids, alkalis and solvents. Mould and other microorganisms attack the fibres. Manila and sisal are comparatively resistant to natural weathering, but hemp loses its strength very quickly unless it is impregnated. As impregnated ropes usually have a strong odour, they must not be used for odour-absorbent or odour-sensitive cargoes. Natural fibres swell when they absorb moisture and shrink on drying. Fluctuations in moisture content can result in considerable changes in the rope length:

- The ropes become shorter when wet; and
- The ropes become longer when dry.

For the above reasons, it is exceedingly difficult to maintain uniform tension in the ropes. Their range of application is therefore limited to use in closed cargo transport units where no great fluctuations in humidity is expected. Natural fibre ropes are also combustible and can lead to spontaneous combustion when used in connection with certain chemicals. One major advantage, however, is that natural fibre ropes are considered environmentally friendly. Since they decay quickly, they can be disposed of without the need for much intervention. Man made, artificial or synthetic material ropes use synthetic fibres, such as:

- Polyester (PES).
- Polyamide (PA).
- Polypropylene (PP).
- Polyethylene (PE); and
- Hempex.

Polyester exhibits exceptionally good resistance to acids, alkalis and solvents and also only loses little strength due to natural weathering; it is therefore very UV-stable. Polyester ropes are not purchased for load securing purposes, however, as they are expensive. Polyamide (known under its trade names nylon, perlon, etc.) exhibits good resistance to alkalis and solvents but is sensitive to acids. It loses around 20% of its strength per year from natural weathering, especially UV radiation. Because it is expensive, polyamide is rarely used for load securing purposes. Polyethylene has no place in load securing on account of its poor mechanical properties. It is even banned for use as slinging ropes. Hempex is a modified polypropylene material which looks so similar to natural fibre hemp that the two are often confused with each other. Although it is UV-resistant, in terms of strength, hempex is not much better than natural fibre ropes. As it is particularly soft, it is sometimes used to fasten tarpaulins or other weight and stress sensitive loads. Polypropylene exhibits particularly good resistance against acids, alkalis and solvents. If it is not stabilised, however, it is overly sensitive to natural weathering. For this reason, polypropylene ropes should always be light, and heat stabilised. Black dyed ropes are best suited for this purpose. Their strength only decreases at around five percent per year from natural weathering. There are two main groups of polypropylene ropes:

- Monofilament ropes; and
- Multifilament fibre ropes.

Monofilament ropes behave exceptionally well under dynamic strain, but because of their material properties, are difficult to knot. This means knots tend to come undone. Monofilament ropes are thus exclusively used as slinging ropes. Despite being stronger than steel-reinforced "Hercules," they are not to be recommended to be used as load securing material. Multifilament ropes exhibit negative properties when they are dynamically stressed. Movement causes the fibres to separate, which quickly produces a rupture. This type of stress rarely occurs with load securing. For this reason, composite ropes are produced for load securing purposes, using multi-filament materials. The type of composite rope that is used for lashing purposes is known as Hercules. This is a rope made of steel wire and natural or synthetic yarns. Hercules made of interwoven polypropylene and wires is sold as lashing material. Ropes of interwoven wire and sisal are no longer commonly used in most industrialised countries. Hercules has the same strength for lashing purposes as ropes made of sisal or polypropylene with the same diameter, as the wire in Hercules does not have a load-bearing function. The individual wrapped wires have an extremely low strength and are only intended to increase the rigidity of the material and improve its twisting properties, e.g., when tightening with a Spanish windlass. The main advantage of Hercules over fibre ropes lies not in its strength but rather in its improved handling. In terms of pure material properties, the same applies as for sisal or polypropylene. Hercules also exhibits suitable properties in terms of resistance to aggressive substances.

Fibre ropes are suitable for securing tarpaulins and light loads, for example, automobiles in or on containers, drum cargoes, fibre drums and similar loads in small batches and light receptacles, fairly light cases, small batches of boxed goods, individual pallets, and so forth. The MSL of fibre ropes is 33% of the breaking load. Since the breaking load of polypropylene can be calculated with a rule of thumb using the formula 12 $\times$ d2, the MSL for the rope shown below (Fig. 19.1) is 1 kN or 100 daN, depending whether the diameter was specified in centimetres or millimetres (Fig. 19.2 and Tables 19.1).

Since a factor of 33% is corresponds to one third, the MSL can also be calculated directly with an easily remembered formula:

$$MSL\ of\ polypropylene\ ropes = d \times d \times 4$$

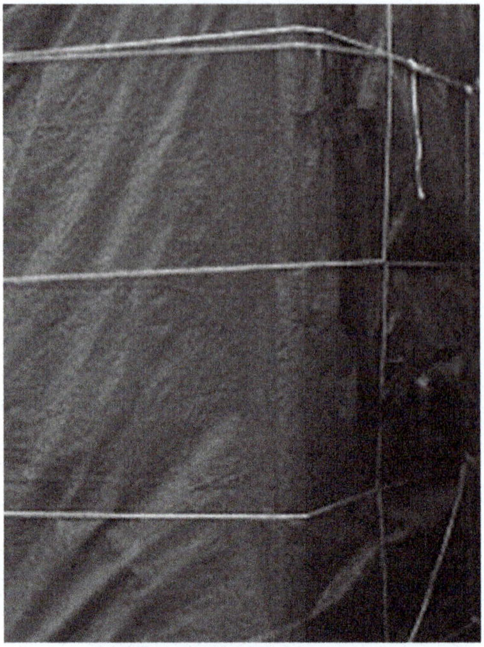

Fig. 19.1 Thin ropes for securing tarpaulins

Fig. 19.2 Vehicle secured with Hercules lashing

Table 19.1 Calculating the diameter of a polypropylene strap—5 mm

Breaking load	MSL factor	MSL	
0.5 × 0.5 × 12 = 3 kN	× 0.33 =	1 kN	(diameter in cm)
5 × 5 × 12 = 300 daN	× 0.33 =	100 daN	(diameter in mm)

If a strap or rope is doubled, the values can also be doubled; if it is used four-fold, the values can be quadrupled. It is not possible to increase the values any further, since there is no guarantee that all the runs of rope would be evenly loaded (Table 19.2).

It would be rather quicker to calculate this using the direct rule of thumb for the MSL:

$$\textit{MSL PP Hercules}\ 1.2\ \text{cm}\ \textit{thick} = 1.2 \times 1.2 \times 4 = 5.76\ \text{kN}$$

$$\textit{MSL PP Hercules}\ 12\ \text{mm}\ \textit{thick} = 12 \times 12 \times 4 = 576\ \text{daN}$$

In this case as well, the values can be doubled or quadrupled respectively for double or (at the most) four-fold use:

To fabricate a direct lashing with a Hercules rope of this type, the necessary rope length is estimated and cut off. The end is threaded through one of the lashing points (A). A loop (B) is knotted into one of the ends using a bowline. The other end is threaded through the second lashing point at (C). The loose end is threaded through the loop of the bowline at (D) and pre-tensioned by hand. The end is knotted at (E) with at least two half hitches. If greater pre-tensioning forces are required than those which can be achieved manually, this can be achieved using a Spanish windlass—i.e., using heavers. If a strap or rope is doubled, the values can also be doubled; if it is used four-fold, the values can be quadrupled. It is not possible to increase the values any further, since there is no guarantee that all the runs of rope would be evenly loaded. The type of steel wire ropes used for lashing are cross-laid stranded ropes laid to the right. Occasionally, left-handed cross-lay used (Figs. 19.3 and 19.4).

$$6 \times 19 = 1\ \text{FC} \qquad 6 \times 37 + 1\ \text{FC}$$

Table 19.2 Determining the diameter of polypropylene-covered Hercules—12 mm

Breaking load	MSL factor	MSL	
1.2 × 1.2 × 12 = 17.28 kN	× 0.33 =	5.76 kN	(diameter in cm)
12 × 12 × 12 = 1,728 daN	× 0.33 =	576 daN	(diameter in mm)

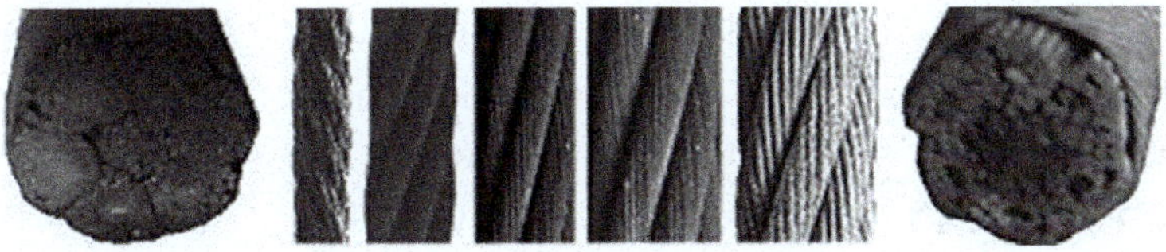

Fig. 19.3 Different rope structures: cross-sections and plan views

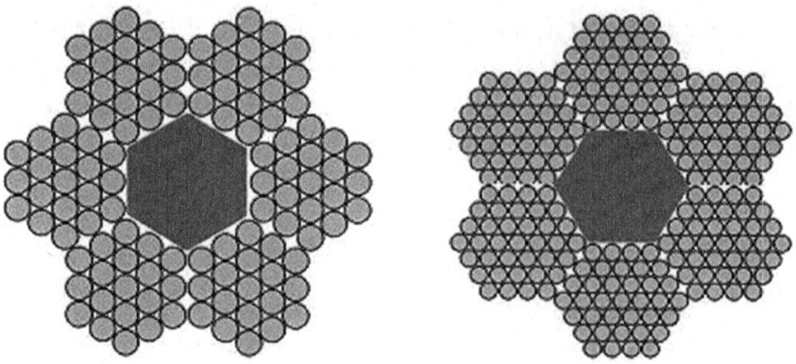

Fig. 19.4 Different rope structures: cross-sections and plan views

Stranded rope is the designation for a wire rope construction where the individual wires are laid into strands, and these are in turn laid into the finished product. The number of individual wires and strands significantly influences the pliability of the wire and its price. The more individual wires there are in a wire rope, the more flexible but also the more expensive it is. A frequently used construction consists of 6 strands each with 19 individual wires per strand, with a fibre core to fill the cavity remaining in the centre. The fill factor for rope constructions of this kind is around 0.455, i.e., the cross-sectional area occupied by metal accounts for around 45.5% of the total cross-section. Wire ropes with the construction $6 \times 37 + 1$ FC are also favoured. These have the same fill factor. Using a rule of thumb, the breaking load and MSL of single-use wire ropes can be calculated as follows (Table 19.3):

Since the 80% of 50 equates to a value of 40, the MSL can also be calculated directly with an easily remembered formula:

$$MSL\ \sin gle - use\ steel\ wire\ ropes\ of\ the\ above\ type = d \times d \times 40$$

It would be a little quicker to calculate the MSL using the direct rule of thumb:

$$MSL\ for\ steel\ wire\ rope,\ 1.6\ \text{cm}\ thick = 1.6 \times 1.6 \times 40 = 102.4\ \text{kN}$$

$$MSL\ for\ steel\ wire,\ 16\ \text{mm}\ thick = 16 \times 16 \times 40 = 10,240\ \text{daN}$$

Using steel ropes of this kind is a waste of materials, since the high MSL value is out of proportion to that of the lashing points on the containers. The strongest lashing rings have an MSL value of 8,000 daN, many are around 5,000 daN and

Table 19.3 Determining the diameter of a steel wire rope—16 mm

Breaking load	MSL factor	MSL	
$1.6 \times 1.6 \times 50 = 128$ kN	$\times$ 0.80 =	102.4 kN	(diameter in cm)
$16 \times 16 \times 50 = 12{,}800$ daN	$\times$ 0.80 =	10,240 daN	(diameter in mm)

most are around 3,000–4,000 daN. For homogeneous lashings, it is usually perfectly sufficient to use wire ropes of a smaller diameter. For instance, 12 mm wires:

$$MSL\ for\ steel\ wire\ rope,\ 1.2\ \text{cm}\ thick = 1.2 \times 1.2 \times 40 = 5.76\ \text{kN}$$
$$MSL\ for\ steel\ wire,\ 12\ \text{mm}\ thick = 12 \times 12 \times 40 = 5,760\ \text{daN}$$

Different MSL values must be applied for reusable wire ropes. Wires of this kind must only be calculated with 30% of the breaking load for the MSL (Table 19.4 and Fig. 19.5):

$$6 \times 12 = 7\ \text{FC} \qquad 6 \times 18 + 7\ \text{FC}$$

Significantly lower fill factors are exhibited by wire ropes with constructions such as 6 × 9 + 7 FC, 6 × 12 + 7 FC, 6 × 15 + 7 FC, 6 × 18 + 7 FC or similar, where, as the designation already indicates, a fibre filler has been incorporated into the centre of each strand. Ropes of this kind are frequently used by Dutch and Belgian companies as lashing wires. The breaking loads and MSLs of wires of this kind are only half as high, because a factor of 25 rather than a factor of 50 has to be used for calculation:

Breaking load MSL factor MSL

d × d × 25 × 0.80 = kN (diameter in cm)

d × d × 25 × 0.80 = daN (diameter in mm)

The same applies for re-usable wire ropes of this type:

Table 19.4 Determining the diameter of a steel wire rope—16 mm

Breaking load	MSL factor	MSL	
1.6 × 1.6 × 50 = 128 kN	× 0.30 =	38.4 kN	(diameter in cm)
16 × 16 × 50 = 12,800 daN	× 0.30 =	3,840 daN	(diameter in mm)

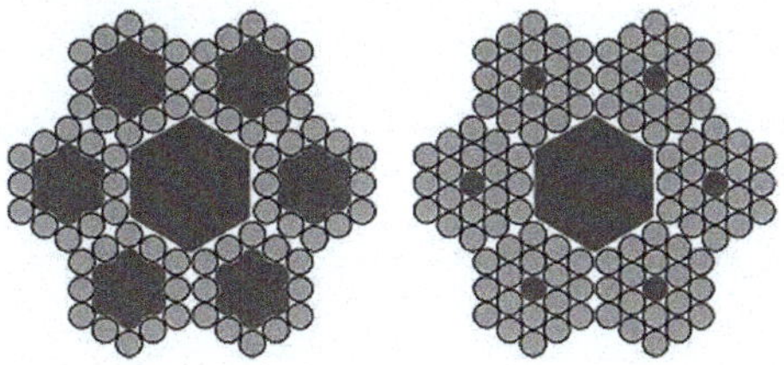

Fig. 19.5 Different rope structures: cross-sections and plan views

Breaking load MSL factor MSL

$d \times d \times 25$	$\times\ 0.30$ =	kN	(diameter in cm)
$d \times d \times 25$	$\times\ 0.30$ =	daN	(diameter in mm)

When ordering wire ropes, the following details should be stipulated to ensure that the correct rope type will be received:

• Diameter in mm or inches	e.g., 16 mm or 5/8"
• Quantity and length	e.g., 20×200 m
• Rope construction	e.g., $6 \times 19 + 1$ FE
• Type and direction of lay	e.g., cross lay, to the right
• Distinctive features (Plain, Galvanised)	e.g., plain
• Nominal strength	e.g., 1570 N/mm^2
• Greasing	e.g., normal
• Termination: swaged/Sharpened	e.g., swaged
• Packaging	e.g., coils

Wire ropes can be cut to length with hydraulic shears or other special tools if larger quantities need to be processed. Hand shears are sufficient on site and with small quantities (Fig. 19.6).

Running wires around sharp edges will lower their MSL. There are precise values, but a number of additional factors should be considered. Rough but nonetheless usable values are provided in this easily remembered table:

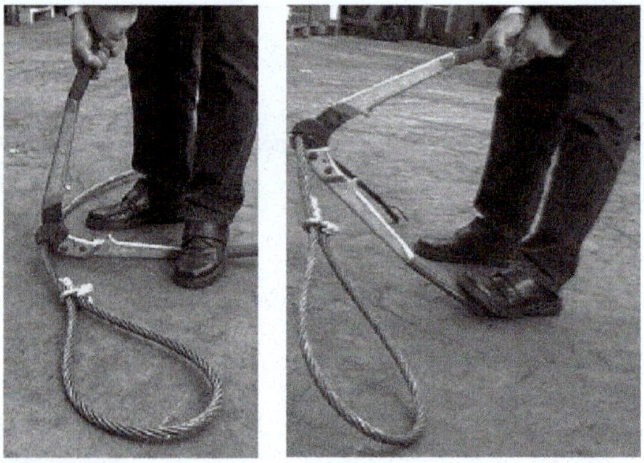

Fig. 19.6 Cutting wire ropes with shears: (l) incorrect—small lever arm; (r) correct—large lever arm

19.2 Wire Clamps (Clips)

Wire ropes with spliced or pressed eyes are rarely used for lashing. Wire cable clamps (clips) are used to connect to other lashing elements, such as turnbuckles, shackles, lashing rings, etc.

There are suitable clips available for every thickness of rope, in both metric and Imperial sizes. For a 12 mm wire, 12 mm or ½″ clips are required. For 16 mm wires, 16 mm or 5/8″ clips must be provided. When the nuts are correctly tightened as described further below, the following figures can be seen as guideline values for the strength of wire cable clips with common rope diameters (Fig. 19.7).

- Up to 25% of the MSL can be achieved with one clip. If there is no tightening torque, the value approaches 0.
- 50% of the MSL of the wire rope is achieved using two wire cable clips. The indentations in the wire indicate a good tightening torque. The following rule of thumb applies: if the free end on which the U-bolts of the wire cable clips should be placed is crushed to half the thickness of the wire, the clips have been tightened to a sufficient torque.
- A maximum MSL of approximately 75% can be achieved with three wire cable clips.

All too often in practice, too few wire cable clips are used, or they are attached incorrectly. Often, the nuts are not tightened sufficiently. The wires slip through the wire cable clips under load. If a wire rope lashing is to be homogeneous, four wire cable clips must be placed with the U-bolt on the free end (Fig. 19.8).

Fig. 19.7 Example of a typical wire cable clamp

Fig. 19.8 Single cable with eye

Fig. 19.9 Sling or double run—unsuitable construction

This type of attachment is advisable if long lashings have to be made. They are usually not necessary for container loads (Fig. 19.9).

This construction is often used for shorter lashings. But it should not be made like this (Fig. 19.10).

This format should always be used for short lashings because it is the most resilient. Inadequacies such as the presence of sharp edges can also be counteracted with this method (Fig. 19.11).

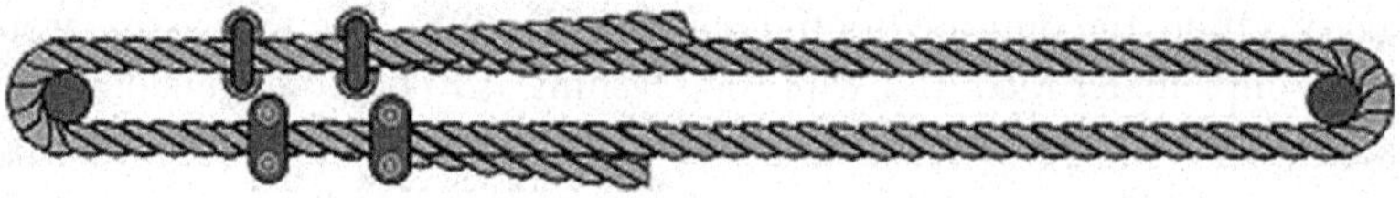

Fig. 19.10 Sling or double run—ideal construction

Fig. 19.11 Example of lashing which is not homogeneous

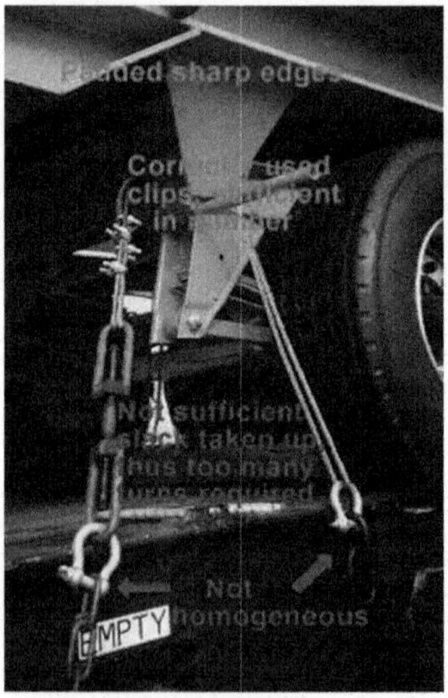

19.3 Anchor Shackles and Chain Shackles

Anchor shackles should be preferred for use over lashing shackles. They can adapt better to the directions in which forces arise (Fig. 19.12).

Using a rule of thumb, the breaking load and MSL of lashing shackles can be calculated as follows:

Breaking load MSL factor MSL

$d \times d \times 20$	$\times 0.50$ =	kN	(diameter in cm)
$d \times d \times 20$	$\times 0.50$ =	daN	(diameter in mm)

Since the factor 50% of 20 is equivalent to 10, the MSL can also be calculated directly with an easy-to-remember formula:

$$MSL\,for\ lashing\ shackle = d \times d \times 10$$

Determining the diameter of a bolt on a lashing shackle—26 mm. The diameter of the bolt is first measured in centimetres or millimetres. The breaking load formula only applies to shackles made of standard strength steel. The value for this shackle is quickly calculated using the direct rule of thumb for the MSL.

Fig. 19.12 Anchor shackles

$$\textit{MSL for lashing shackle bolt of diameter } 2.6\,\text{cm}$$
$$= 2.6 \times 2.6 \times 10 = 67.6\ \text{kN}$$
$$\textit{MSL for lashing shackle bolt of diameter } 26\,\text{mm}$$
$$= 26 \times 26 \times 10 = 6,760\,\text{daN}$$

19.4 Turnbuckles

The same values apply to turnbuckles as to lashing shackles. The diameter of the thread is first measured in centimetres or millimetres. The breaking load formula only applies to turnbuckles made of standard strength steel. The value for this turnbuckle is quickly calculated using the direct rule of thumb for the MSL, if the thread was measured at 1 1/8 inch = 28.575 mm.

$$\textit{MSL for lashing turnbuckle of thread diameter } 2.86\ \text{cm}$$
$$= 2.86 \times 2.86 \times 10 = 81.796\ \text{kN}$$
$$\textit{MSL for lashing turnbuckle of thread diameter } 28.6\ \text{mm}$$
$$= 2.86 \times 2.86 \times 10 = 8,179.6\ \text{daN}$$

When purchasing turnbuckles, it is important that only turnbuckles with sufficiently wide nuts and weld seams are purchased. Spot welded turnbuckles are available on the market. These should not be used. The turnbuckle shown above is available in three sizes: 1 1/8″, 1¼″ and 1½″ thread diameter. Note: one inch is equivalent to 25.4 mm (Figs. 19.13 and 19.14).

Fig. 19.13 Lashing turnbuckle

Fig. 19.14 Turnbuckle made of higher strength steel for round steel chains with short links

The manufacturer's specifications must be observed for turnbuckles made of higher strength steel. 50% of the specified breaking load is assumed as the MSL. Unless, as with this turnbuckle, a WLL (Working Load Limit) has already been specified. This then corresponds to the MSL. The same rules of thumb apply for lashing points made of normal strength steel as for lashing shackles and lashing turnbuckles. The diameter of the ring material or the bars is critical.

Breaking load MSL factor MSL

$d \times d \times 20$	$\underset{=}{\times 0.50}$	kN	(diameter in cm)
$d \times d \times 20$	$\underset{=}{\times 0.50}$	daN	(diameter in mm)

19.4.1 Determining the Diameter of a Lashing Point: 20 mm

$$\begin{aligned} &MSL\ lashing\ point\ material\ diameter\ 2.0\,\text{cm} \\ &= 2.0 \times 2.0 \times 10 = 40.0kN \\ &MSL\ lashing\ point\ material\ diameter\ 20\,\text{mm} \\ &\ \ = 20 \times 20 \times 10 = 4,000\,\text{daN} \end{aligned}$$

Even if it is not entirely clear what is meant by a strength specification, the MSL should still be calculated using the rule of thumb. Loads may only be applied to lashing points in a permissible manner. If a number of lashings are attached to a single lashing point, the total load must be considered. If lashings attached to the same lashing point are alternately subjected to load and relieved of load, every lashing can have the full MSL. If the lashings subject the lashing point to load simultaneously, the MSLs of the lashings must be distributed accordingly. Steel strapping can be applied quickly and easily and is therefore suitable for securing purposes on many loads. The following values can be used for calculations for untreated steel strapping (Table 19.5).

Blued straps have a higher strength. It can be calculated using these values (Table 19.6).

A blued steel strap 32 mm wide and 1.6 mm thick would produce the following value for the MSL:

$$MSL\ for\ blued\ steel\ strap\ 36\ \text{mm}\ wide,\ 1.6\ \text{mm}\ thick$$
$$= 3.2 \times 0.16 \times 85 \times 0.7 = 30.464\ \text{kN}$$
$$MSL\ for\ blued\ steel\ strap\ 36\ \text{mm}\ wide,\ 1.6\ \text{mm}\ thick$$
$$= 32 \times 1.6 \times 85 \times 0.7 = 3,046.4\ \text{kN}$$

The dimensions of steel straps are often coded without specifying the dimension units. It is usual practice to encode the width in mm and the thickness in 0.1 mm. The strap referred to above would have the designation 32/16 steel strap. Strength values calculated by rule of thumb must not be used if the steel strap is used incorrectly. Another important requirement is that the strapping seals are applied correctly. Steel straps have an extremely high breaking strength, and this is a common selling point. The disadvantage of steel straps, however, is their low elasticity. This ranges from around 0.25% to a maximum of 0.5%. Steel straps are therefore not really suitable for use as tie-down lashings. For instance, if a lashing is 7 m long, an ordinary commercial steel strap would have an elasticity of 7 m $\times$ 0.0025 = 0.0175 m or 17.5 mm. About half of this is exhausted for pre-tensioning, which means that just 8–9 mm remains. If the circumference of the lashing is reduced by this amount due to the strap cutting into the wood of cases, shifting of goods, etc., the steel strapping will become loose and no longer able to fulfil its function as a securing aid. Further jolts during transportation then produce greater forces on account of the resulting

Table 19.5 Calculating the breaking load and MSL for steel strapping

Breaking load		MSL factor	MSL	
Strap width w × strap thickness t	w × t × 70	× 0.70 =	kN	(diameter in cm)
Strap width w × strap thickness t	t × w × 70	× 0.70 =	daN	(diameter in mm)

Table 19.6 Calculating the breaking load and MSL for blue strapping

Breaking load		MSL factor	MSL	
Strap width w × strap thickness t	w × t × 85	× 0.70 =	kN	(diameter in cm)
Strap width w × strap thickness t	t × w × 85	× 0.70 =	daN	(diameter in mm)

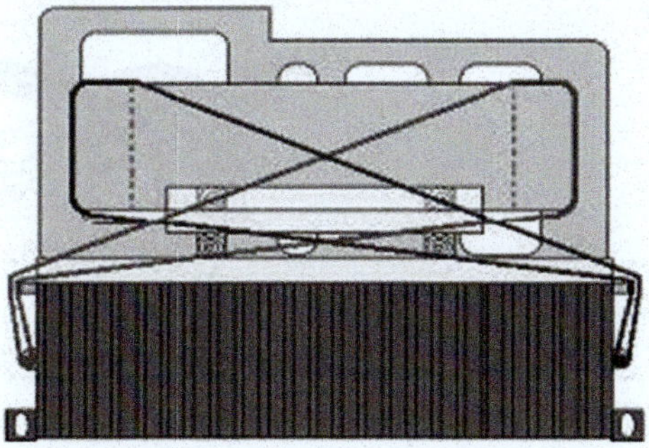

Fig. 19.15 Loads ideally suited for securing with steel straps

Fig. 19.16 Lashing chain with long links

Fig. 19.17 Lashing chain with short links

acceleration forces which can cause the steel strapping to tear. The low elasticity of steel straps can be compensated for by padding with materials with a high degree of elasticity and recovery. It is more convenient to use steel straps for direct lashings. They are extremely suitable for this (Figs. 19.15, 19.16 and 19.17).

Lashing chains are only used to secure container cargoes if the chains will be returned. On valuable cargoes, however, it may also be worthwhile to use them as a "lost load securing aid," because of the benefits offered by their high strength and speed with which they can be deployed. Since lashing chains are made of high strength steel, there are no rules of thumb for calculating the breaking loads. The manufacturer's specifications should be observed. The MSL may be assumed to be fifty percent of the breaking load. A range of different tensioning mechanisms are used with chains. Errors are frequently made when using these aids.

Tightening the chain using the hooks (Figs. 19.18 and 19.19).

Tightening the chain with the tension lever (Figs. 19.20 and 19.21).

The tension levers must not remain at an angle of 90°. They must be adjusted to an angle of at least 45°. Otherwise, the tongues could buckle or shear off (Figs. 19.22, 19.23 and 19.24).

No rules of thumb are provided in the CSS code for high strength chains, because the diversity of the products is extremely high. There are chains made to all kinds of

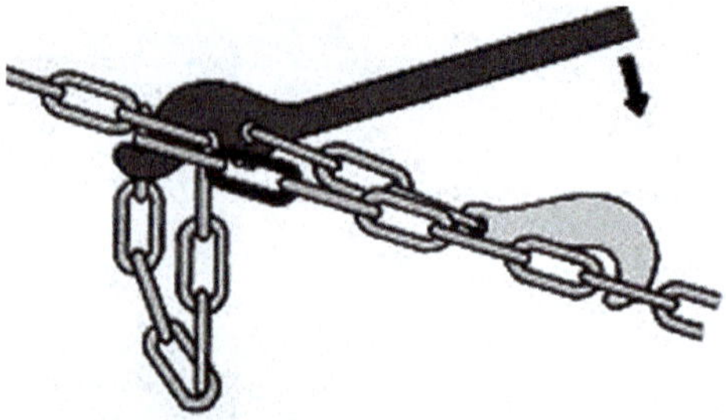

Fig. 19.18 Inserting the hook

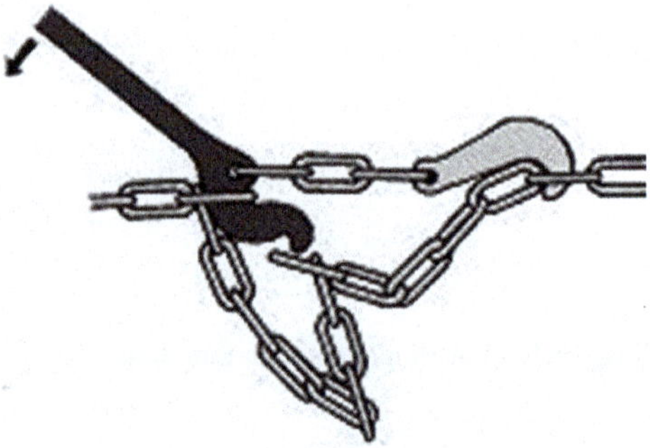

Fig. 19.19 Shortening the chain

Fig. 19.20 Inserting the tensioning lever

different standards on the market. If the stamp on the chain and its accessories clearly identifies it as quality class 8, the rule of thumb d2 × 120 can be used to calculate the breaking load. The MSL is to be assumed to be fifty percent of the breaking load. This produces the following figures for the following diameters of chain steel (Table 19.7).

Fig. 19.21 Tightening and securing the lever (l) the incorrect way (r) the correct way

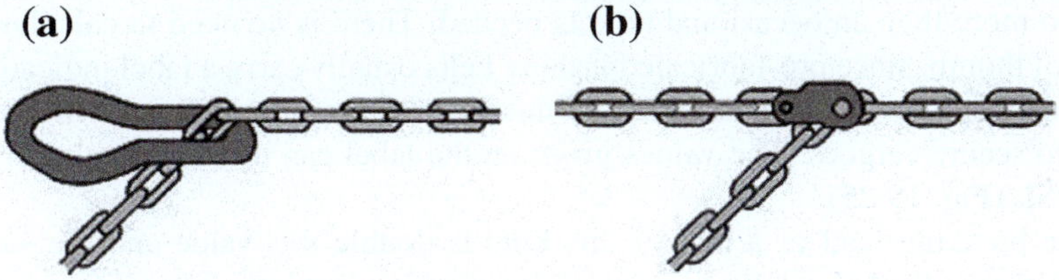

Fig. 19.22 Permissible shortening of chains

Fig. 19.23 Correct shortening of a chain—no chain link is trapped

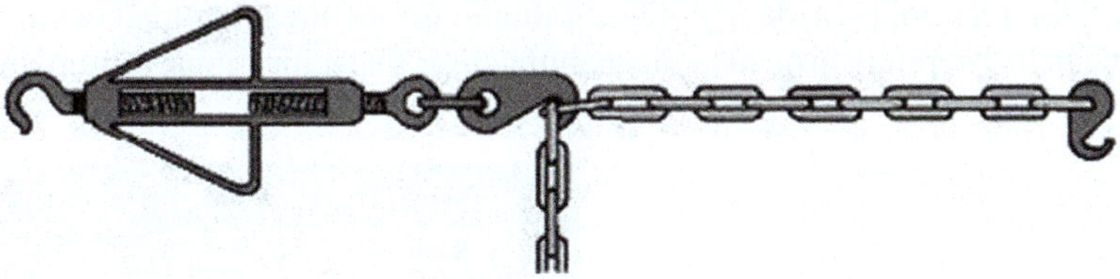

Fig. 19.24 Incorrect shortening of a chain—a chain link is trapped

Table 19.7 Diameters of chain steel

Diameter of steel	Breaking load	MSL factor	MSL
0.6 cm	43.2 kN	0.5	21.6 kN
6 mm	4,320 daN	0.5	2,160 daN
0.8 cm	76.8 kN	0.5	38.4 kN
8 mm	7,680 daN	0.5	3,840 daN
1.0 cm	120 kN	0.5	60 kN
10 mm	12,000 daN	0.5	6,000 daN
1.3 cm	202.8 kN	0.5	101.4 kN
13 mm	20,280 daN	0.5	10,140 daN

19.5 Pre-fabricated Lashing Belts

It is assumed that staff will be familiar with how to use systems of this kind and that no more than an occasional note is needed. There is no need to calculate with rules of thumb, since pre-fabricated lashing belts usually carry a label indicating the Lashing Capacity (LC). This is the permissible lashing force (F_{max}) which can be used to secure cargoes. The values given on the label can therefore be accepted as the MSL (Fig. 19.25).

The breaking load of a belt of this kind is double this value and the standard therefore sets the LC at fifty percent of the breaking strength. The CSS also names "50% of breaking strength" as the permitted value. In relation to all lashings used, the proportion of pre-fabricated lashing belts is small, since the belt systems can only be returned with specific shipping companies and cargo types. Flexibility suffers in day-to-day lashing operations as a result of fixed lengths and sewn-in end fittings. Due to the speed with which they can be applied, however, the proportion of these lashings being used is constantly growing. If lashing of this kind is used, however, it should be applied out correctly and professionally. Belt systems permanently installed in containers must also be included among the pre-fabricated systems. The lashings below used to secure automobiles must also be classed as pre-fabricated goods (Fig. 19.26).

The CSS does not provide any rules of thumb for the breaking load of single-use webbing belts but refers instead to the manufacturer's specifications. Seventy percent

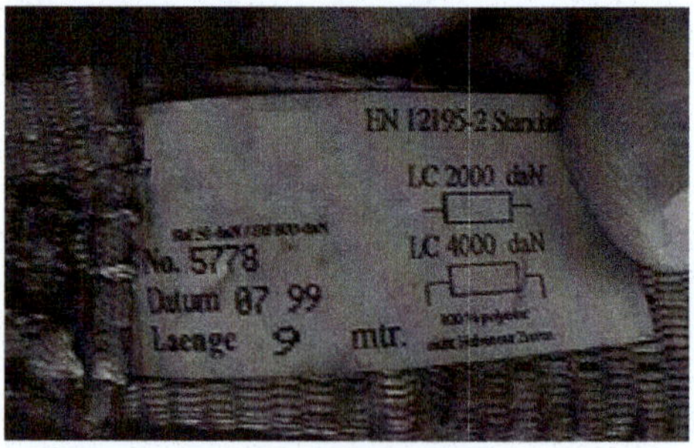

Fig. 19.25 Textile lashing belt

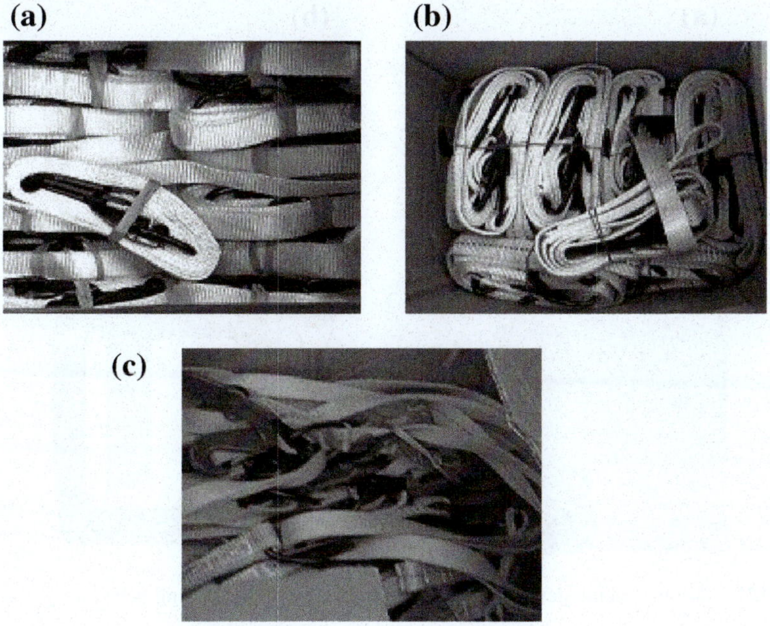

Fig. 19.26 Various automobile lashings

is given as a factor for the MSL.

$$MSL\,for\ synthetic\,fibre\ lashing\ belts \\ = breaking\ load\ as\,per\ manufacturer's\,specification \times 0.7$$

$$\text{Size : 40 mm}$$
$$\text{BS : 3,500 kg}$$

In the example above, the breaking strength is specified as 3,500 kg on the label, i.e., 3,433.5 daN. The MSL would then be seventy percent of this, or 2,403 daN. In the opinion of the author, this is far too high. It is well-known that in defining this factor the national interests of some IMO members played a greater role than the known problems associated with single-use lashing belts, where the quality of production differs. They are often very elastic. Some materials exhibit permanent deformation at threshold loads above fifty percent of the breaking load. This means that the lashings made from them will then become slack. Since they are also identical in many other ways to synthetic fibre ropes, the author recommends that at least the fifty percent factor of the European Standards is assumed for pre-fabricated lashing belts. For the material discussed above, this would be a value of 1,717 daN. If the author were personally responsible for lashing work, he would only assume the value for synthetic fibre ropes, i.e., 33% of the breaking load. For a belt with a breaking

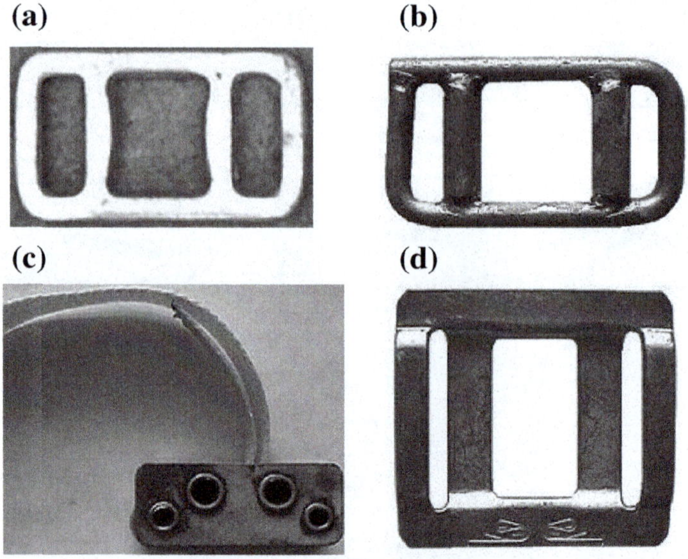

Fig. 19.27 Various locking devices for use with single-use textile lashing belts

load of 3,433 daN (3,600 kg), this would be 1,133 daN. Since belts are always used doubled, the value would double accordingly (Fig. 19.27).

It is useful and practical to have operating instructions stamped onto the locking devices.

19.6 Packing Straps

Packaging straps of this kind are only suitable for securing individual, light packages on pallets and similar supports as well as for unitisation. There are a wide variety of end fittings and other connecting elements available. It is important to select the one which best suits the lashing points on the particular cargo transport unit. Hooks must where possible be hooked in from the bottom to the top or from the inside to the outside, so that they cannot loosen if jolted during transport, should the lashing become slack. It is essential that belts are protected against sharp edges whether this be by means of belt protectors or special sleeves. If this is not possible or items like this are not available, edge protectors are to be used (Figs. 19.28, 19.29 and 19.30).

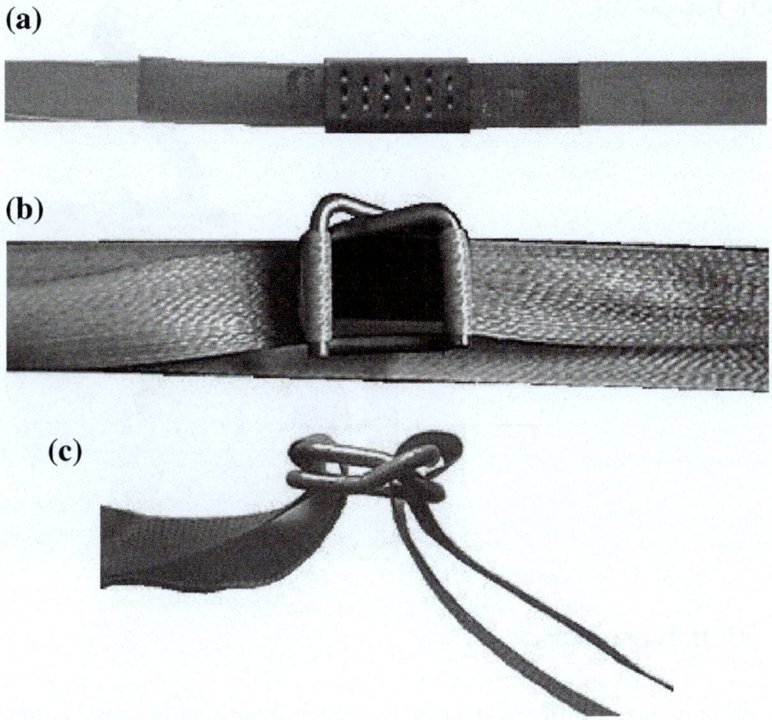

Fig. 19.28 Different types of packing straps

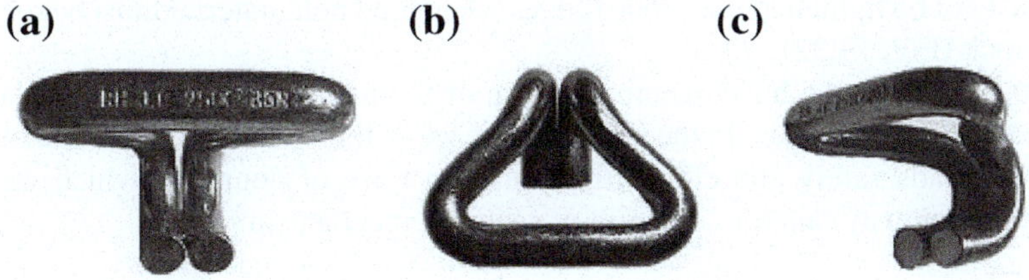

Fig. 19.29 Different types of end fittings

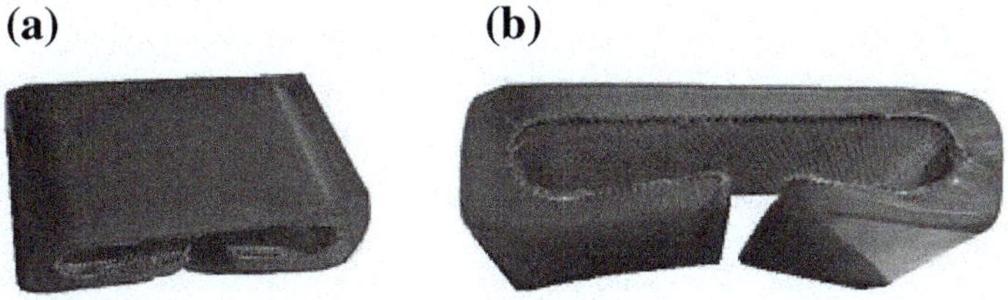

Fig. 19.30 Belt protectors

Fig. 19.31 Belt tensioner

Fig. 19.32 Knotted textile belts

19.7 Belt Tensioners

There are a number of different devices available for tensioning textile lashing belts (Fig. 19.31).

Belts and straps must not be knotted and twisted. They must rest on a wide surface and must not be tightened over sharp edges. Damaged belt material must be removed from stock (Fig. 19.32).

In this chapter, we have examined some of the most common forms of lashing materials and accessories. In the next chapter, we will look at how wood can be used to secure loads safely and efficiently, both as a form of dunnage (which we have covered previously) and or as a form of securing mechanism.

Glossary

Absorption Refers to the taking up of gases or vapours by liquids and/or solids where the gas or vapour fully penetrates the liquid or solid (adsorption is limited to the surface).

Acceleration of shipment Negative and positive accelerations are dynamic, mechanical stresses which occur in two main types during the transportation of goods:

- regular acceleration forces; and
- irregular acceleration forces.

Regular acceleration forces primarily occur in maritime transport. Acceleration of up to one g (g = 9.81 [m/s^2]) and, in extreme cases, even more, may occur due to rolling and pitching in rough seas. Such regular acceleration forces have an impact on the effort involved in load securing. Irregular acceleration forces occur during cornering or when a train passes over switches, during braking, starting up, hoisting and lowering. Such acceleration forces are not repeated, but they may occur several times at varying intensities during transport. These are the typical stresses of land transport and transport, handling and storage operations.

Acid value The acid value specifies how many milligrams of potassium hydroxide (KOH) are necessary to neutralise the free fatty acids in a single gram of fat.

Across the grain The cross-section that is the result of cutting a tree trunk perpendicular to its length (as opposed to "with the grain").

Active behaviour Applies to goods that have an active influence on other goods and their environment.

Additives Substances added to other substances in order to modify their properties in a particular manner or to improve processing.

Adsorption The uptake of gases, vapour or dissolved substances (this is limited to the surface of solid objects).

Adsorption delay Delayed take up of water vapour by the cargo. In this book, this refers exclusively to water vapour and describes the fact that during the day the

A. A. Olsen and F. Karkori, *Containerized Cargo Handling and Stowage*, Springer Series on Naval Architecture, Marine Engineering, Shipbuilding and Shipping 22, https://doi.org/10.1007/978-3-031-63329-4

water vapour is given off by the cargo to the air in the container (desorption) faster than the cooling of the container air is introduces water vapour to the cargo (adsorption). The water vapour condenses on the walls of the container thus leading to damage.

Adsorption isotherms As used in this book, the graphical representation of the uptake of water vapour by a substance at a specified temperature. For product information purposes, the sorption isotherms are usually taken for a temperature of 20 °C (68 °F).

Aerobe (Greek: aer—air). Microorganisms that require atmospheric oxygen.

Aerobic respiration (Greek: aer—air). Metabolic process of goods of vegetable origin, during which glucose and atmospheric oxygen are consumed by respiration processes to form carbon dioxide, water vapour and heat.

Aflatoxins Short form of Aspergillus—flavus—toxin. A mycotoxin (mould toxin) that is produced by the fungus Aspergillus—flavus, among others.

A-frame Frame shaped like the letter "A" used to transport sheet loads, similar to the type of frame used when transporting panes of glass.

Agglomeration Lump formation.

Airbag Airbags can be positioned in the gaps between the cargo in order to achieve a tight fit during loading.

Allelopathy (Greek: allelon—mutual; pathe—influence). The mutual influence of goods of vegetable origin when stored together, caused as a result of gaseous metabolic products, for example, ethylene or carbon dioxide.

American Rust Standards Used to classify corrosion damage. The *American Rust Standard* is recognised worldwide and widely used. It is often used in reports produced by surveyors.

Amorphous Without shape, without a fixed form.

Anaerobe (Greek: an—without, and aer—air). A microorganism that survives without atmospheric oxygen.

Anaerobic respiration (Greek: an—without; and aer—air). Occurs with goods of vegetable origin if the permitted carbon dioxide content is exceeded.

Animal diseases Animal diseases are diseases which are caused by bacteria, viruses, protozoa or fungi and may be transmitted to animals and humans.

Animal goods Goods of animal origin.

Anthrax Notifiable febrile disease caused by the anthrax bacillus.

Anthropogenic influence Influence caused by human activity.

Apparent density Density of lumber including hollow space.

Arrhenius equation An equation describing the dependency of reaction speed on temperature. A simplified rule of thumb for transportation purposes is that when the temperature is lowered by 10 °C (50 °F), degradation processes are reduced to between half and a third, i.e., the storage life can be doubled or even trebled by lowering the temperature by 10 °C (50 °F).

Articulated train/double train An articulated train is a combination of an articulated truck and one or more trailers, i.e., truck tractor plus semitrailer plus trailer.

Articulated truck (lorry) An articulated truck is a combination of a truck tractor and a semitrailer.

Athwartships stowage Load stowed across the beam of the container or ship (as opposed to fore and aft stowage). In the context of load securing, it is of utmost importance whether a container is stowed athwartships or fore and aft on a ship. In the case of athwartships stowage, the greatest acceleration forces act on the actual container longitudinally rather than transversely. Load securing measures must then be taken with this in mind.

ATP Agreement Agreement on the *International Carriage of Perishable Foodstuffs and on the Special Equipment to be used for such Carriage (Agreement on the Transport of Perishables); UN agreement for the cross-border transportation of foodstuffs.*

Autolysis (Greek: autos—self; lucin—separation). Self-digestion.

Autoxidation Oxidation caused atmospheric oxygen and taking place at normal temperatures.

Baud Unit of speed for data transfer; 1 baud = 1 bit/s.

Bay A row of containers running from one side of the ship to the other. To be distinguished from a hold.

Bay plan Plan for the stowage of containers, in which each bay on a vessel is represented in a cross-sectional plan of the ship from bow to stern. A bay plan shows all positions for stowage on the vessel.

Bay-row-tier system Numbering system for the arrangement of containers on a vessel. In this numbering system, the bay is specified first, then the container row, which runs the length of the ship, and finally the tier (vertical layer).

Bay-tier-row system Numbering system for the arrangement of containers on a vessel. In this numbering system, the bay is specified first, then the tier (vertical layer) and finally the container row, which runs the length of the ship.

Bevelling If a board or lumber beam is used diagonally for load securing purposes in a container, the ends must be bevelled to ensure a tight fit into the container, thus allowing forces to be transferred to the container.

Bilge sump Sump at the lowest point of the hold, where water can collect and be pumped off.

Billets Semi-finished products (steel products) with a square or rectangular cross-section, edge lengths of 50 to 125 mm, rounded edges and a minimum length of approximately 1000 mm. This are the used to produce for bars, wire rod and special profiles or for forged or drop-forged items.

Bill of lading, bill of lading number Consignment notes issued by the consignor as part of ocean-going and inland transportation processes which regulates the legal relationships between the loader, the consignor and the receiver of the goods that are being transported. This note may be confirmation from the consignor that the goods have been received for shipping (Received for Shipment B/L) or a confirmation that the goods have been put on board (On Board B/L, Shipped on Board B/L). This document also covers the duty of the consignor to deliver the goods to their port of destination and to hand the goods over to the legitimate owner of the bill of lading in return for the bill of lading.

Biotic activity Refers to activities evident in products of vegetable or animal origin that during transportation exert an influence on the products themselves and/or on other animal or vegetable products. Biotic activity covers biochemical, microbial and other decomposition processes (e.g., allelopathy).

Bird's mouth A bird's mouth is formed when a very flat V-shape is cut into the end of a piece of wood.

Bitter ends The ends of rope after a change of direction of any kind. The ends are usually joined by means of a knot, wire cable clamps, turnbuckles, shackles etc.

Black rot Rot attack on fruit. Black rot can always occur if damage to the fruit extends as far as the flesh. This is then easily transferred from an infected piece of fruit to the other fruits. Black rot is a frequent type of chilling damage occurring in pineapples.

Block stowage A method of loading: in block stowage, the goods are stowed as a single block, without leaving space of the flow of air. Thus, air can only pass around the block itself.

Blowing Bulging observed on the lids and bases of cans caused by the formation of gas after the sterilisation process and leading to bursting. Such bulging can be caused by high temperatures (see *heat-induced blowing*) or low temperatures (see cold-induced) erfolgen.

Blue discolouration Blue discolouration of wood caused by blue stain fungi.

Blued steel strap A normal steel strap can be refined by heat treatment (heating, quenching, tempering). It undergoes a structural change which gives it optimal properties in terms of hardness and elasticity. Heat treatment leaves an oxide layer on the surface of the steel which is a blueish colour (steel blue), hence the name.

Botulism Dangerous food poisoning caused by the bacterium clostridium botulinum occurring for instance in insufficiently sterilised preserved meat or vegetables.

Box container Alternative colloquial term for the standard ISO container.

Bracing/bracing beams Items used to secure general cargo loads. Bracing is usually made from squared lumber beams that are fitted between the load and the load-bearing sections of the container in order to take up the horizontal and vertical forces acting on the load. The bracing is subject to compressive stress.

Brine circuit Intermediate circuit of a cooling system that uses a saline solution (brine) as a coolant to reduce the freezing point. This method is often used to reduce the amount of chemical-based coolant used in the cooling system.

Brittle fracture Caused by mechanical stress placed on plastic films at freezing temperatures.

Buckling Drying and consequential bulging of the exterior of paper packages on pallets.

Buffer stows A so-called "buffer" is required when transporting a heavy load with a high mass and small surfaces for distributing the forces acting upon the object. Lumber beams are usually used for this purpose. For certain types of loads, the effort involved in this load securing method can be reduced by using part of the load as a "buffer." Round bars are a good example, as they are (1) capable of bearing high loads, and (2) suffer no damage as a result of being used as a buffer. For this method, a round bar is first loaded across the end wall of the container.

This then acts as a buffer and, in the event of jolts during transport, cushions or distributes the forces evenly across the end wall. Since the buffer is created the cargo is loaded, this method is referred to as "buffer stow."

CA, CA atmosphere CA = controlled atmosphere. For example, an artificial reduction of the level of oxygen and increase in the level of carbon dioxide to extend the lifetime of goods.

CA container Container with controlled atmosphere.

Cakes Cuboid semi-finished copper product that is used to make other semi-finished or finished products.

Caking Hardening of bulk cargo caused by a release of water vapour; Causes the formation of lumps if moisture is present in salts or sugars.

Candying Hardening caused by an excessively dry atmosphere, e.g., on dried fruits.

Capacitive network Data network with capacitive coupling to the power network for remote monitoring of refrigerated containers.

Capillary condensation Condensation of vapours in the fine pores of adsorption media.

Cargo information This is an online system for transport professionals containing transport-related risk and damage prevention information.

Cargo sweat Condensation on the surface of the cargo: Condensation can form on the cargo as a result of climatic conditions. Cargo sweat can always occur when the temperature of the cargo is lower than or equal to the dew point. It can also occur during transportation from temperate latitudes, e.g., from northern-hemisphere winter to the tropics.

Cargo temperature The temperature of the cargo within the container. Can vary according to place and time (e.g., during a cooling or warming process).

Catalysis (Greek: katalysis—dissolution, decomposition). Change, usually the acceleration, of a chemical reaction by means of substances (catalysts) which remain unchanged during the reaction process.

CBU Capacitive bridge unit. Component of a remote refrigerated container monitoring system where data is transferred via the power cable. A CBU is used to capacitively modulate the data signal (using a capacitor) over the power cable.

Cell lumen Pores in the wood structure.

CFC Chloro Fluoro Carbons. CFCs are artificially produced gases or liquids. They are used as coolants, propellants or detergents. CFCs can remain in the atmosphere for decades before being broken down. They are a source of chlorine radicals, which react with ozone and contribute significantly toward the destruction of the ozone layer. CFCs are now widely around the world.

Chafe Become damaged as a result of abrasion.

Chilled bananas Bananas are chilled when the temperature has dropped below the acceptable limit. This process means that the fruit suffers irreparable metabolic damage which renders it incapable of ripening.

Chilled meat Fresh meat is transported chilled.

Chilled operation/chilled mode Operating mode of a refrigerated container. During chilled operation, the circulating fans run at maximum rpm and the intake air temperature is regulated constantly.

Chill haze Change in consistency due to the deposition of dissolved components as a result of the cooling of aqueous solutions.

Chilling Chill damage is caused when the storage temperature goods of vegetable origin drop below the specified limits. This process means that fruit suffers irreparable metabolic damage which renders it incapable of ripening.

Chilling damage Occurs when the load is subject to temperatures below the acceptable limits. This process means that fruit suffers irreparable metabolic damage which renders it incapable of ripening.

Chipping This a mechanical defect of enamel goods.

Circulation bypass Occurs when air in a space does not circulate throughout the entire space, but rather seeks the path of least resistance in parts of the space.

CKD (Completely knocked down cars). For the purposes of customs, cars are not transported fully assembled, or are partially dismantled after being assembled. Final assembly of the vehicles is subsequently carried out once they reach their destination. CKD transportation makes it possible to avoid high import duties (100% and more).

Climacteric (Latin: climactericus—turning point, critical time). Ready-to-eat or ripe for consumption.

Climacteric fruits (Latin climactericus—turning point, critical time). For bananas, the "climacteric" is the point at which the fruits turn from green to yellow. Starch breaks down to form sugar. Fruits that reach the climacteric are ready-to-eat or ripe for consumption.

Climacteric rise (Latin climactericus—turning point, critical time). The final increase in carbon dioxide excretion which marks the beginning of maturation of a fruit. After the climacteric rise, fruits are ready-to-eat.

Clip-on-units Refrigeration units used to cool porthole containers on deck, in port or during road/rail transportation.

Coagulation (Latin: coagulare—to congeal) Clotting of a colloid solution, for example, protoplasma.

Coefficient of expansion, thermal, cubic Increase in volume of a material caused by an increase in temperature.

Coefficient of sliding friction Dimensionless factor indicated by the Greek letter μ. This factor is used to determine the force required to overcome the friction which is produced by a specific material combination constantly sliding under the influence of a pulling force. The coefficient of sliding friction is important in load securing, since the sliding friction of a load with a specific material combination (load/loading area) must be overcome before the load slides. The greater the friction, the less effort is required in securing the load.

Cohesive resistance Resistance of an object against slipping on a surface (a function of cohesion and friction).

Coil box container Special container for transporting coils. They handle greater line loads and are equipped with coil wells and special load securing features.

Coil containers are built like flats or flat racks, i.e., they consist of a container floor and flat or frame-like end walls. The container floor has cargo troughs for accommodating coils/rolls of steel sheet.

Coils Products such as steel sheets, wide strip, steel piping or wire which are coiled into rolls. Coils weighing up to 35 t are transported.

Cold chain Unbroken chain from manufacture through to the consumer in which the prescribed cooling temperature is constantly maintained.

Cold-induced blowing Bulging on cans caused by the canned product increasing in volume when the temperature falls below freezing point.

Cold-rolled strips Flat cold-rolled products in widths of up to 650 mm and thicknesses of 0.1–6 mm. Can be made from any type of steel and is available with any required surface treatment. Cold-rolled strips are available in the form of rings or strips (bars). The advantage of cold-rolled strips and sheets over hot-rolled and cold rerolled sheets is the greater degree of accuracy of dimensions.

Collapsible flat rack container Flat with end walls that can be folded in when empty. The end walls are unlocked and folded down onto the loading area. In this state, the flats can be stacked. Three of the "folded" flats take up about the same space as a standard container.

Collo Pl. colli, internationally used word meaning an item of freight (crate, box, bale, bag, bundle, etc.).

Compatibility characteristics Interrelationships between the transport properties of products in relation to each other and in relation to the environment.

Conair container Refrigerated containers without their own refrigeration unit. Conair refrigerated containers, also called insulated or porthole containers, do not have their own refrigeration unit. They are thus reliant on an external supply of cold air. This is achieved by refrigeration units of several types, permanently installed on the ship, permanently installed in the terminal or clip-on units for individual containers.

Porthole containers are thermally insulated and have two sealable openings on the end walls (the portholes) through which cold air can be blown into the container and warm air can be extracted. The cold air is forced through the lower opening into the container, then distributed throughout the load via the T-bar floor, and subsequently flows through the load to the top of the container and is extracted through the upper opening.

Consolidated packages Group of cargo items fastened together with straps or similar devices.

Contact insecticide Contact poison that kills pests on contact.

Contact poison (contact insecticide) Poison that kills pests on contact.

Container bolster A container floor without sides or end walls used for Ro/Ro operations.

Container dry Water content of goods which would not negatively impact the quality of the goods being transported in the container. For any given load, this water content will be dependent on the duration of the journey and the route and may vary over different routes.

Containerisable Goods (and their packaging) must have the physical properties which ensure that they will survive transport in a container without damage as a result of climatic, mechanical and biotic conditions (depending on the route and duration of transport). These properties include the humidity, temperature and ventilation requirements of the load.

Container packing certificate Certificate indicating correct loading of a dangerous goods container and the observance of the regulations set out in the IMDG Code (*International Maritime Dangerous Goods Code*). The container packing certificate is issued by the person responsible for packing the container.

Container section A packed section of a container.

Container sweat Condensation which forms on the surfaces of the container.

Container template A moveable frame that maps the size and shape of the inside of a container. If the exact dimensions of a cargo or its packing are unknown, the 'template' can be pushed over the load in order to make sure that the correct size of container has been selected.

Contamination Damage to goods caused by dust, dirt or staining substances.

Contraction (Latin: contrahere—shrink). Reduction in volume.

Cooked bananas If upper temperature limits are exceeded, the bananas can ripen so quickly that they burst. This is usually referred to as "cooked" bananas.

Corner castings Corner fittings located at all eight corners of the container. They are used to handle the container. Lifting gear, such as container gantries and cranes, is attached to the corner castings of the containers in order to lift them. The corner castings are used in conjunction with twist locks to secure containers when traveling on road vehicles or stack and fasten them securely to the deck of ocean-going vessel. The twist locks are placed in the corner castings and engage with the corner castings (on the floor) of the next container.

Corner posts Corner posts of a container. These connect the upper corner casting of a container to the lower one and form one of the most stable parts of the container.

Corrosion (Latin: corrodere). Destruction of materials. When applied to metals this is an electrochemical process that can, in particular, be promoted by a good electrolyte (sea water).

Corrugated board carton Carton made of corrugated board.

Corrugations Metal sheets are deliberately corrugated in order to increase their stability. Standard containers, as far as they are made of steel sheet are usually manufactured using sheet sheets with trapezoidal corrugations. Corrugations can also be used when making steel drums, again for the purposes of stabilisation.

Cradle A cradle is a frame that is used to secure loads with a tendency to roll, such as boats or cylindrical goods (boilers, large-diameter pipes, etc.). The upper side of the cradle is shaped to fit the product it is designed to carry.

Crate A wooden receptacle formed like a case or box, with the difference that the external walls and the lid are not fully closed.

Crate restraint shoe Steel bracket specially made to suit a specific crate, where the upper side of the shoe takes the crate, and the underside of the shoe is designed to fit tightly to the transport medium or container.

Crispness Property of a product (e.g., bakery products).

Critical water content Water content of cargo at which a change in quality is expected.

Cryptoclimate (Greek: kryptos—concealed, hidden) Microclimate in a closed space such as a container.

CTU/CTU [container] packing guidelines Cargo Transport Unit. Guidelines for the packing of loads except bulk goods either in or on containers for transport by means of any method of land-based or water-based transportation. In this book, CTUs are referred to generically as 'containers.'

Dan Deca Newton.

Data logger Device used for automatic (and continuous) recording of data. The data can then be read out at a later time, for instance, using a laptop.

Dead air zone Area in a cargo stack with insufficient ventilation.

Degree of rusting Assessment of the visible amount of rusting, i.e., the degree of damage to iron and steel cargoes as a result of corrosion. The *American Rust Standard* is frequently used in international trade to describe the degree of rusting.

Deodourisation (Latin: odour—odour; de—negation = to render odourless). Elimination of odours.

Dermatitis (Latin: derma—skin; -itis—suffix denoting 'inflammation'). Inflammation of the skin.

Desiccant Substance for absorbing moisture.

Desorption Release of the water vapour from a substance.

Desorption isotherm Graphical representation of the release of water vapour from a substance in relation to its water content at a specified temperature. For product information purposes, the isotherms are usually taken for a temperature of 20 °C (68 °F).

Dew point temperature The temperature at which water vapour in the air reaches saturation point and condenses ("dew" appears).

Diffusion law (Latin: diffundere—to pour out, spread). Water vapour always flows from the higher to the lower partial pressure.

Dilatation, thermal (Latin: dilatare—to extend). Increase in the volume of a substance as a result of increased temperature.

Display packaging Packaging designed for a specific product (e.g., pears) which are also suitable for presentation at the point of sale.

Dormancy temperature The temperatures of cargo of vegetable origin are regulated in order to prevent loss of quality.

Dose (French: dose—appropriate measure). Quantity of a toxic substance in the body.

Dressed lumber Wood from which the sapwood has been removed.

Drip loss Loss of juice as the result of recrystallisation of frozen meat. If the meat is frozen rapidly at low temperatures (e.g., −27 °C (−16.6 °F), small ice crystals are produced. If the meat rises to a higher temperature (e.g., −10 °C (14 °F), the ice crystals grow larger, i.e., they recrystallise. As of a certain size, they can rupture the cell walls of the meat. These "perforated" cell walls lose more cell juices than cell walls which are not damaged. The result is a loss of juice, also known as "drip loss."

Driving wedges See *tapered blocks*.

Dry bulk Bulk cargo.

Dunnage Dunnage is used to protect the contents of the container, for instance against sweat. A distinction is drawn between top dunnage side dunnage and floor dunnage. Paper (see paper dunnage), wood, plastic or many other materials can be used as dunnage.

Dunnage material Materials which are not fastened firmly to the means of transport or transport container and are used to protect the goods from sweat, dirt, or mechanical stress, for instance, or as an aid in stowage (interlayer dunnage). Depending on its application, it is referred to as floor dunnage, interlayer dunnage, lateral dunnage or top dunnage.

Eating ripeness Ready-to-eat or ripe for consumption (climacteric). Ideal degree of ripeness for consumption.

EN standard European standards.

Enzymes (Greek: enzymos—fermented). These are catalytic systems which break down nutrients.

Equilibrium moisture content The water content of the goods is in equilibrium with the quantity of water vapour in the ambient air (relative humidity).

Ergot Resting body (sclerotium) of the ergot fungus. Meal contaminated by ergot causes poisoning; ergot alkaloids produce uterine contractions.

Essential oil Constituent part of spices that have a strong seasoning action and odour.

EUROSAL service Container line running between the west coast of South America, the Caribbean and Europe. Members of the EUROSAL service include Hapag-Lloyd, Hamburg-Süd and CSAV.

Eutectic point (E.P.) Temperature (approx. −62 °C (−79.6 °F) at which all the water in the cells of the product is completely frozen, preventing microbial biotic activity and therefore preventing any loss of quality to the product. The product has an infinite storage life. Water and dissolved substances crystallise out together.

Fat-cleaving enzymes Lipases.

FCL container "Full Container Load." Container packed by the shipper for door-to-door delivery without any transshipment operations.

Ferment (Latin: fermentum—yeast, fermentation). See also *Enzyme*: Certain organic substances induce decomposition, particularly fermentation, in other organic substances. The term "enzyme" has replaced the term "ferment."

Fermentation Biochemical process in which energy is released from carbohydrates by enzymes.

FEU The abbreviation "FEU" is occasionally used for "Forty-foot Equivalent Unit" and refers to a 40ft container. A 40ft container comprises 2 × TEUs (Twenty-foot Equivalent Unit, 20ft container). Therefore: 1 FEU = 2 × TEU.

Fibre drum Cylindrical transport container (similar to a large can or a barrel). It is made of wound kraft paper and has a sheet steel base and lid. It is sealed by means of closing rings and a locking lever.

Field bus Part of a network solution that based on a four-tier model (actuator-sensor level, field-level, cell-level, management-level). The "field-level" will comprise spatially separated machine lines and local machines linked to each other or to the superordinated level.

Fit-for-purpose packaging Packaging that is able to withstand any stresses that can be expected during the voyage and that will afford appropriate protection to the products. The stresses could vary significantly depending on the method of transport (road, rail, air, sea, combined methods of transport, repeated transshipment, etc.).

Flaking Flaking is a common fault during the production of enamel goods. In a general sense, it refers to the separation of areas of the surface due to strain and mechanical stress.

Flange A ring welded onto the end of a pipe. It is used with its counterpart to join pipes. The rings are provided with drilled holes through which the pipes can be bolted together.

Flap Collapsible rail/ramp for moving loads into the belly of the ship.

Flat rack (flat) Container without side walls and roof, with fixed or collapsible end walls.

Flexural strength The ability of a material or a construction to withstand bending.

Flowability Ability of crystalline, pulverulent or granular goods to flow freely.

Flow moisture point (FMP) Point at which the goods change from solid to liquid state.

Food trays Preformatted small packages which are also used for sales presentations.

Fore and aft stowage Load stowed along the length of the container or the ship. Contrast athwartships stowage. In the context of load securing, it is of utmost importance whether a container is stowed fore and aft or athwartships on a ship. In the case of athwartships stowage, the greatest acceleration forces act on the actual container longitudinally rather than transversely. Load securing measures must then be taken with this in mind.

Freezer burns Drying out of frozen goods.

Freezer container This is a special type of refrigerated container which can maintain the temperature of the cargo at approximately −65 °C (−85 °F). At approximately −62 °C (−79.6 °F), the "eutectic point" is reached, i.e., as of this temperature, all the water in a product is completely frozen. No microbial activity can occur when there is no free water, and therefore the product does not suffer any further loss of quality. At these temperatures, the product has an infinite storage life.

Freezing chain Unbroken chain from manufacture through to the consumer in which the prescribed freezing temperature is constantly maintained.

Friction-enhancing mat Mat made of material which hinders slipping.

Fruit respiration Exchange of gases between fruits and the ambient air. Apples, for instance, constantly take in oxygen when they respire and release the same quantity of carbon dioxide. If you change the concentration of the gases which are involved in respiratory metabolism, you can permanently affect the intensity of the fruit respiration. This will delay the maturity and degradation processes and thus increase the duration of storage.

Fumigant A gas that acts as a respiratory poison against pests.

Fungal film Film of fungal growth, e.g., on microscope lenses.

Fungal hyphae Fungal filaments.

Fungicide (Latin: fungus—fungus, cide—inhibiting). A means used to combat mould.

General cargo Cargo, consisting of goods, unpacked or packed, for example in cartons, crates, bags or bales, often palletised.

Generator set (gen-set) A generator which can be attached to a container or a truck chassis and which generates power for supplying electricity to a refrigerated container.

Grating Floor of a refrigerated hold/container which is designed so that air can flow along the floor and into the space where the floor is not covered by a load. T-bar gratings comprising aluminium tee profiles are used in refrigerated containers. Hole gratings (braced boards with holes) are normally used on refrigerated cargo ships.

Grey-mould rot Storage disease frequently occurring in carrots and tomatoes.

Ground conveyor General term for conveying equipment which runs on wheels along the ground, can be freely steered and is used to convey, pull or push loads, such as forklift trucks or side loaders.

Half-height open-top container A half-height container with no roof, which is particularly suitable for heavy and compact cargo. The container can be loaded through the open top using a crane. Due to the reduced height of this container, the rates on container ships can sometimes be more favourable.

Handling symbol Handling symbols are an essential part of the marking of packages and ensure greater care is taken during cargo handling.

Hardboard This refers to hardboards, around 2–3 cm thick, that are used for load securing and for interlayer dunnage or top dunnage. Since these hardboards only has minimal material strength and, in addition, one side of the board is exceptionally smooth, they are not suitable for use as dunnage and cannot be recommended. The exception is use as lateral load distribution or to protected neighbouring cargo from low physical loads.

Hard-top open-top container Container that opens at the top and can be closed with a solid roof (steel sheet). Like all containers that open at the top, this is suitable for crane loading.

Heat capacity Physical property of a material indicating its ability to retain heat; measured in kJ/(kg K); typical values: 4.182 kJ/(kg °C/°F)) for water, 2.1 kJ/(kg °C/°F) for ice, 1.6 kJ/(kg °C/°F) for many frozen foodstuffs.

Heat expansion rupture Damage to a product or its packaging caused by an increase in volume when the temperature rises above a critical point.

Heat-induced blowing Bulging on cans caused by dilation (expansion) of the canned product.

Heat transition coefficient Also known as the k-value; physical value that specifies the degree of insulation provided by a wall or similar. For containers, heat transition takes account of the heat transfer from the air inside the container to the wall, the heat transmitted by the wall, and the heat transferred to the air outside the container; unit of measurement: W/(m^2K); Typical value for refrigerated containers: 0.3 W/(m^2K) (new), 0.4 W/(m^2K) (old).

Heavy-lift steel corner pieces These are angled steel plates attached to the bottom and/or top of cargo items to prevent damage by slinging equipment.

Heavy plate Hot-rolled thick steel sheets—sometimes several centimetres—further processed or used in the "raw" steel industry (shipbuilding).

Heeling Heel, heel angle: Temporary inclination of a ship to one side along its fore-and-aft axis brought about by external forces working on the ship. Heeling can be caused by a rough sea, winds or the rudder position. The heel angle describes the inclination of the ship. The heel angle is the deviation from the perpendicular.

High cube container Large volume container (many cubic meters), as it is 9½ feet high (as opposed to 8ft or 8½ft).

Highly perishable foodstuffs Foodstuffs that are at risk of rapid quality degradation.

Hoist operation Operation with lifting gear (cranes etc.).

Hold meteorology study group The objective of the container meteorology study group at the Warnemünde-Wustrow University of Seafaring, established by Professor Ulrich Scharnow in 1970, who has headed the group since that time, was to investigate the complex thermodynamic processes occurring in containers. The investigations were carried out on two standard containers, each of which was equipped with an air lock to prevent disturbing the cryptoclimate when monitoring and making measurements and a weather station. The results of the investigations were able to provide information to both container manufacturers and consignors, particularly to consignors of hygroscopic goods, about the potential for climatological stress to which the goods stored in containers are subjected under specific climatological conditions, particularly in ambient conditions exposed to radiation.

Homeothermic pests (Greek: homoios—identical; thermos—heat). Warm-blooded pests.

Hot rolled (wide) strip Hot rolled steel product with a rectangular cross-section of at least 600 mm in width. It is immediately wound into coils after rolling with the edges aligned as closely as possible (like a watch spring). It is subsequently used to produce thin steel sheet and heavy plate as well as cold rolled strip/cold rolled wide strip.

Humidity motor Circulation produced when the temperature gradient between the cargo and the container wall is too steep. This increases the transport of humidity to the surfaces of the container.

Hydrolytic class The degree of water resistance for glass products.

Hydrolytic/enzymatic fat cleavage (Greek: hydro—water; lysis—solution; enzymos—fermented). Fat cleavage due to moisture and fat-cleaving enzymes (lipases).

Hydrometer Measuring equipment used to determine the density of liquids, also areometer.

Hygrophilic microorganisms (Greek: hygros—moist, philos—love). Microorganisms that require a relative humidity of > 90%.

Hygroscopicity (Greek hygros—moist). A property of products that react to the air humidity.

IBC Intermediate Bulk Container. A special type of bulk container for hazardous materials.

IEC International Electrotechnical Commission.

Imago (plural: imagoes). Fully developed insects.

IMDG Code *International Maritime Dangerous Goods Code.* Regulations concerning the international transport of dangerous goods by sea.

Infection (Latin: inficere—to poison). Invasion of the body by pathogenic microorganisms causing inflammation.

Ingot Ingots are blocks of metal which are cast into a particular shape. They can vary from metal to metal and from production site to production site. Aluminium, lead, zinc and tin are very often transported in this form.

In-package desiccation Drying out that occurs within the package.

Insecticide Pesticide designed specifically to combat insects.

Intake air temperature Temperature of the air as it leaves the cooling mechanism and is blown into the refrigerated hold/container. The intake air temperature is regulated during chilled operation.

Integral refrigerated container Insulated refrigerated container with an integrated refrigeration unit. Requires a three-phase power supply for operation.

Integrated unit Refrigerated container with integrated refrigeration unit.

Interfacial problems Transfer of heat and water vapour at interfaces.

Internal breakdown Chilling damage in poaceous fruit. If, for instance, apples are stored at too cold a temperature, they "freeze" and suffer irreparable metabolic damage. The flesh of the fruit usually becomes soft and brown from the core. Such damage cannot be detected externally.

Invasion In the context of the Container Handbook: Insect attack of goods.

Iodine value Measure of the degree of unsaturated hydrocarbons in oils and fats.

Isomerisation (Greek: isos—same; meros—part). Intra-molecular restructuring. The atoms in the molecule are reorganised in such a way that the actual composition does not change. This usually has an effect on the chemical and physical properties.

ISO standard Standard compliant with the "*International Organisation for Standardisation.*"

Kraftliner Kraft paper (see below) made of bleached/unbleached kraft pulp and used to make the outer layers of corrugated board and millboard.

Kraft Paper This is made of at least 90% fresh, usually unbleached, sulphate pulp (kraft pulp). It is characterised by high strength and resistance.

Label Labels are, in accordance with the dangerous goods stipulations of the IMDG code, adhesive labels that indicate to all parties concerned what type of dangerous goods are being transported. They differ from placards in terms of size, although placards have the same purpose. Labels with dimensions of 10 × 10 cm are designed for use on each of the items to be transported. Placards of at least 25 × 25 cm must be attached to the outside of the container or vehicle.

Lashing Lash: attaching or fixing/securing a load to a means of transport/or securing a load inside a container in order to prevent it from sliding or falling. Lashing is a widely used international term. The term can be modified to indicate the type of lashing that is meant (direct lashing, diagonal lashing, tie-down lashing) The lashing equipment may be prefabricated lashing belts, chains, wire rope, as well as ropes made of artificial or natural fibres, or a combination of such.

Lashing capacity The permissible lashing capacity (LC) is the maximum force that may be applied to a lashing. This lashing capacity can be found on a label that is attached to the lashing equipment.

Lashing crosspiece A construction similar in appearance to a girder, manufactured primarily from steel and offering attachment points for lashing equipment.

Lashing equipment Devices or aids for securing loads against slippage, tipping and falling. It may comprise chains, wires, ropes, straps or rods and is used in conjunction with a tensioning device (e.g., turnbuckle, twist lock or ratchet tensioner). Depending on the way in which the load is secured, it may be referred to as cylinder lashing, cross lashing, wire-rope lashing or round-turn lashing. Verb: to lash

Lashing rod Metal rod for securing (lashing) containers on deck.

Lateral dunnage Covering the sides of the container with dunnage provides protection against dripping sweat.

Lateral tier Lateral layer of cargo. Generally used to stabilise the stow or to close off the face of the load.

Lattice Grid-like structure, wooden, used to secure the load.

Lethal (Latin: letalis—fatal). Fatal

Lgjs container car Series of rail cars for container transportation. The letter j in the name indicates that the car is fitted with buffers.

Lift-on-lift-off operation (Lo-Lo). Loading and unloading of Intermodal Transport Units (ITUs) using lifting gear.

Lighter A lighter is an inland water vessel or barge usually without an engine that was originally used for reducing the draft of an ocean-going vessel by lightening the load thus allowing it to enter the port.

Line load Weight force which can be withstood by a container floor per running meter.

Liner bag Plastic fabric fixed inside a standard container with bulk cargo.

Lipases (Greek: lipos—fat, oil). Fat-cleaving enzymes.

Liquid bulk Liquid cargo

Load limit line A red line in refrigerated containers that indicates the maximum stowage height of the load to allow sufficient air circulation below the roof.

Load-securing foam Foams which are used as fillers, made from PE or PU. They are used to fill gaps in the container, and thus to ensure tight fit and firm securing of the load.

Locking coil Coils (rolls of steel sheet) are loaded athwartships with their winding axis fore and aft. They are stowed from the side walls inwards, so that one or two spaces are left in the centre that are too small for a further coil. The locking coil(s) is (are) placed in this (these) space(s). The opening angle leads to significant lateral pressure. This pressure forces the other coils towards the ship's sides, thus stabilising the stow/row of coils. Locking coils must not protrude down into the cant lines of the lower tier by more than one-third of the diameter of the locking coil, since if they protrude any further, the increased lateral pressure can lead to damage to the coils.

Lower temperature limit Temperatures below this limit lead to quality degradation, e.g., chilling.

Low temperature flexibility Increased kink resistance achieved by adding plasticiser to synthetic plastics.

Low-temperature refrigerated containers Refrigerated containers that enable goods to be transported at temperatures as low as −60 °C (−76 °F).

Lumen (Latin: lumen—cavity). Pores in the structure of wood, cavities in vegetable fibres.

Maximum cargo height Frozen goods are packed using the block stowage method merely for protection against the effects of external heat, so it must be ensured that subzero temperatures are maintained in the floor, side and ceiling areas. A free space of at least 8 -10 cm must be left for the return air between the top of the cargo block and the container roof. This maximum cargo height is indicated in the containers.

Maximum securing load (MSL) The strength of a material to resist a tensile load without suffering plastic/permanent deformation. The MSL is extremely important for cargo securing, since materials which are liable to suffer permanent deformation must not be used for securing cargoes. Cargo securing materials must therefore be dimensioned to ensure that they are loaded no further than their MSL.

Mediterranean fruit fly (Ceratitis Capitata) Quarantine pest which particularly attacks citrus fruits.

Medulla (Latin: medulla—marrow). The medulla is responsible for feeding the wool fibre.

Mesophilic microorganisms (humidity) (Greek: mesos—middle; philos—love). Microorganisms thriving above 86% relative humidity.

Mesophilic microorganisms (temperature) (Greek: mesos—middle; philos—love). Microorganisms thriving at medium temperatures.

mm H_2O Millimetres of water. Unit for small differences in pressure, measured as the difference in height in mm between the two surfaces of a water column in a siphon. 1 mm H_2O = 9.81 Pa.

Modular unit load dimensions Packages or unit loads are suitable for modular container stuffing if their dimensions or multiples of their dimensions are equal to the internal dimensions of the container. By using appropriately dimensioned load units it is possible to tightly pack the container making good use of the "natural" securing elements of the container, namely the side and end walls. If the cargo modular unit load dimensions and the overall load is of appropriate dimensions, there may be no need for additional load securing measures to be implemented.

Modular unit load packaging Packaging whose dimensions permit the items in a consignment to be grouped together to form modules. Modules are predefined basic sizes defined for cargo units which enable them to fit together, to be combined and to be transported with different means of transport.

Moisture sensitivity Degree to which the quality of a product is impacted even by minimal uptake of water vapour.

Mould growth threshold Relative humidity in excess of 75% promotes rapid multiplication of the majority of mould species.

Monophagic (Greek: monos—one; phagein—to eat). Associated with a particular food source.

MSL Maximum Securing Load (MSL). is the strength of a material to resist a tensile load without suffering plastic/permanent deformation. The MSL is extremely important for cargo securing, since materials which are liable to suffer permanent deformation must not be used for securing cargoes. Cargo securing materials must therefore be dimensioned to ensure that they are loaded no further than their MSL.

Mycotoxins Poisonous metabolic products produced by mould.

Narrowband Frequency band for data transfer via the power cable. The frequency on narrowband is 55 kHz.

Neutral behaviour Goods which are mutually compatible and compatible with their environment display neutral behaviour.

Nominal breaking load Breaking load (BL). Specifies the load at which a material will fail.

Non-climacteric fruits (Latin: climactericus—turning point, critical time). Fruits which are not (yet) ready to eat or ripe for consumption.

Non-respiring goods Goods which do not cause a change in the ambient atmosphere (except a release of water). These include all frozen goods and goods sealed in gas-tight packaging, steel products and all products which are not subject to decomposition processes. Caution! Degradation processes occur even in scrap lumber/wooden chips, shredded scrap metal, wastepaper, etc. Although these cannot be described as respiration, they nevertheless consume O_2 and release CO_2 to the extent that this may lead to the build-up of life-threatening concentrations in closed rooms.

Nonwoven fabric Water absorbent cloth underneath the container roof.

Odour sensitivity Sensitivity of the goods to foreign odours.

Odour tainting Taking up of odours from foreign substances or gases in the environment.

Onion neck rot Rot of the neck of onions caused by the mould Botrytis allii.

Onward-carriage The colloquial meaning is the transport of goods from the seaport/airport to the domestic port of destination. In accordance with the German Commercial Code, onward-carriage is the delivery of goods from the store of a receiving agent to the final recipient. During this phase of the transport chain, the freight regulations of the German Commercial Code apply.

Open-sided container Container with sides which can be opened, usually closed with tarpaulin and bows.

Open-top container Container open at the top (enables the container to be loaded by crane).

Oxidative fat cleavage Fat cleavage produced by atmospheric oxygen.

Pail A container made from plastic or steel with a firmly closed lid, for transportation of liquids, in particular paints and oils.

Pallet box Transport container/box with integrated pallet.

Palletisation Loading of uniform packages onto pallets in accordance with a predefined packing plan. For reasons of stability this should be done in such a way that the load items interlock.

Palletised Cargo which has been grouped into a single cargo unit. Better rates are available for packing containers with palletised loads in the case of LCL (Less than Container Load), as a pallet load can be better handled with ground conveyors (q.v.) than individual items.

Paper dunnage Covering the floor, sides and cargo surface with paper dunnage provides protection against dripping sweat.

Parasitoses These are diseases caused by parasites.

Partial pressure difference In a mixture of gases, each component has a partial pressure. The sum total of all partial pressures is the absolute pressure. The partial pressure difference is the difference between the partial pressures of the gases either side of a gas separation membrane.

Particulate air contaminants Air pollution caused by dust.

Passive behaviour Defines the sensitivity of cargo to other cargo and to its environment (e.g., sensitivity to foreign odours).

Payload The revenue-producing load carried by a means of transport.

PCT Power Cable Transmission: Transfer of digital data by means of the high-voltage power cable.

Peel discolouration Chilling damage in citrus fruit.

Penetration Penetration of packages occurs as a result of insect activity (holes eaten out by insects).

Permeation rate Property of a material which measures the rate at which gases pass through the material as a result of solution diffusion processes.

Phase change Change to the state of an aggregate as a result of temperatures rising above or below the solidification and melting temperatures, e.g., with oils and fats

Ph value Hydrogen ion concentration. This indicates the acidity or alkalinity of a solution.

Physiological ripeness Post-climacteric: This is the start of over ripeness. The fruits start to decompose, the taste suffer, and the fruits can easily be attacked by mould or rot.

Phytosanitary certificate Certificate of health under the *International Plant Protection Convention.*

Picking or industrial ripeness Pre-climacteric, the stage of ripeness prior to the ready-to-eat/ripe for consumption stage.

Pileworm/shipworm Teredo which destroys wood.

Pipe clamps These are used to protect cylindrical goods, to unitise them and to ensure that they can be transported safely. They comprise pairs of squared lumber, steel profiles or similar materials which are placed above or below the goods and are connected to each other with threaded rods at the sides.

Pitching Movements of a ship around its transverse axis.

Placard Placards are, in accordance with the dangerous goods stipulations of the IMDG code, adhesive placards that indicate to all parties concerned what type of dangerous goods are being transported. They differ from labels in terms of size, although labels have the same purpose.

Placards of at least 25 × 25 cm must be attached to the outside of the container or vehicle.

Labels of a size of 10 × 10 cm must be attached to each package to be transported.

Poikilothermic pests (Greek: poecil—multicoloured; thermos—heat). Cold-blooded pests.

Polymerisation (Greek: poly—many; meros—part). Gelling of oils caused by exposure to heat, in particular in the context of plastics manufacture.

Polyphagic (Greek: poly—many). Associated with many food sources.

Polypropylene wrapped Hercules, Hercules cordage A particularly strong type of rope: Hercules rope is a combination of fibre and wire rope; the individual strands have a wire core. This construction means that Hercules rope is easier to handle than ordinary wire rope.

Polyurethane foam Used to insulate refrigerated containers. It is also known as construction foam.

Porthole container Refrigerated containers without a separate refrigeration unit:

Porthole refrigerated containers, also called insulated or CONAIR containers, do not have their own refrigeration unit. They are thus reliant on an external supply of cold air. This is achieved by refrigeration units of various types, permanently installed on the ship, permanently installed in the terminal or clip-on units for individual containers.

Porthole containers are thermally insulated and have two sealable openings on the end walls (the portholes) through which cold air can be blown into the container and warm air can be extracted. The cold air is forced through the lower opening into the container, then distributed throughout the load via the T-bar floor, and subsequently flows through the load to the top of the container and is extracted through the upper opening.

Post-climacteric (Latin: post—after; climactericus—turning point, critical time) Physiological ripeness.

Post-fermentation Heat and moisture can cause products such as coffee, cocoa, tobacco to ferment again, which in turn can make them unfit for consumption.

Postmortem (Latin: post—after; mors—death). After death.

Pre-carriage The carriage of goods (containers) by any mode of transport from the place of receipt to the port (place) of loading into the ocean vessel (main means of transport).

Pre-climacteric (Latin: climactericus—turning point, critical time). Picking, harvesting, shipping or industrial ripeness.

Pre-fabricated lashing belts These are ready-made lashing belts with hooks and tensioning elements sewn in.

Preserved foods Semi-preserved foodstuffs with limited shelf life.

Preserved foodstuffs Foodstuffs preserved by sterilisation and stored in cans/jars.

Pre-slung cargo units In order to accelerate unloading, the strops used on loading are left on the cargo units. This then makes slinging unnecessary when discharging the cargo.

Pre-trip inspection (PTI) Test to examine correct operation of a refrigerated container. The test should be carried out before transporting refrigerated goods.

Protection of stored commodities Protection of goods against spoilage by animal pests.

Psychrophilic microorganisms (Greek: psychros—cold, philos—love). Microorganisms that thrive at low temperatures.

Psychrophilic spoilage Spoilage on the surface of meat/fish.

Putrefaction Autolysis due to enzymes within the meat.

Pycnometer (Greek: pyknos—sealed). Measuring device used to determine the density of liquids.

Quarantinable diseases (French: quarante—forty). Infectious and notifiable diseases that carry a mandatory isolation period (quarantine) for infected parties. Such diseases can be spread as a result of trade and transportation.

Quarantine Precaution against contagion and infection: Isolation, (French: quarante—forty) originally lasting forty days, to prevent the introduction of diseases and pests.

Quarantine pests (French: quarante—forty). Pests that carry or spread quarantinable diseases (see above), or which can themselves be termed quarantinable diseases.

Radiation class Radiation conditions calculated on the basis of the measured duration of sunshine and solar altitude for 10 day measurement periods.

Rancidity Oils and fats spoil by becoming rancid. This causes changes in odour and taste, which can make edible oils and fats inedible.

Recovery Capacity of a material or construction to return to its original position (elastic deformation).

Recrystallisation If meat is frozen rapidly at low temperatures (e.g., −27 °C (−16.6 °F), small ice crystals are produced. If the meat rises to a higher temperature (e.g., −10 °C (14 °F), the ice crystals grow larger, i.e., they recrystallise. As of a certain size, they can rupture the cell walls of the meat. These "perforated" cell walls lose more cell juices than cell walls which are not damaged. The result is a loss of juice, also known as "drip loss."

Relative humidity Relative humidity is derived from the ratio of absolute moisture content to saturation content. Depending on its temperature, air can absorb different quantities of water vapour until it is saturated (100% relative humidity). Absolute humidity (moisture content) is the quantity of water actually present in the air and is measured in grams per cubic meter (g/m^3). Relative humidity expresses as a percentage the quantity of water vapour the air has at a specific temperature, relative to its saturation content. If, for instance, air at 20 °C (68 °F) has an absolute humidity of 12.1 it will have a relative humidity of 70%. If air has reached its saturation content at 20 °C (68 °F), it has absorbed 17.3 g/m^3 of water.

Residual oil, feedstuffs containing Pressing residues arising from oil extraction from oil-bearing seeds/fruits.

Respiration Breathing. Even after it has been harvested, fruit undergoes ripening processes. Bananas, for instance, absorb O_2 from the ambient atmosphere, convert starch to sugar and release CO_2, ethylene, water vapour and heat. This ripening process is also known as respiration.

Respiration heat Heat which is released during respiration.

Respiring goods Goods of vegetable origin, for example fruits, for which ripening, or degradation processes are taking place. These take air (or gases) from the atmosphere surrounding them and give off CO_2, water vapour and heat to their environment.

Return air temperature Temperature of the air which is drawn off from the refrigerated hold/container. The return air temperature is controlled during refrigeration.

Risk factors Required information on risks for the transport, handling and storage processes.

Risk of larceny/theft Risk of the illegal removal or appropriation of goods or objects. A distinction is made between simple larceny (i.e., without the use of force) and compound larceny (i.e., with forced entry by breaking and entering or by copied keys.

Roll angle Angle at which a ship moves around its longitudinal axis.

Roll oscillation Oscillation due to rolling. This is the name given to the movement of a ship around its longitudinal axis.

Roof bows Steel or aluminium frame construction in conjunction with removable wooden/aluminium gates that hold the tarpaulin of a truck, trailer, semi-trailer or a swap body. This type of open-sided construction is sometimes known as a tarpaulin truck.

Ro/Ro ship Roll-on/roll-off. Cargo ship on which the mobile cargo can be rolled directly on and off board (by lorry, trailer, etc.) by means of bow, stern and side doors (ramps). In contrast to this, cargo loaded and discharged on other cargo ships by means of their own or external loading tackle (vertical handling).

Round turn Hawsers, ropes or belts wound around an item. On the one hand as a basis for forming a knot, on the other hand as a method of lashing. Unfortunately, round turn lashing is a widespread "securing method." It is absolutely unsuitable for securing a cargo as the cargo can move "freely" in the loop and no securing effect is thus achieved.

Safe to transport Cargo is secured appropriately if it is loaded, stowed and secured in such a way that it cannot be damaged by events that can be expected during carriage.

Salmonellosis Infectious, febrile intestinal and general diseases caused by salmonellae.

San-José scale Quarantine pest that attacks stone fruit, poaceous fruit and berries, as well as decorative woods und structural lumber.

Saturation content Depending on its temperature, air can absorb different quantities of water vapour until it is saturated (100% relative humidity).

Sealed package Packaging impervious to water vapour used to store goods at risk of corrosion. Desiccants are used in sealed packages to maintain an artificial atmosphere with a relative humidity of $< 40\%$. No corrosion is possible below this threshold.

Secondary infection Contamination caused as a result of microorganisms (durable spores) from previous contaminated loads that are able to attack a subsequent cargo.

Self-heating Temperature increase within an organic solid without external heat input.

Semifinished product (semi) Steel products that have already been shaped by hot-rolling or continuous casting processes and which are then subjected to further shaping to produce finished products. These include: Rough-rolled blocks (blooms with round or square cross-sections), slabs, billets, boards and ingots.

Semitrailer (trailer) Trailer with a rear axle assembly that is pulled by a towing tractor or truck tractor. There are two different kinds of trailer: Semitrailers that are also permitted in road traffic, and roll trailers (cargo trailers) that are pulled by towing tractors and are used only in ro-ro operations. These towing tractors are permitted only in terminal operations, and not allowed to be used on public roads.

Separation Chill haze.

Shipping Mark Part of the marking on packages. The shipping mark comprises:

- Identification mark: e.g., initial letters of recipient or shipper.
- Identification number: e.g., the recipient's order number.
- Total number of items in the complete consignment.
- Sequential number of the package in the consignment.
- Final destination and port of destination.

Shock and/or tilt indicators These indicators are attached to sensitive loads for monitoring purposes. They indicate whether specified threshold values have been exceeded either in terms of acceleration or tilt.

Shrinkage in mass Loss of volume, for instance with leather goods as a result of drying out.

Shrinkage/shortage Loss in volume, reduction (weight loss) of goods during transport or storage.

Shrink wrapping Shrink wrapping involves enclosing the package contents in shrink film (flat or tubular film), heat sealing any unsealed portions and separating the package from the film web or covering the package contents with a shrink cover. Depending upon the shape and weight of the package contents, the shrink material used should be PE or plasticised PVC film of a thickness of 0.01 to 0.2 mm, with PE films being particularly suitable for heavy items. The film is heated from the outside in a shrink-wrap oven or with hand-held heat gun, thus releasing the captive tension in the film. Shrink films are produced in either monoaxial (in a single direction) or biaxial (in two directions) orientation. As the film cools down, it shrinks around the package contents, applying an extremely low pressure per unit area.

Sintering Physical changes in materials caused by high temperatures; applied in particular to metals, ceramics and minerals/rock.

Sirex wasp (Urocerus Gigas L.) Giant wood wasp. When using packaging made of wood for marine transport to Australia and New Zealand, the appropriate treatment and certification of the wood must be observed.

SKD SKD (semi knocked-down cars): For export to countries which apply high levels of customs duty to luxury goods, automobiles are shipped in completely knocked down (CKD) or semi/part knocked down (SKD, PKD) form. In this case, the package sizes for the individual components are adapted precisely to the container in which they are transported, which means that the cargo is secured by tight fit.

Skids Skids are sled-like bases used to facilitate ground conveyance of/securing of, for example, heavy machinery and machine parts

Slab Cuboid half-finished steel products that are hot-rolled into sheets at the rolling mill.

Slim wedges See tapered blocks

Slinging Describes the process of attaching slinging equipment (for example, ropes, chains or webbing slings) to a load in order for it to be lifted/trans-shipped safely using lifting gear, for example, a crane.

Slit strip Rolled sheets are produced in wide strips (e.g., > 600 mm), also called wide strip. In order to be able to offer narrower strips, the strips are cut to slit strip in a slitting unit.

Slot position Specifies the position of the container on board the ship, for example using the Bay Row Tier system.

Slot rate Rate used to calculate the cost of renting a slot for transporting a single container on board a ship.

Smart reefer A type of refrigerated container (trade name).

Sorption behaviour (Latin: adsorbere – take up). Absorption or release of water vapour by a hygroscopic cargo until equilibrium is reached.

Sorption isotherms Graphical representation of the sorption behaviour of a substance at a specified temperature.

Specially-shaped stanchions These are collapsible stanchions that can be inserted and locked in place or unlocked and folded down. Stanchions are suitable for providing tight fitting lateral load securing for many types of loads.

Spontaneous combustion A rise in temperature of an organic solid without external heat to a point where the solid ignites spontaneously.

Spreader Containers were traditionally handled using four wire ropes and cranes. However, as the opening angle of the wire ropes placed considerable horizontal pressure on the tops of the containers, steel framed constructions (spreaders) were introduced in order reduce these forces. These spreaders have undergone constant development and are still used on container gantries today.

Stacking load The force acting on lower packing units as a result of the units stacked on top of them. Standardised DIN and ISO symbols are used to identify the permitted stacking loads.

Stalk position Different positions of tobacco leaves on the stalk which are harvested at different times.

Stanchion Rods or bars on the sides and corners of an open means of transport or container, used to secure the cargo. Removable stanchions are inserted into the stanchion pockets provided.

Stomach poison Poison which enters the body of the insect on food intake (ingestion). Stomach poisons are usually administered using artificial bait.

Storage climate conditions Refers to the specific external conditions (humidity, temperature, ventilation) in which particular products can be stored without risk of a reduction in value or of damage to the products.

Stowage factor The stowage factor specifies how many cbm or cbf of hold space in a ship will be taken up by a metric tonne, a US ton, or an Imperial ton (1,000 kg/ 907.18 kg/1016 kg) of a cargo considering gaps, shape of the hold, number of 'tween decks, supports and the use of dunnage.

Straddle carrier/van carrier Special self-propelled vehicle for moving containers within the terminal.

Stretch wrap Stretch film used to wrap cargo to form bales or to wrap cargo to pallets.

Strip To unpack/unload a container.

Strop Slinging equipment from various materials. Strops are slung round the load and attached to the hook of the crane.

Stuff To pack or load a container.

Sublimation Transition from the solid to gaseous state without passing through the liquid state.

Swap body Swap bodies are vehicle bodies that are not permanently fixed to the carrying vehicle. They are similar to containers and are locked to the carrying vehicles in the same way using twist locks. At the loading bays, the swap bodies can be stored standing on their own legs. They are suitable for use for multimodal transport by road and rail and have grappler pockets which allow them to be moved using gantry cranes.

Sweat class Classification of sweat water according to intensity. Grades range from 0 (no sweat water) through to 6 (greatest formation of sweat water). The classification is made by observing the volume of water produced per m^2 surface, thus how many grams of sweat water appear for each square meter for each of the grades.

Swellability Increase in volume of solid substances, particularly hygroscopic substances, (paper, natural fibres) as a result of exposure to moisture.

Syneresis Congregation of molecules leads to "sweating" in cereals.

Syrup formation Formation of viscous liquids as a result of humidity and warmth, for example, in dried fruits.

Tank container A large container comprising an ISO frame in which liquid containers are mounted. They must meet special requirements with respect to pressure-resistance as a result of the pressures they have to withstand. For this reason, the actual containers are cylindrical or spherical in form. Several containers can be mounted either horizontally or vertically in a single frame. Only specific containers must be used for foodstuffs and must be marked "Potable liquids only." Tank containers for hazardous goods must meet the requirements of the IMDG code. Half-height tank containers serve to carry high-density liquids which cannot be carried in normal tank containers because they cannot be filled to a high enough level and are therefore subject to surging.

Tapered blocks These are wedges cut at a narrow angle to allow small gaps between items of cargo to be braced. They are also referred to as slim wedges or driving wedges.

TBU Transformer Bypass Unit. Component of a remote refrigerated container monitoring system with data transfer via the power cable. A TBU allows a transformer to be bridged, that would otherwise not let the data signal pass.

Telescope carton The top of telescope cartons is fabricated as a separate part and overlaps all four sides of the carton (e.g., banana boxes).

Temperature-controlled transport Transportation during which the optimum temperature for chilled and frozen products is maintained.

Temperature/dew point difference The difference between the temperature of the goods and the dew point temperature of the air surrounding the goods.

Temperature gradient Differences in temperature within a spatial distance.

TEU Twenty-foot Equivalent Unit: Corresponds to a 20-foot container, where the container height (8 ft, 8½ ft odour 9½ ft) is not considered; unit for counting containers.

Textured coated board Plywood board with a phenolic laminate backing to enhance durability. The laminate is textured with a screen-printing process.

Theoretical density Pure solid lumber without cavities.

Thermophilic microorganisms (Greek: thermos—heat, philos—love). Microorganisms that thrive on heat.

Thigmotaxis (Greek: thigma—contact, touch; taxis—arrangement, sequence). Stimulus by touch.

Tier (container) Layer in a container that is numbered from bottom to top.

Tight fit A method of securing the load by direct or indirect physical contact with load-bearing components of the means of transport or the transport container.

TLV Threshold Limit Value (referring to harmful substances in the workplace).

Tower units Cooling equipment for supplying of cool air to porthole containers in terminals.

Toxic (Greek: toxikon—poison). Poisonous

Toxicity (Greek: toxikon—poison). The degree to which a substance is poisonous.

Trade imbalance Discrepancy between the amount a country imports and the amount it exports.

Transpiration losses Evaporation losses in vegetable products as a result of respiration (transpiration).

Transport-related acceleration forces Negative and positive accelerations are dynamic, mechanical stresses which occur in two main types during the transportation of goods:

- regular acceleration forces and
- irregular acceleration forces.

Regular acceleration forces primarily occur in maritime transport. Acceleration of up to one g (g = 9.81 [m/s^2]) and, in extreme cases, even more, may occur due to rolling and pitching in rough seas. Such regular acceleration forces have an impact on the effort involved in load securing.

Irregular acceleration forces occur during cornering or when a train passes over switches, during braking, starting up, hoisting and lowering. Such acceleration forces are not repeated, but they may occur several times at varying intensities during transport. These are the typical stresses of land transport and transport, handling and storage operations.

Transport-related properties of products Product properties required for transport, handling and storage processes.

Transport temperatures Loading, travel and pulp temperatures, e.g., for sweet oils.

Travel or transport temperature Optimum storage temperature of a product.

Tray Trays are preformed containers made from plastic to take individual fruits that are particularly sensitive to pressure.

Trichinosis Human disease caused by triciniellae (trichinella spiralis). Household pets, for example may be secondary hosts. Meat inspections are statutory preventive measures.

Turgor (Latin: turgescere—to swell). Inner pressure, distension of cells.

Turnbuckle Tightening elements made up of two threaded rods, a threaded barrel and a lever, usually in the form of a ratchet. Turnbuckles are the preferred tensioning method when securing loads with chains.

Turner Banana that is becoming ripe to eat.

Twist lock A securing mechanism for securing containers in a vertical direction; twist locks are placed between the containers and fastened in the oblong holes on the container corners.

Unitisation Grouping together of items for transport into transport units (e.g., in containers or on pallets).

Unit loads Unit loads are parts of a shipment that can be treated as a single unit during cargo handling and transportation. The size or dimensions of the unit load can vary according to requirements and to the means of transport and packaging container available. To optimise the cargo handling, transport and storage processes, standardisation of unit loads is desirable. The most common unit loads are pallets and components with bases that resemble pallets, such as pallet boxes, IBCs, etc. The most important properties of a unit load are that it can be loaded to ensure a tight fit, its modularity and its stability, the last of which ensures that it can be stowed safely and will not be damaged by load securing measures.

Upper temperature limit Temperatures above this limit lead to quality degradation.

Upright blocking In the context of load securing, upright blocking is understood to be wooden elements inserted vertically in order to take up and transfer loads.

USDA United States Department of Agriculture; U.S. governmental department responsible for the regulations concerning the import of foodstuffs.

VCI (Volatile Corrosion Inhibitor) method A method used to inhibit corrosion.

Viscosity (Latin: viscum—tenacity). Degree of resistance to flow in a fluid or semifluid for gas, caused by internal friction.

Volume-to-payload ratio The volume-to-payload ratio of a means of transport specifies the volume available per unit weight payload. In the metric system, it is usual to specify the ration in cubic meters per metric ton (m^3/t). In English-speaking countries, it is usual to use cubic feet per ton (cu.ft./t).

Walking board Plywood board, around 2 cm thick, often used for load securing and for interlayer dunnage or top dunnage.

Water content The percentage of total mass of a substance that is constituted by water.

Water content class Classification of goods in accordance with their water content.

Water vapour release Desorption.

Water vapour uptake Adsorption.

Waviness Damage as a result of moisture affecting, for instance paper plies, veneered boards.

Wideband Frequency band for data transfer via the power cable. The frequencies used for wide band range from 140 to 400 kHz.

Wireless LAN Wireless Local Area Network: a wireless (radio) network used for transferring data; is also used in offices.

Wire rod coil Steel product made in rod wire rolling mills with varying cross-sections. German standards refer to wire rod as a product that can be wound immediately into coils from the rolling mill whilst still hot.

Coils of wire rod. Depending on the nature of the goods and any intended subsequent processing these coils are corrosion-sensitive and sensitive to mechanical damage, for example, kinks forming in the wire. The length of the winding axis is a critical factor in determining the way they are stowed.

Wood treatment certificate Certificate of treatment against insect infestation.

Xerophilous microorganisms (Greek: xeros—dry, philos—love). Microorganisms that survive at a low relative humidity of (approximately 75%).

SPRINGER NATURE

GPSR Compliance

The European Union's (EU) General Product Safety Regulation (GPSR) is a set of rules that requires consumer products to be safe and our obligations to ensure this.

If you have any concerns about our products, you can contact us on ProductSafety@springernature.com

In case Publisher is established outside the EU, the EU authorized representative is:

Springer Nature Customer Service Center GmbH
Europaplatz 3
69115 Heidelberg, Germany

The manufacturer's authorised representative in the EU is Springer Nature Customer Service Centre GmbH, Europaplatz 3, 69115 Heidelberg, Germany. If you have any concerns regarding our products, please contact ProductSafety@springernature.com

Printed and bound by CPI Group (UK) Ltd, Croydon, CR0 4YY
30/09/2026
02225586-0016